YS

ertin

F

TAXONOMY OF FLOWERING PLANTS

A SERIES OF BOOKS IN BIOLOGY

A flowering branch of the Tulip-tree, or Yellow Poplar, *Liriodendron tulipifera* L. The Magnolia Family, of which this plant is a member, is often regarded as one of the most primitive of all families of flowering plants. The scientific name, meaning "the lily tree that bears tulips," is highly descriptive.

Second Edition

Taxonomy of Flowering Plants

C. L. PORTER *University of Wyoming*

W. H. Freeman and Company
SAN FRANCISCO AND LONDON

Printed in the United States of America.

Library of Congress Catalog Card Number: 66-19914

TO

all who have ever wondered
"What plant is that?"
and especially to Marj,
this book is fondly dedicated.

PREFACE

ALTHOUGH most branches of botany have long been adequately served by an array of suitable texts, plant taxonomy, the oldest branch, has not been so fortunate. Until quite recently it has been necessary to improvise text material for classwork; and even now we are faced, for the most part, with a choice between texts that are really reference books for advanced students and much abbreviated texts that have had much of the meat of the subject deleted from them. It is my hope that this book will help to fill that gap.

To this end I have attempted to put together such a factual and simplified account of basic principles as is needed by students in a beginning course in taxonomy, together with illustrated descriptions of more than a hundred families of flowering plants representative of the North American flora. Only the flowering plants, or angiosperms, are included, for the vascular cryptogams (ferns and fern allies) and the gymnosperms are often dealt with in separate courses in morphology and dendrology, and for those groups excellent texts are available. In my experience, a study of the basic principles and of selected angiosperm families alone provides ample material for the average beginning course in taxonomy, which is given for a minimum of one semester but preferably extends throughout one school year.

The content of the text is based on the assumption that the students have had at least some introduction to plant science, such as that provided by the usual beginning course in botany. It is particularly aimed at undergraduate students in such practical fields of study as agronomy, range management, forestry, and wildlife management and conservation. This has necessitated a short, concise treatment and the omission of the detailed discussion and elaboration that are found in the published research of many taxonomic experts of the past century. Perhaps I have been guilty of oversimplification in attempting to get at the essence of a complex subject. Those who feel that this is true, as well as more advanced students, will find numerous references to pertinent literature. By delving into at least some of it, the student will gain a better and more detailed knowledge of some of the classical work accomplished in this field.

The text has been divided into three major parts: Part I, dealing with historical and theoretical aspects and with terminology and morphology; Part II, dealing with orders and families of monocotyledons; and Part III, dealing with orders and families of dicotyledons. In practice I have used Parts I and II as the basis of one semester's work and Part III as the basis of a second semester's work. The instructor may or may not wish to follow this sequence.

One method of learning families, proved by a goodly number of classes over the years, is the study of floral diagrams and flower sections, together with other illustrative material, such as habit sketches or photographs and illustrations of fruits. This has called for the development of a system of diagrams not usually found in texts of this sort. These pictorial aids impress the minds of students with the varied floral morphology of plant families much better than conventional descriptions do. It is generally possible to identify a family by its floral diagram alone. In attempting to become acquainted with the flora of any region, one must first learn to recognize *families* rather than a great diversity of genera and species. These may come later with added field experience.

These pages have had a slow evolution. In the beginning they were merely outlines that I used as guides when conducting classes in taxonomy some thirty years ago; but at the insistence of the suffering students, who had difficulty in taking down notes from the blackboard, they gradually took the form of mimeographed notes distributed to classes in lieu of a textbook. These notes have been revised repeatedly as need arose until they now form the basis for this text. In drafting the final manuscript I have incorporated numerous suggestions of my colleagues. The drawings have been prepared by Mr. Evan Gillespie from sketches I have furnished, and I have added some of my own photographs to supplement the drawings.

In this second edition the text has been updated by taking into account some of the newer aspects of taxonomy, and additional references to some of the recently published literature have been added. A few illustrations have been improved or changed, and a few have been added.

I am indeed grateful to the many persons who have offered helpful suggestions, and I shall welcome further comments from students and fellow teachers.

July 1966 C. L. PORTER

CONTENTS

Part I

HISTORY, PRINCIPLES, AND METHODS

Part II

SELECTED ORDERS AND FAMILIES OF MONOCOTYLEDONS

Part III

SELECTED ORDERS AND FAMILIES OF DICOTYLEDONS

Part I

HISTORY, PRINCIPLES AND METHODS

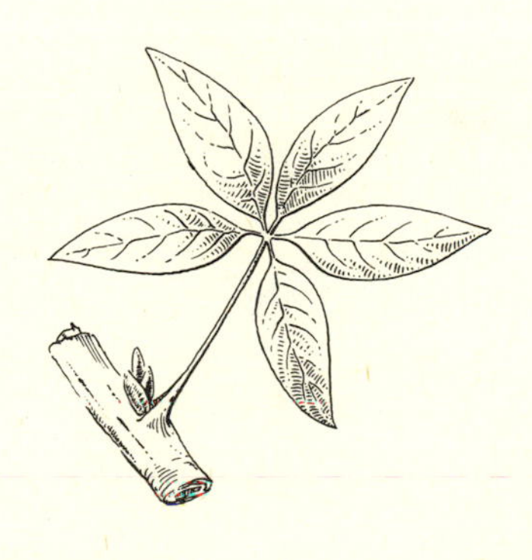

CHAPTER 1

The Aims of Taxonomy

"It has been already suggested, and forcibly enough, that plant taxonomy was not invented in any school, or by any philosopher; that it is everywhere as old as language; that no plant name is the name of an individual plant, but is always the name of some group of individuals, and that all grouping is classifying."—Edward Lee Greene in *Landmarks of Botanical History* (1909), page 106.

Plant taxonomy has two aims: (1) to identify all the kinds of plants; (2) to arrange the kinds of plants into a scheme of classification that will show their true relationships.

The first aim requires us to make a complete inventory of all the plants on the face of the earth. This is not an insurmountable task, but—largely because our knowledge of many tropical regions is far from complete—it is still a long way from accomplishment. But even the fully explored and more civilized parts of the globe still present problems in plant identification, particularly in such perplexing groups as the genus *Lupinus* (Lupines) of western North America.

Anyone can learn, with a little practice, to identify a goodly number of things, be they people, dogs, rocks, or plants. The ancient civilizations produced people who could recognize and name many hundreds of stars and even arrange them into constellations. This was pure taxonomy. People have always needed to name things in order to have a means of communication. In scientific work it is essential that we be able to apply names with precision, for the validity of much research hinges on the identification of the materials involved. For botanists it is often necessary to identify materials beyond the species level, for minor differences in the kinds of plant under investigation may mean a major difference in the results. By furnishing such identification to others, the taxonomist serves other branches of science in a basic way. *All specialists and all botanical laboratory scientists should realize the importance of accurate identification of the materials with which they work.*

The second aim of the taxonomist is to seek out the evidence that will enable him to understand the relationships among groups of plants —among the lesser groups, or taxa, such as species and their subdivisions, and among the larger groups such as genera, families, and orders. To do this effectively, the taxonomist must utilize the methods and resources of all the major fields of botanical investigation. The *morphologist* gives him an understanding of form and structure, including such refinements as comparative anatomy and embryology. The *physiologist* can point out requirements for existence (such things as "physiological species" seem to occur in plants—groups that appear to be identical but differ in their requirements). The *ecologist* can furnish information about the relationships between plants and their environment, about how environment may affect form and structure, and about the selective action of the environment in determining which plants will survive. The *geneticist* and the *cytologist* contribute information concerning inheritance and reproduction as well as chromosome number and morphology, enabling the taxonomist to judge better whether he is dealing with distinct species or with lesser categories. *Biochemistry* is used effectively to solve taxonomic riddles, while some workers use *statistics* and *computers* in dealing with plant characters.

In cultivating these fields of investigation, the taxonomist must be able to call on others for assistance. *Geology* (particularly historical geology) furnishes information about past life, climates, and changing land forms, and this enables the taxonomist to interpret plant distribution and to understand something of the long history and evolution of plant life on the earth. A knowledge of *physical geography* is useful when he studies plant distribution and migration, whether local or world-wide, and it also points out what physical barriers to interbreeding may be present.

All this does not imply that the taxonomist is always an expert in all the fields mentioned; but it indicates the diversified knowledge that may be useful to the student.

Taxonomists are sometimes criticized for continually changing the names of plants. It should be remembered that taxonomy, though it is the oldest of the botanical sciences, is still in the learning process, and that we are continually adding to the great fund of information that will eventually enable us to name and classify all plants to the general satisfaction of everyone. That day is still a long way off. New techniques, new discoveries, and a continually expanding fund of information all demand revisions in taxonomic treatment, even though these revisions may inconvenience the users of our product. It should also be remembered that we are dealing with living, dynamic, and often fluctuating populations, which will not always be interpreted by all workers, however expert, in exactly the same manner.

The science of taxonomy may be thought of as a synthesis of the four interrelated fields[1] outlined below:

1. *Systematic botany,* the fact-finding field, which includes genetic and cytological studies as well as any other techniques applicable to the problem.
2. *The taxonomic system,* based on the facts that were found, and including:
   (a) Taxonomic concepts of plant groups, or taxa.
   (b) Concepts of the evolutionary sequence of characters.
   (c) Classification and arrangement of taxa.
   (d) Description of taxa, or phytography.
3. *Nomenclature,* a method of naming plants based on international rules that botanists have agreed upon in order to promote a uniform and reasonably stable system. This permits only a single valid scientific name for each kind of plant, the discarded names being known as *synonyms.*
4. *Documentation,* which includes the preservation of living or fossil floras in a museum or herbarium, including *type specimens* (those on which names and concepts of species and lesser taxa were originally based) and *illustrations* (which may sometimes be used in lieu of type materials).

By utilizing all four fields of taxonomy we are building up a mass of information that has already thrown much light on plant populations, their units, their means of perpetuation and dispersal, their interrelationships, distribution, and evolutionary tendencies. There remains, however, much more still to be done.

[1] See Mason, H. L., in the list of references at the end of this chapter.

REFERENCES

ARNOLD, CHESTER A. 1950. "Fossil dicotyledons," in A. Gunderson, *Families of Dicotyledons,* pp. 3–6. Ronald Press, Waltham, Mass.

CAMP, W. H. 1950. "Plant geography," in A. Gunderson, *Families of Dicotyledons,* pp. 25–30. Ronald Press, Waltham, Mass.

CAMP, W. H. 1951. "Biosystematy," *Brittonia* **7**:113–127.

CARLQUIST, SHERWIN. 1961. *Comparative Plant Anatomy,* Holt, Rinehart and Winston, New York.

CONSTANCE, LINCOLN. 1951. "The versatile taxonomist," *Brittonia* **7**:225–231.

CONSTANCE, LINCOLN. 1964. "Systematic botany—an unending synthesis," *Taxon* **13**:257–273.

COPELAND, H. F. 1950. "Embryology," in A. Gunderson, *Families of Dicotyledons,* pp. 18–19. Ronald Press, Waltham, Mass.

FERRIS, C. F. 1923. "The place of the systematist in modern biology," *Sci. Monthly* **16**:514–520.

FERRIS, G. F. 1925. "The content of systematic biology," *Sci. Monthly* **20**:653–658.

GATES, R. R. 1951. "The taxonomic units in relation to cytogenetics and gene ecology," *Am. Nat.* **85**:31–50.

GILMOUR, J. S. L., and W. B. TURRILL. 1941. "The aim and scope of taxonomy," *Chron. Bot.* **6**:217–219.

GLEASON, H. A. 1952. "Some fundamental concepts in taxonomy," *Phytologia* **4**:1–20.

GOLDSCHMIDT, R. B. 1952. "Evolution, as viewed by one geneticist," *Am. Scientist* **40**:84–98.

JUST, THEODOR. 1950. "Carpels and ovules," in A. Gunderson, *Families of Dicotyledons,* pp. 12–17. Ronald Press, Waltham, Mass.

MASON, H. L. 1950. "Taxonomy, systematic botany, and biosystematics," *Madroño* **10**:193–208.

MERRILL, E. D. 1943. "Some economic aspects of taxonomy," *Torreya* **43**:50–64.

ROLLINS, R. C. 1957. "Taxonomy of the higher plants," *Am. J. Bot.* **44**:188–196.

SHARP, A. J. 1964. "The compleat botanist," *Science* **146**:745–748.

TIPPO, OSWALD. 1950. "Wood anatomy," in A. Gunderson, *Families of Dicotyledons,* pp. 6–11. Ronald Press, Waltham, Mass.

CHAPTER 2

Historical Summary

The beginnings of taxonomy undoubtedly antedate recorded history. It has always been one of man's characteristics that he likes to have names for things and to arrange things in a somewhat orderly manner. To say that taxonomy started with Aristotle, Dioscorides, or Pliny, or with anyone else now known, is erroneous. The early cultures of China, Egypt, and Assyria were based, to a certain degree, on cultivated plants, and there is evidence that such plants were studied and described by the scholars of those times. Much interest was also shown in plants reputed to have medicinal value. The pre-Columbian Aztec culture of Central America was likewise based on cultivated plants, and to this culture, and probably to people with a knowledge of plant breeding, we may owe the origin of certain modern crop plants, such as Maize. Even among uncivilized races of the present day we find evidence that a sort of taxonomy of plants is practiced, for such people have what are equivalent to concepts of species, genera, and families.

The attempts to classify all known plants have gradually shifted from purely *artificial systems* to *natural,* or *phylogenetic, systems.* An artificial system of classification is based on mere ease of identification; it uses convenient and readily observed characters of plants regardless of their evolutionary or genetic significance—as, for example, the grouping into herbs, undershrubs, shrubs, and trees (see under Theophrastus below) or the sexual system of Linnaeus, which was based on the number of stamens and styles. A natural, or phylogenetic, system, on the other hand, attempts to classify plants according to their evolutionary sequence and relationships inferred from indirect evidence or proved by genetic experimentation. The gradual growth of botanical terminology and a better understanding of the structure and function of reproductive parts resulted in greater precision in the definition of the various categories, or taxa, such as the species and its subdivisions, the genus, the family, and the order. Though great advances are being made in these fields of study at the present time, we are still a long way from achieving the ultimate goal of a completely phylogenetic system of classification and a complete understanding of all the parts of plants.

The Ancient Greeks and Romans

Among the first persons to leave us a written record of attempts at classification of the plants known to them were the early philosophers and medical practitioners of the Greek and Roman civilizations.

Theophrastus (*c.* 370–287 B.C.), an outstanding Greek naturalist, was a pupil of Plato and Aristotle. In his *History of Plants,* the oldest botanical work in existence, he described some five hundred species of plants, mostly cultivated. These he classified into four groups, herbs, undershrubs, shrubs, and trees, considering trees to be the most highly developed of all. He also laid the foundations for the study of floral morphology.

Pliny the Elder (23–79 A.D.), a Roman naturalist and scholar, mentioned nearly a thousand plants in his *Historia Naturalis.* This series of thirty-seven books, sixteen of which dealt largely with plants, treated such topics as medicinal properties, timber trees, plant anatomy, and the practice of horticulture. Pliny read, or had read to him, most of the available literature of his time. His publications were therefore encyclopedic in scope; though they perpetuated many of the errors of his predecessors, they also included much factual information. His must be considered one of the most significant contributions to early botanical knowledge. It is unfortunate that his studies were cut short by his death during an eruption of Mount Vesuvius.

Dioscorides (first century A.D.) was a military physician under Emperor Nero of Rome. His chief contribution, *Materia Medica,* was a description of about six hundred species, mainly Mediterranean, used for medicinal purposes. Another manuscript, the *Anicia Juliana Codex,* was prepared for the daughter of a Byzantine emperor about 512 A.D. from material originally compiled by Dioscorides. It contained colored illustrations of plants and is still in existence.

The Herbalists

During the Middle Ages, following the decline of the Greek and Roman civilizations, little significant botanical progress was made. The early herbals, such as the *Codex* of Dioscorides, were copied and recopied for centuries with few additions or improvements. In the first half of the sixteenth century, however, a botanical renaissance developed, and it was greatly stimulated by the still young art of printing. Woodcuts were employed in printing illustrations of plants, and these often show a remarkable degree of fidelity, originality, and skill in execution. The herbals of **Brunfels, Bock, Fuchs** (see Fig. 1), **Cordus,** and others, men sometimes referred to as the "German fathers of botany," are representative of this period. A significant contribution to taxonomy was made at this time by **Kaspar Bauhin**, a pupil of Fuchs. In Bauhin's *Prodromus Theatri Botanici* (1620) and *Pinax Theatri Botanici* (1623) we find one of the first attempts to utilize a *binomial system of nomenclature*—which, however, Bauhin did not use exclusively.

Handed down from the ancients, and elaborated on by the herbalists, was the *doctrine of signatures,* which was based upon the belief that plants or plant parts that resembled portions of the human body must have been so created for the purpose of furnishing remedies for the ailments of those portions. Many plants were given common names that referred to supposed remedial properties, and many scientific names that are still in use today trace their origin to this doctrine—for example, the generic name *Hapatica,* the shape of the leaves in that genus being thought to resemble the shape of the liver and the leaves therefore to be a remedy for diseases of that organ.

The Transition Period

The transition from the Renaissance to the modern period produced many notable workers and much literature, which can be only hinted at in this brief discussion. Botanists gradually broke away from the traditional doctrines of the ancients and developed new systems of classification, a new terminology of description, and a system of nomenclature that was to become a permanent part of taxonomy. Since the concept of evolution had not yet been developed, the arrangement of plant groups in the various systems of classification was still more artificial than phylogenetic. We can, however, sense a groping toward a more natural arrangement of plant groups.

The geographical explorations of this period brought a great influx

1 Page from the herbal *De historia stirpium* of Leonhard Fuchs, published in 1542. The illustration is of *Zea mays* L., or what Fuchs called *Turcicum Frumentum* or "Türckisch korn," stating that it was brought into Germany from Asia by the Turks. Actually maize is of American origin and was unknown in the Old World until after the time of Columbus. The original size of the plate was 8 by 13 inches. [*Ann. Mo. Bot. Gard.* **35**:158 (1948); used by permission.].

of new species and genera, which soon became burdensome to the struggling taxonomist.

Andrea Cesalpino (1519–1603), an Italian physician, became director of the botanical garden at Bologna and later professor of botany. His herbarium, assembled in 1563 and still preserved in the Museum of Natural History at Florence, contains 768 dried and mounted plants with Latin or Italian names. Cesalpino's contribution to literature was *De Plantis,* consisting of sixteen books: the first a general exposition of the whole of theoretical botany, the others concerned with descriptions, without synonymy or illustration, of some 1,520 species of plants. He recognized a system of phyllotaxy in plants. His classification was based on the ancient grouping into herbs and trees, but within these groups he recognized the significance of fruit and seed characters. He had a good concept of genera, and he thereby exerted considerable influence on later botanists, such as Tournefort and Linnaeus. A copy of *De Plantis* once owned and annotated by Linnaeus is preserved in the library of the Linnaean Society. Cesalpino is commemorated by the name *Caesalpinia,* a genus of handsome legumes including the Poinciana Tree.

Joachim Jung (1587–1657), trained in both mathematics and medicine, and a brilliant teacher in Germany, is sometimes referred to as "the first terminologist." He left us no publications of his own writing, but two of his pupils left a record of his teachings that clearly reveals his concepts. He succeeded in clearing up some of the confusion that had existed between homologous and analogous structures. He believed that the old classification into trees, shrubs, and herbs was not fundamental.

Among the terms that he clearly defined for the first time are the following: *nodes* and *internodes;* the leaf consisting of *blade* and *petiole,* the blade being *simple* or *compound,* and compound leaves *pinnate* or *digitate; perianth* for the calyx instead of for the combined calyx and corolla, as now; *stamens* and *styles* as understood today, although Jung was not aware of their functions; *capitulum,* or head of flowers of the *Asteraceae,* as composed of *disk* and *ray florets.*

John Ray (1627–1705), the son of a British blacksmith, was a graduate of Trinity College. In his *Historia Plantarum* (1686–1704) we find some of the first indications of a natural system of classification. This work was an odd mixture of the old and the new. The old grouping into herbs and trees was retained as a primary significance, but within these groups we find a division into *dicotyledons* and *monocotyledons,* which represented a notable advance in thinking. His classification of the major groups was as follows:

I. Herbae (herbs)
  A. Imperfectae (essentially the cryptogams)
  B. Perfectae (flowering plants)
    Dicotyledones
    Monocotyledones
II. Arborae (trees and shrubs)
  A. Monocotyledones
  B. Dicotyledones

Pierre Magnol (1638–1715) was a French contemporary of John Ray. Finding Ray's system too difficult, he divided plants into what he called *families*, listing some seventy-six of these in his *Prodromus historiae generalis, in qua familiae per tabulas disponuntur* (1689), and construing them as incorporating certain overall striking characteristic of roots, stems, flowers, and seeds. It is to Magnol, then, that we owe the first concept of modern families, although not all of the families he listed were groups original with him. His name is commemorated by the generic name *Magnolia.*

Joseph Pitton de Tournefort (1656–1708), a pupil of Magnol, was professor of botany at the Jardin du Roi under Louis XIV. Before his time most botanical descriptions were confined to the level of the species, and genera were merely listed. From Magnol Tournefort must have acquired a concept of natural groups, for he reversed this procedure and for the first time gave a *characterization of genera,* merely listing the species under each. He recognized petaliferous and apetalous flowers, corollas with separate and with united petals, and regular and irregular corollas; but he refused to recognize sexuality in plants, and he also retained the ancient basic classification into trees and herbs. Among his contributions to the literature are *Éléments de botanique* (1694), illustrated with 450 copper plates and including 698 genera and 10,146 species, which he enlarged and published in Latin as *Institutiones Rei Herbariae* (1700), and a *Flora of the Environs of Paris* (1698).

Tournefort traveled widely in Europe, and in 1700–1703 he journeyed to Asia Minor, where he succeeded in climbing Mount Ararat. He died of injuries sustained when he was struck in the chest by the tongue of a carriage while crossing the street near the garden where he worked.

Rudolf Camerarius (1665–1721), though not strictly a taxonomist, influenced later taxonomists. He was made professor extraordinary and director of the botanical garden at Tübingen, Germany, in 1688. He experimented with certain plants in which the stamens and pistils were produced in separate flowers, reporting his findings in letters to scientists in other universities. One such letter, entitled *De sexu plantarum epistola,* dated August 25, 1694, and addressed to a professor at Giessen, described his experiments and the failure of the pistillate flowers to set seed in the absence of the staminate flowers. He concluded that the stamens were the male sex organs, that pollen was necessary for seed formation, and that the style and ovary constituted the female parts of the flower. Camerarius thus *established the fact of sexuality in flowering plants.*

Carolus Linnaeus, latinized form of Carl Linné (1707–1778), great Swedish naturalist and classifier, was born at Råshult, Sweden, where his father was a clergyman. Young Linnaeus started out to become a churchman also, but he gave that up at an early age to study natural history under the guidance of a local physician, who befriended him and introduced him to the writings of Tournefort. He attended the university at Uppsala, where he encountered the literature that first suggested to him the artificial *sexual system of classification,* which he later elaborated. His obvious interest in the university garden endeared him to Olof Rudbeck, an elderly professor, who helped him financially and used him as an assistant.

Under the sponsorship of the Academy of Sciences of Uppsala, and at a cost of about $125, Linnaeus undertook in 1732 a botanical exploration of Lapland. He covered about 4,800 miles in five months, and returned with 537 specimens, among them one he named *Campanula serpyllifolia,* which turned out to be not a *Campanula* at all and was renamed *Linnaea borealis* by the Dutch botanist Gronovius in honor of Linnaeus. The results of the Lapland expedition were later published as the *Flora Lapponica* (1737).

Linnaeus, having been urged to study medicine abroad, traveled in 1735 to the Netherlands, where after some time he received his degree in medicine at Harderwijk. While in the Netherlands he met the two eminent Dutch botanists J. F. Gronovius and Hermann Boerhaave, and it was Gronovius who financed the publication of his *Systema Naturae* (1735), a tabular outline consisting of eight folio sheets. He continued his stay in the Netherlands, with short visits to England and France, until 1742, when, at the age of thirty-five, and because of the death of his former teacher, Rudbeck, he returned to Uppsala to fill the chair of medicine and later that of botany.

As a tribute to his success as a teacher, the enrollment at Uppsala soon rose from 500 to 1,500; and through his industry the number of species of plants cultivated in the botanical garden was augmented by more than a thousand. Among his students were such notables as Peter

Forskål, Fredrik Hasselquist, Peter Kalm, Carl Thunberg, and Daniel Solander, all explorers of renown who brought back plants from distant lands.

With the publication of his *Species Plantarum* (1753) we have the starting point for the system of priority that is used in our present-day nomenclature of the higher plants (see Figs. 2 and 3). Some 7,300 species were described therein, synonyms and countries of origin were included, and the groups were arranged according to the *sexual system* proposed earlier in the *Systema Naturae.* Here, for the first time, we find a consistent use of the binomial system of nomenclature, and this was one of Linnaeus' greatest contributions even though he was not the first to suggest its use. Linnaeus accepted many of Tournefort's genera, citing his descriptions and figures.

The system of classification divided all plants into twenty-four classes as follows:

1. *Monandria*—stamen 1.
2. *Diandria*—stamens 2.
3. *Triandria*—stamens 3.
4. *Tetrandria*—stamens 4.
5. *Pentandria*—stamens 5.
6. *Hexandria*—stamens 6.
7. *Heptandria*—stamens 7.
8. *Octandria*—stamens 8.
9. *Enneandria*—stamens 9.
10. *Decandria*—stamens 10.
11. *Dodecandria*—stamens 12.
12. *Icosandria*—stamens more than 12, attached to the calyx.
13. *Polyandria*—stamens more than 12, attached to the receptacle.
14. *Didynamia*—stamens didynamous (2 long, 2 short).
15. *Tetradynamia*—stamens tetradynamous (4 long, 2 short).
16. *Monadelphia*—stamens monadelphous (in 1 bundle).
17. *Diadelphia*—stamens diadelphous (in 2 bundles).
18. *Polyadelphia*—stamens polyadelphous (in several bundles).
19. *Syngenesia*—stamens syngenesious (with united anthers).
20. *Gynandria*—stamens gynandrous (adnate to the pistil).
21. *Monoecia*—plant monoecious (unisexual flowers on the same plant).
22. *Dioecia*—plant dioecious (unisexual flowers on separate plants).
23. *Polygamia*—plants polygamous.
24. *Cryptogamia*—flowers concealed (the cryptogams).

These primary divisions, or classes, were subdivided by Linnaeus into orders, which were usually based on the number of styles of the

CAROLI LINNÆI

Sıæ Rigiæ Mitis Sveciæ Archiatri; Medic. & Botan.
Profess. Upsal. Equitis aur. de Stella Polari,
nec non Acad. Imper. Monspel. Berol. Tolos.
Upsal. Stockh. Soc. & Paris. Coresp.

SPECIES PLANTARUM.

EXHIBENTES

PLANTAS RITE COGNITAS.

AD

GENERA RELATAS.

CUM

Differentiis Specificis,
Nominibus Trivialibus,
Synonymis Selectis,
Locis Natalibus,

Secundum

SYSTEMA SEXUALE

DIGESTAS.

Tomus I.

Cum Privilegio S. R. M:tis Sveciæ & S. R. M:tis Poloniæ ac Electoris Saxon.

HOLMIÆ,
Impensis LAURENTII SALVII.
1753.

2 Title page of Linnaeus' *Species Plantarum.*

DECANDRIA MONOGYNIA. 373

Classis X.

DECANDRIA

MONOGYNIA.

SOPHORA.

1. SOPHORA foliis pinnatis. foliolis numerosis villosis oblongis. *alopecuroides.*
   Sophora. *Hort. cliff.* 156.
   Ervum orientale alopecuroides perenne, fructu longissimo. *Tournef. cor.* 27. *Dill. elth.* 136.
   Glycyrrhiza siliquis nodosis quasi articulatis. *Buxb. cent.* 3. *p.* 25. *t.* 46
   Habitat in Oriente. ♃

2. SOPHORA foliis pinnatis: foliolis numerosis subrotundis. *tomentosa*
   Sophora tomentosa, foliolis subrotundis *Fl. zeyl.* 163. *
   Indigophora foliis tomentosis. *Hort. cliff.* 487.
   Colutea zeylanica argentea tota. *Herm. lugdb.* 169. *t.* 171. *Raj. hist.* 1720.
   Habitat in Zeylona. ♄

3. SOPHORA foliis pinnatis, foliolis septenis glabris. *heptaphylla.*
   Sophora glabra, foliolis septenis. *Fl. zeyl.* 104.
   Fruticulus sinensis, sennæ sylvestris folio angustiore, nodosa siliqua rostro longiore donata *Pluk. amalth.* 18. *t.* 451. *f.* 10.?
   Habitat in India. ♄.

4. SOPHORA foliis ternatis sessilibus: foliolis linearibus. *Genistoides.*
   Genistra africana, foliis galii. *Old. afr.* 31.
   Habitat ad Cap. b. Spei.

5. SOPHORA foliis ternatis subsessilibus: foliolis subrotundis glabris. *tinctoria.*
   Cytisus foliis fere sessilibus, calycibus bractea triplici auctis. *Gron. virg.* 82.
   Cytisus procumbens americanus, flore luteo, ramosissimus, qui Anil suppeditat. *Pluk. alm.* 129. *t.* 86. *f.* 2 *Ehret. t.* 1. *f.* 3.
   Habitat in Barbados, Virginia.

A a 3 6. SO-

3 Sample page from Linnaeus' *Species Plantarum.* This is the beginning of the treatment of the legumes.

pistil, the ordinal names being *Monogynia, Digynia, Trigynia,* etc. The tremendous increase, during this period, in the number of known plants rendered such a simple classification very popular with the botanists who were called upon to deal with all the new material; but Linnaeus himself realized the limitations of the system and knew that it was more convenient than natural. He suggested that an increased knowledge of relationships among plants would some day permit a more natural grouping.

In 1753 Linnaeus was made a Knight of the Polar Star, the first scientist in Sweden to be knighted. In 1761 he was granted a patent of nobility, and from this date he became known as Carl von Linné.

His personal herbarium and library were sold by his widow, after his death, to Dr. James Edward Smith, the founder and first president of the Linnaean Society of London, for the sum of £1,000. After Smith's death they were purchased by subscription among the Fellows of the Society for presentation to the Linnaean Society, where they are preserved today. The Society has recently made available a set of microfiche reproductions of the more than 14,000 specimens in the Linnaean Herbarium.

The Modern Period

There is no sharp line of demarcation between the transition period, marked by various attempts at classification, all of which were more or less artificial, and the modern period, which progressed steadily in the development of a system based on natural affinities. Progress is still being made in this direction, the ultimate and as yet distant goal being an arrangement that will reflect the true phylogenetic relationships of the whole plant kingdom.

Bernard de Jussieu (1699–1777), a French contemporary of Linnaeus, attempting to lay out the Royal Gardens at Versailles according to a natural system and finding that the Linnaean system was unsatisfactory, proceeded gradually to modify it into a more natural arrangement. Never being completely satisfied with his groupings, he did not publish his results; but his nephew, **Antoine Laurent de Jussieu** (1748–1836), published his uncle's plan along with his own in *Genera plantarum secundum ordines naturales disposita* (1789). This publication is the starting point for the list of conserved family names for flowering plants in the *International Code of Botanical Nomenclature.*

This first attempt at a natural classification included a hundred orders (now mostly recognized as families), all described and classified in fifteen classes as follows:

1. *Acotyledones.*
2. *Monocotyledones*—stamens hypogynous.
3. *Monocotyledones*—stamens perigynous.
4. *Monocotyledones*—stamens epigynous.
5. *Dicotyledones, Apetalae*—stamens epigynous.
6. *Dicotyledones, Apetalae*—stamens perigynous.
7. *Dicotyledones, Apetalae*—stamens hypogynous.
8. *Dicotyledones, Monopetalae*—corolla hypogynous.
9. *Dicotyledones, Monopetalae*—corolla perigynous.
10. *Dicotyledones, Monopetalae*—corolla epigynous, anthers united.
11. *Dicotyledones, Monopetalae*—corolla epigynous, anthers distinct.
12. *Dicotyledones, Polypetalae*—stamens epigynous.
13. *Dicotyledones, Polypetalae*—stamens hypogynous.
14. *Dicotyledones, Polypetalae*—stamens perigynous.
15. *Diclines irregulares.*

According to this system, the *Acotyledones* included mainly the cryptogams, as we know them today, but also included some aquatic flowering plants whose reproduction was not understood at that time. The *Monocotyledones* and *Dicotyledones,* as defined, represented essentially the modern concepts of those groups. (The *Apetalae,* however, were construed as related groups; they are now considered to represent widely divergent lines of evolution and to be reduced forms of the *Polypetalae,* and they are still something of a puzzle to taxonomists.) The term "perigynous," as used by de Jussieu, is likewise at variance with our interpretation of that term; and his *Diclines irregulares* were an odd assortment of gymnosperms, *Amentiferae,* Nettles, and Euphorbias.

Augustin Pyrame de Candolle (1778–1841), the senior member of a notable French-Swiss family, following in the footsteps of de Jussieu, further developed the morphological approach to classification and recognized the significance of vestigial organs. His views were expressed in his *Théorie élémentaire de la botanique* (1813), in which 135 orders (or families as now treated) were delimited, that number being raised to 213 in a revision edited by his son **Alphonse** in 1844. The de Candolles, father and son, and later the grandson **Casimir,** collaborated on the preparation of the monumental *Prodromus systematis naturalis regni vegetabilis* (1824–1873). This series of volumes included nearly all the dicotyledons and probably would have included other groups had the complexity of the problem and the number of known species not increased to such an extent that the project as originally conceived got somewhat out of hand. A later series of monographs resulted in the revision of a number of families. The

Candollean herbarium and library are now located at Geneva, Switzerland, where the members of the family did much of their work.

Robert Brown (1773–1858), a Scottish botanist, broke with the British tradition of following the Linnaean system and followed that of the de Candolles. He published no system of his own, but through his studies of seeds he *proved conclusively that the gymnosperms were a discrete group with naked ovules and seeds.* He established several families, among them the *Asclepiadaceae* and *Santalaceae;* and he proposed that the orders then recognized (families as we know them) should be grouped into larger categories (which would be equivalent to the modern orders).

Adolphe Théodore Brongniart (1801–1876), a French botanist, proposed a system of classification in which the *Apetalae* were treated as reduced members of the *Polypetalae.* This represents a distinctly modern trend, which was to be taken up later by Bessey.

George Bentham (1800–1884) and **Sir Joseph Hooker** (1817–1911), British botanists associated with the Royal Botanic Gardens at Kew, contributed an outstanding system of classification in their *Genera Plantarum* (1862–1883), which described some 202 orders (families as now understood) grouped into cohorts (orders as now understood). A combination of the systems of Bentham and Hooker and of the de Candolles was adopted by **Asa Gray** (1810–1888), Harvard botanist, during the formative period of American taxonomy.

The Bentham-Hooker system may be summarized as follows:

Dicotyledones

Polypetalae. Petals separate.

Series 1. *Thalamiflorae.* Petals and stamens hypogynous. (Includes 6 cohorts, and families now known as the *Ranunculaceae, Brassicaceae, Caryophyllaceae,* and *Malvaceae.*)

Series 2. *Disciflorae.* Stamens usually definite, inserted on or near a disk surrounding the base of the ovary. Ovary superior. (Includes 4 cohorts, and families now known as the *Geraniaceae, Rutaceae, Rhamnaceae, Sapindaceae,* and *Anacardiaceae.*)

Series 3. *Calyciflorae.* Petals and stamens perigynous, and the gynoecium often enclosed by the development of the floral axis, the ovary sometimes inferior. (Includes 5 cohorts, and families such as the *Rosaceae, Leguminosae, Onagraceae, Loasaceae, Cactaceae,* and *Apiaceae* or *Umbelliferae.*)

Gamopetalae. Petals united.

Series 1. *Inferae.* Ovary inferior, stamens the same number as the lobes of the corolla or fewer. (Includes 3 cohorts, and families such as the *Caprifoliaceae, Rubiaceae, Campanulaceae,* and *Asteraceae* or *Compositae.*)

Series 2. *Heteromerae.* Ovary usually superior; stamens as many as the lobes of the corolla, or more, on the corolla tube or free; carpels more than 2. (Includes 3 cohorts, and families such as the *Ericaceae* and *Primulaceae.*)

Series 3. *Bicarpellatae.* Ovary usually superior; stamens the same number as the lobes of the corolla or fewer, on the corolla tube; carpels usually 2. (Includes 4 cohorts, and families such as the *Gentianaceae, Asclepiadaceae, Polemoniaceae, Boraginaceae, Solanaceae, Scrophulariaceae,* and *Menthaceae* or *Labiatae.*)

Monochlamydeae. Petals lacking. (No subdivision into cohorts.)

Series 1. *Curvembryeae.* Embryo curved round the endosperm; ovules usually single; flowers usually perfect; stamens the same number as the sepals or fewer. (Includes *Chenopodiaceae, Polygonaceae,* and *Nyctaginaceae.*)

Series 2. *Multiovulatae aquaticae.* Aquatics with syncarpous ovary and numerous ovules. (Includes *Podostemaceae.*)

Series 3. *Multiovulatae terrestres.* Terrestrial herbs or shrubs with syncarpous ovary and numerous ovules. (Includes *Nepenthaceae, Cytinaceae,* and *Aristolochiaceae.*)

Series 4. *Micrembryeae* Embryo very small in copious endosperm. Ovary apocarpous or syncarpous, ovules usually single. (Includes *Piperaceae, Chloranthaceae, Myristicaceae,* and *Monimiaceae.*)

Series 5. *Daphanales.* Ovary usually with 1 carpel, ovules single or paired; perianth sepaloid, in 1 or 2 series; woody plants; flowers perfect. (Includes *Lauraceae, Elaeagnaceae,* etc.)

Series 6. *Achlamydosporeae.* Ovary 1-celled, 1–3-ovuled; seeds without a seed coat; perianth sepaloid or petaloid; plants often parasitic. (Includes *Santalaceae, Loranthaceae,* etc.)

Series 7. *Unisexuales.* Flowers unisexual; ovary syncarpous or of 1 carpel; ovules solitary or in pairs; perianth sometimes lacking. (Includes *Euphorbiaceae, Urticaceae, Juglandaceae,* etc.)

Series 8. *Ordines anomali.* Flowers often unisexual; not closely related to other orders. (Includes *Salicaceae,* etc.)

Gymnospermae. (Includes *Gnetaceae, Pinaceae,* etc., and *Cycadaceae.*)

Monocotyledones

Series 1. *Microspermae.* At least the inner perianth petaloid; ovary inferior, 1-celled, with 3 parietal placentae; seeds very small, numerous, and without endosperm. (Includes *Orchidaceae, Burmanniaceae,* etc.)

Series 2. *Epigynae.* At least the inner perianth petaloid; ovary usually inferior; seeds with endosperm. (Includes *Bromeliaceae, Iridaceae, Amaryllidaceae,* etc.)

Series 3. *Coronarieae.* At least the inner perianth petaloid; ovary superior; endosperm present. (Includes *Liliaceae, Xyridaceae, Commelinaceae,* etc.)

Series 4. *Calycinae.* Perianth small, sepaloid, stiff or herbaceous; ovary superior; endosperm present. (Includes *Juncaceae, Arecaceae,* etc.)

Series 5. *Nudiflorae.* Perianth lacking or reduced to scales or bristles; ovary superior; carpels united or single, with from 1 to many ovules; endosperm usually present. (Includes *Typhaceae, Araceae, Lemnaceae,* etc.)

Series 6. *Apocarpeae.* Perianth in 1 or 2 series, or lacking; ovary superior; carpels 1 or else free; endosperm lacking. (Includes *Alismaceae, Najadaceae,* etc.)

Series 7. *Glumaceae.* Flowers in heads or spikelets, subtended by usually imbricated bracts; perianth small, scale-like, or chaffy, or lacking; ovary with 1 ovule in each cell or 1-celled and 1-seeded; endosperm present. (Includes *Poaceae* or *Gramineae, Cyperaceae,* etc.)

During this period in the development of botanical science, much of the confusion in the various systems of classification was due to a lack of understanding of the method of reproduction in the lower plants. **Wilhelm Hofmeister** (1824–1877), a German botanist and music dealer of Leipzig, clearly established the fact of *alternation of generations* and demonstrated the essential similarities running through various plant groups. His research focused attention on *resemblances,* whereas earlier workers had emphasized arbitrary *differences.* Hofmeister's findings, coupled with the revolutionary *Origin of Species* (1859) by **Charles Darwin** (1809–1882), stimulated much thought and many publications and controversies, the result of which was a better understanding of relationships. Gradually there emerged a concept of continuity in the plant kingdom, from the lowest forms to the highest, with numerous gaps in the record remaining to be filled. Darwin's theory of evolution was based on three principles: variation and overproduction, a resulting competition or struggle for existence, and the survival of the fittest.

It is worth mentioning in passing that Darwin, realizing the difficulties and complexities involved in nomenclature, provided in his will for the cost of compiling an index to the names, authorities, and countries of origin of all known flowering plants. This work, carried out under the direction of Joseph Hooker, produced the famous and useful *Index Kewensis,* which has been kept up to date by supplements to the present time.

August Wilhelm Eichler (1839–1887), professor of botany at the University of Kiel, Germany, modified previous systems to place the gymnosperms in their proper sequence as follows:

Cryptogamae
    Thallophyta
    Bryophyta
    Pteridophyta
Phanaerogamae
    Gymnospermae
    Angiospermae
        Monocotyleae
        Dicotyleae
            Choripetalae (petals separate or none)
            Sympetalae (petals united)

Adolf Engler (1844–1930), professor of bontany at the University of Berlin, adopted the main features of the Eichler classification, and in collaboration with **Karl Prantl** (1849–1893) and others published *Die natürlichen Pflanzenfamilien,* a comprehensive world treatment of the plant kingdom, the first edition being completed in 1909 and some revisions being added more recently. Herein were given details concerning the morphology, anatomy, and economic aspects of the various families, profuse illustrations, and keys to the genera, which were dealt with in synoptical fashion. Engler's system is still followed in the arrangement of families of flowering plants in nearly all manuals and floras and in most herbaria. For details of the system the student is referred to Engler and Diels' *Syllabus der Pflanzenfamilien,* eleventh edition (Berlin, 1936). The significant feature of the system is that it places the monocotyledons ahead of the dicotyledons, and the apetalous and catkin-bearing dicotyledons ahead of the others, indicating that the authors considered these last to be primitive rather than derived groups.

Richard von Wettstein (1862–1931), an Austrian systematist, produced a classification that was similar in most respects to that of Engler. He also considered the unisexual, naked flower to be primitive and derived perfect flowers with a perianth from such a source. But he took into account more background information than Engler did, and this resulted in some realignment and a more nearly phylogenetic arrangement of groups, for he considered the dicotyledons to be primitive and the monocotyledons to be derived from them through the *Ranales.*

Charles Edwin Bessey (1845–1915), a student of Asa Gray and thereafter professor of botany at the University of Nebraska, introduced some new ideas concerning primitive and advanced characters in plants. After trial flights into the field of classification in 1894 and 1897, at which time he followed rather closely the Bentham-Hooker system, and again in 1907 and 1909, he finally evolved what is now known as the Besseyan System under the title "The Phylogenetic Taxonomy of Flowering Plants" (*Ann. Mo. Bot. Gard.* **2**:108–164, 1915). Portions of this are given below through the courtesy of the Missouri Botanical Garden.

Bessey's system was based on a series of "dicta," or statements of the guiding principles he used in determining the degree of primitiveness or evolutionary advancement of a plant group. According to these principles he constructed a chart (Fig. 4) showing the relationships as well as the approximate sizes of the relatively few orders of flowering plants he recognized. He took the *Ranales* (or Buttercup order) as the basal group—a feature of the Bentham-Hooker system—from which both monocotyledons and other groups of dicotyledons have evolved. He also believed that all flowering plants originated from strobiliferous cycad ancestors, probably from the *Bennettitales.*

Following are Bessey's dicta:

A. GENERAL DICTA

1. Evolution is not always upward, but often it involves degradation and degeneration.
2. In general, homogeneous structures (with many and similar parts) are lower, and heterogeneous structures (with fewer and dissimilar parts) are higher.
3. Evolution does not necessarily involve all organs of the plant equally in any particular period, and one organ may be advancing while another is retrograding.
4. Upward development is sometimes through an increase in complexity, and sometimes by a simplification of an organ or a set of organs.
5. Evolution has generally been consistent, and when a particular progression or retrogression has set in, it is persisted in to the end of the phylum.
6. In any phylum the holophytic (chlorophyll-green) plants precede the colorless (hysterophytic) plants, and the latter are derived from the former.
7. Plant relationships are *up and down* the genetic lines, and must constitute the framework of phylogenetic taxonomy.

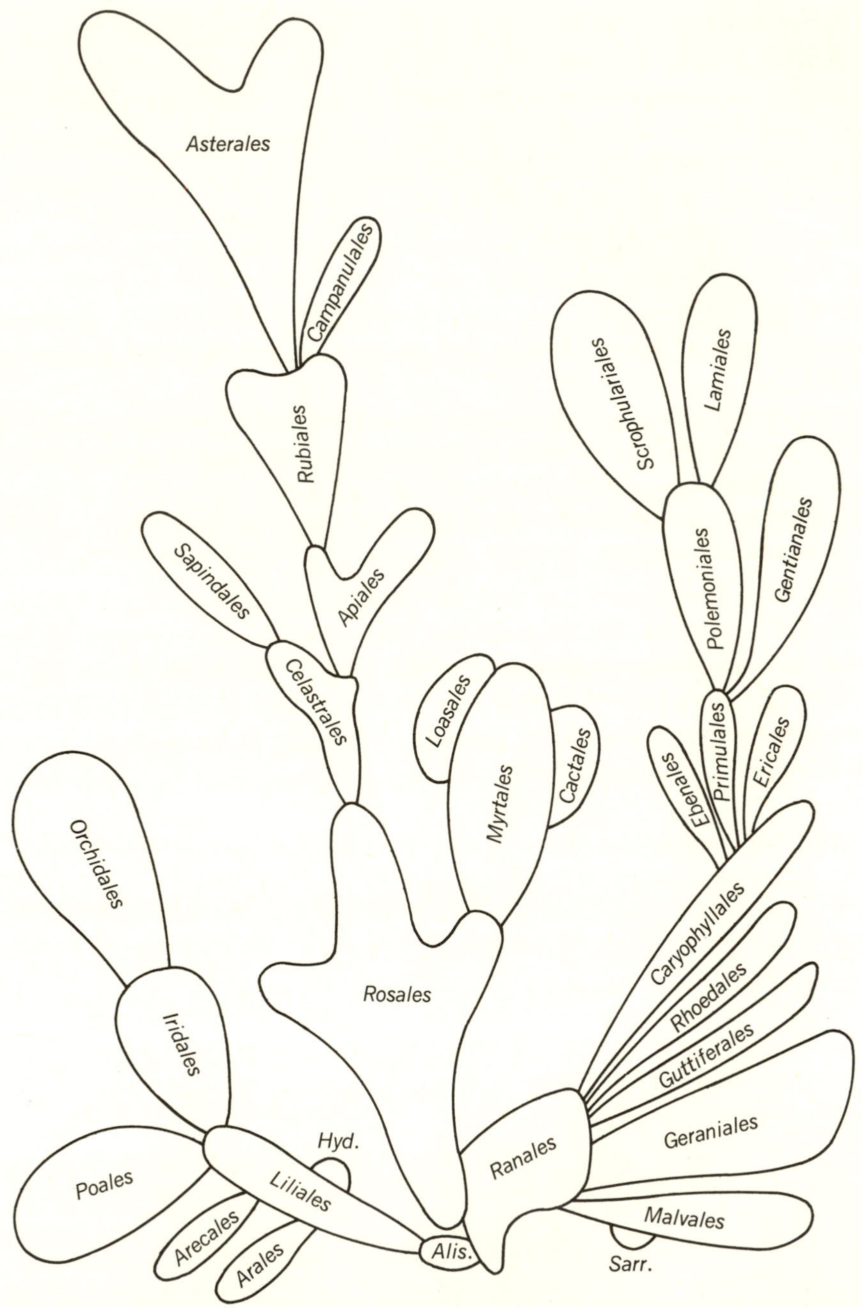

4 Bessey's chart showing relationships of orders he recognized. The areas are approximately proportional to the number of species in the orders. To the left are the monocotyledons, in the center are perigynous and epigynous dicotyledons, and to the right are hypogynous dicotyledons. [*Ann. Mo. Bot. Gard.* **2**:118 (1915); used by permission, and slightly modified.]

B. DICTA HAVING SPECIAL REFERENCE TO THE GENERAL STRUCTURE OF THE FLOWERING PLANTS

8. The stem structure with collateral vascular bundles arranged in a cylinder is more primitive than that with scattered bundles, and the latter are to be regarded as derived from the former.
9. Woody stems (as of trees) are more primitive than herbaceous stems, and herbs are held to have been derived from trees.
10. The simple, unbranched stem is an earlier type, from which branching stems have been derived.
11. Historically the arrangement of leaves in pairs on the stem is held to have preceded the spiral arrangement in which the leaves are solitary at the nodes.
12. Historically simple leaves preceded branched ("compound") leaves.
13. Historically leaves were first persistent ("evergreen") and later deciduous.
14. The reticulated venation of leaves is the normal structure, and the parallel venation of some leaves is a special modification derived from it.

C. DICTA HAVING REFERENCE TO THE FLOWERS OF FLOWERING PLANTS

15. The polymerous flower structure precedes, and the oligomerous structure follows from it, and this is accompanied by a progressive sterilization of sporophylls.
16. Petaly is the normal perianth structure, and apetaly is the result of perianth reduction (aphanisis).
17. The apochlamydeous perianth is earlier and the gamochlamydeous perianth is derived from it by symphysis of the members of perianth whorls.
18. Actinomorphy is an earlier structure than zygomorphy, and the latter results from a change from similar to a dissimilar growth of the members of the perianth whorls.
19. Hypogyny is the more primitive structure, and from it epigyny was derived later.
20. Apocarpy is the primitive structure, and from it syncarpy was derived later.
21. Polycarpy is the earlier condition, and oligocarpy was derived from it later.
22. The endospermous seed is primitive and lower, while the seed without endosperm is derived and higher.
23. Consequently, the seed with a small embryo (in endosperm) is more primitive than the seed with a large embryo (in scanty or no endosperm).

24. In earlier (primitive) flowers there are many stamens (polystemonous) while in later flowers there are fewer stamens (oligostemonous).
25. The stamens of primitive flowers are separate (apostemonous), while those of derived flowers are often united (synstemonous).
26. The condition of powdery pollen is more primitive than that with coherent or massed pollen.
27. Flowers with both stamens and carpels (monoclinous) precede those in which these occur on separate flowers (diclinous).
28. In diclinous plants the monoecious condition is the earlier, and the dioecious later.

Hans Hallier (1868–1932) was a German systematist who attempted to develop a phylogenetic system. It is significant that, although the two developed independently, Hallier's system closely resembled that of Bessey. Hallier made much more use of ovule structure and position than other workers had.

John Hutchinson (born in 1884), a British botanist associated with the Royal Botanic Gardens at Kew, has proposed a system somewhat like that of Bessey in his *Families of Flowering Plants:* Vol. I, *Dicotyledons* (1926), and Vol. II, *Monocotyledons* (1934). Hutchinson, however, considering the woody versus the herbaceous habit as of fundamental importance, maintained that the predominantly woody orders stemmed from the Magnolias and their relatives in the *Ranales* as we know them, and that the chiefly herbaceous orders originated from the Buttercups and their relatives, these two lines developing independently but in a more or less parallel fashion. He also proposed many more than the generally accepted number of orders, with fewer families in each. In the following treatment, Hutchinson's scheme has been the basis for the arrangement of monocotyledons, and Bessey's scheme has been the basis for the arrangement of the dicotyledons.

Recent Developments

Much new information has come to light since the above attempts to produce a phylogenetic classification. This information is derived mainly from additional knowledge of certain plant groups found in the Pacific area, from comparative anatomical studies of various organs, particularly of flowers and embryos, and from cytological studies of many aberrant groups.

I. W. Bailey has pointed out that new genera, which are apparently relics of an ancient, woody, dicotyledonous flora, and most of which are represented by one or only a few species, are being discovered in the southwest Pacific area (Fiji, New Caledonia, New Guinea, northern Australia, and northward to China and India). Investigations centering on these plants may in time unravel the great mystery of the origin of angiosperms.

Arthur J. Eames and others have studied a number of families of flowering plants with particular attention to the comparative anatomy of the flowers and secondary wood. Such research has indicated both that a number of genera are misplaced in the traditional treatments of families and also that certain genera and species, despite various unusual characteristics, properly belong where they have been placed by conservative taxonomists. In the future it will probably be necessary to establish several additional families to take care of aberrant genera.

The detailed studies of comparative embryology made by **John R. Reeder** and others substantiate some elements of our traditional system of classification but also point out places where it is probably at fault.

Many cytogenetic studies of a wide range of plant groups have contributed to a better understanding of relationships among the higher plants. At first this research was aimed mainly at infraspecific problems, but as it broadened it began to take on significance in higher levels of classification, even at the family level.

Since evolution has resulted in a branching system of relationships, perhaps even in a reticulate one, any linear sequence, such as must be followed in a book, or any numbered sequence of families, such as that used in many herbaria, must of necessity be artificial. Furthermore, though a broad general pattern of evolution in the flowering plants has emerged from the various studies made of them through the years, many unsolved riddles still exist. As someone has said, "We stand on the wrong side of the tapestry—a confusion of colors, knots, and loose ends. But, be assured, on the other side there is the pattern." It has even been suggested that a science of "systematic physiology" should be developed as an aid in the formulation of phylogenetic systems. In the meantime the "family tree" type of arrangement serves a useful purpose in that it helps us visualize possible relationships—even though some eminent workers have ridiculed the diagrams as "deformed hatracks." Such diagrams, of course, are subject to the modifications that new information may suggest. And it should be remembered that an order or a family, having been clearly differentiated by the evolution of its members, is not likely to have given rise to other orders or families as pictured in these diagrams. But the more plastic and primitive progenitors of such groups could have had this possibility in their germ plasm.

REFERENCES

ARBER, AGNES. 1938. *Herbals: Their Origin and Evolution.* Cambridge University Press, Cambridge.

BAILEY, I. W. 1953. "The anatomical approach to the study of genera," *Chron. Bot.* **14**:121–125.

BLAKE, S. F. 1935. "Systems of plant classification: A review," *J. Heredity* **26**:463–467.

EAMES, A. J. 1953. "Floral anatomy as an aid in generic limitation," *Chron. Bot.* **14**:126–132.

FERNALD, M. L. 1942. "Some historical aspects of plant taxonomy," *Rhodora* **44**:21–43.

FINAN, JOHN, J. 1948. "Maize in the great herbals," *Ann. Mo. Bot. Gard.* **35**:149–191.

GREENE, E. L. 1909. "Landmarks of botanical history," *Smithsonian Inst., Wash., Misc. Coll.* **54**:1–329.

GUNDERSON, ALFRED. 1918. "A sketch of plant classification from Theophrastus to the present," *Torreya* **18**:213–219, 231–239.

GUNDERSON, ALFRED. 1954. "World families for angiosperms," *Bull. Torr. Club* **81**:210–214.

HALLIER, H. 1905. "Provisional scheme of the natural (phylogenetic) system of flowering plants," *New Phytol.* **4**:151–162.

HARSHBERGER, J. W. 1907. "Taxonomic charts of the monocotyledons and the dicotyledons," *Proc. Am. Philos. Soc.* **46**:313–321.

HARVEY-GIBSON, R. J. 1919. *Outlines of the History of Botany.* New York.

HAWKS, ELLISON, and G. S. BOULGER. 1928. *Pioneers of Plant Study.* London.

HUMPHREY, H. B. 1961. *Makers of North American Botany.* Ronald Press, New York.

JACKSON, B. D. 1923. *Linnaeus: The Story of His Life,* adapted from the Swedish of T. M. Fries (London).

JESSEN, KARL F. W. 1948. *Botanik der Gegenwart und Vorzeit in culturhistorischer Entwickelung* (1864). Reprinted by Hafner Publishing Company, New York.

MAXON, W. R. 1931. "Systematic botany: Its development and contacts," *Old and New Plant Lore,* Smithsonian Scientific Series **11**:133–164.

REED, HOWARD S. 1942. *A Short History of the Plant Sciences.* Waltham, Mass.

REEDER, J. R. 1957. "The embryo in grass systematics," *Am. J. Bot.* **44**:756–768.

RENDLE, A. B. 1904. *The Classification of Flowering Plants,* Vol. I, pp. 1–31. Cambridge University Press, Cambridge.

ROLLINS, R. C. 1953. "Cytogenetical approaches to the study of genera," *Chron. Bot.* **14**:133–139.

SACHS, J. VON. 1890. *History of Botany.* Garnsey and Balfour, Oxford.

SINGER, CHARLES. 1959. *A History of Biology* (rev. ed.). Abelard-Schuman, New York.

SINGER, CHARLES. 1960. *From Magic to Science.* Peter Smith, Gloucester, Mass.

STAUFFER, R. C. 1959. "On the Origin of Species: an unpublished version," *Science* **130**:1449–1452.

TIPPO, O. 1942. "A modern classification of the plant kingdom," *Chron. Bot.* **7**:203–206.

TIPPO, O. 1946. "The role of wood anatomy in phylogeny," *Am. Midl. Nat.* **36**:367–372.

TRELEASE, WILLIAM. 1924. "Four generations of memorable botanists," *Sci. Monthly* **19**:53–62.

TURRILL, W. B. 1942. "Taxonomy and phylogeny," II, *Bot. Rev.* **8**:473–532.

WETTSTEIN, R. 1935. *Handbuch der systematischen Botanik* (4th ed.). Leipzig and Wien.

CHAPTER 3

Taxonomic Literature

The literature of taxonomy is tremendous, varied in scope, and diverse in purpose. It runs the gamut from ponderous volumes to obscure notes in periodicals and even letters of correspondence between workers. Because of this great volume and diversity, one of the vexing problems of the research worker and teacher is that of keeping abreast of recent publications and developing a knowledge of past publications in his field of special interest. To facilitate this task he may consult the lists of publications that have been issued for many years in the *Bulletin of the Torrey Botanical Club* and in the *Taxonomic Index*, the latter recently published as a supplement to issues of *Brittonia*, the publication of the American Society of Plant Taxonomists. Abstracts of published papers and texts can also be found in *Biological Abstracts*, published by the University of Pennsylvania and including works under the heading "General and Systematic Botany." Other fertile sources of pertinent literature will be found at the end of research papers in the form of bibliographies. Most serious workers gradually accumulate such information by the simple expedient of making card files, each card keyed to subject matter or author, or variously cross-indexed. A very satisfactory file of this sort, used by the author, is based on an arrangement of subject matter by families, the breakdown within the family being an alphabetical sequence by genus. Within the genus, the file is alphabetical by author and chronological by date. Other major files, besides the one by family, are used for more general subjects. Such a file, kept up to date, will yield desired information with a minimum of time and effort, because the system is much the same as that employed by many herbaria for the filing of plant specimens.

For the advanced student and the research taxonomist there are three texts that will be very helpful: *Plant Taxonomy: Methods and Principles* by Lyman Benson (Ronald Press, New York, 1962); *Principles of Angiosperm Taxonomy* by P. H. Davis and V. H. Heywood (Van Nostrand, New York, 1963); and *Taxonomy of Vascular Plants* by G. H. M. Lawrence (Macmillan, New York, 1951).

It is difficult to classify satisfactorily the major kinds of taxonomic literature, but the following will give some indication of the variety of publications that is to be found in most large libraries.

Comprehensive Publications

These are wide, even world-wide, in scope and often comprise a series of large volumes. Because of the tremendous amount of labor involved in their preparation, the cost of publication, and their limited sale, such publications are almost a thing of the past. But they were formerly the mainstay of the taxonomist, and they are still invaluable to the serious student. In Linnaeus' *Species Plantarum* (1753), de Candolle's *Prodromus Systematis Naturalis Regni Vegetabilis* (1824–1887), and the more recent Engler and Prantl's *Die natürlichen Pflanzenfamilien* (1887–1909, a second edition being unfinished), we have some of the most monumental taxonomic work of all time.

Manuals and Floras

These regional treatments include keys, and usually descriptions, of families, genera, and species. Early publications of this type usually included large areas, but the increasing complexity of the problems involved has led to floras of smaller regions. A modern manual for a single state in the United States may be just as voluminous as one for the whole of North America published many years ago.

Manuals and floras may be illustrated or not; they may vary in the amount of descriptive matter; and some give citations to original publications of taxonomic groups and fairly complete synonymy, but others do not. Here, again, the high cost of publication, and the limited sale, have limited the scope.

In addition to the larger, regional floras, many *local floras* have been published. These are generally treatments of small areas such as valleys, islands, counties, or other more or less natural physiographic provinces. They serve a very useful purpose and make a very worthwhile project for the local botanist.

Following is a list of the more commonly consulted major floras of North America north of Mexico:

ABRAMS, LEROY. 1923, 1944, 1951. *An Illustrated Flora of the Pacific States,* Vols. I–III. Stanford Uniersity Press, Stanford, Calif. Dr. Abrams having died in 1956, Vol. IV (1960) was prepared by Roxana S. Ferris.

ANDERSON, J. P. 1959. *The Flora of Alaska and Adjacent Parts of Canada,* Iowa State University Press, Ames, Iowa.

ARCHER, W. ANDREW, ed. 1940– *Contributions toward a Flora of Nevada.* National Arboretum and U.S. Department of Agriculture. Issued as mimeographed parts. By 1965 49 parts had been issued.

BAILEY, L. H. 1949. *Manual of Cultivated Plants* (2nd ed.). Macmillan, New York.

BAILEY, V. L., and H. E. BAILEY. 1949. *Woody Plants of the Western National Parks.* Notre Dame, Ind.

BENSON, LYMAN, and R. A. DARROW. 1944. *A Manual of Southwestern Desert Trees and Shrubs,* Univ. Ariz. Biol. Sci. Bull. No. 6.

CHAPMAN, A. W. 1897. *Flora of the Southern United States* (3rd ed.). Cambridge, Mass.

COULTER, J. M., and AVEN NELSON. 1909. *New Manual of Botany of the Central Rocky Mountains.* New York.

DAVIS, R. J. 1952. *Flora of Idaho.* Wm. C. Brown, Dubuque, Iowa.

DEAM, C. C. 1940. *Flora of Indiana.* Indianapolis, Ind.

FERNALD, M. L. 1950. *Gray's Manual of Botany* (8th ed.). New York.

GLEASON, H. A. 1963. *New Britton and Brown Illustrated Flora of the Northeastern States and Adjacent Canada,* 3 vols. Hafner, New York.

GLEASON, H. A., and ARTHUR CRONQUIST. 1963. *Manual of the Vascular Plants of Northeastern United States and Adjacent Canada.* Van Nostrand, Princeton, N.J.

GRAY, ASA. 1886. *Synoptical Flora of North America:* Vol. 2, *Gamopetalae,* 2 parts (2nd ed.). New York.

GRAY, ASA, SERENO WATSON, and B. L. ROBINSON. 1895–1897. *Synoptical Flora of North America:* Vol. 1, *Polypetalae,* 2 parts. New York.

HARRINGTON, H. D. 1954. *Manual of the Plants of Colorado.* Alan Swallow, Denver, Colo.

HITCHCOCK, C. L., *et al.* 1955–1965. *Vascular Plants of the Pacific Northwest* (a projected 5-part illustrated flora). Parts 2–5. University of Washington Press, Seattle, Wash.

HOOKER, W. J. 1829–1834. *Flora Boreali-Americana.* London. (An account of the plants of northern British America, including collections of such notable explorers as Richardson, Drummond, Franklin, and Douglas.)

HULTÉN, ERIC. 1937. *Flora of the Aleutian Islands.* Stockholm.

HULTÉN, ERIC. 1941–1950. *Flora of Alaska and Yukon.* Lund.

JEPSON, W. L. 1960. *A Manual of the Flowering Plants of California.* University of California Press, Berkeley, Calif.

JEPSON, W. L. 1909–1943. *A Flora of California* (3 vols., incomplete). San Francisco and Berkeley, Calif.

JONES, G. N. 1950. *Flora of Illinois* (2nd ed.). Notre Dame, Ind.

JONES, G. N., and G. D. FULLER. 1955. *Vascular Plants of Illinois.* Urbana, Ill.

KEARNEY, T. H., and R. H. PEEBLES. 1942. *Flowering Plants and Ferns of Arizona.* U.S. Dept. Agr. Misc. Publ. 423.

KEARNEY, T. H., and R. H. PEEBLES. 1960. *Arizona Flora* (2nd ed.). University of California Press, Berkeley and Los Angeles, Calif.

MOSS, E. H. 1959. *Flora of Alberta.* University of Toronto Press.

MUNZ, P. A. 1935. *A Manual of Southern California Botany,* Claremont, Calif.

MUNZ, P. A., and D. D. KECK. 1959. *A California Flora.* University of California Press, Berkeley and Los Angeles, Calif.

NEW YORK BOTANICAL GARDEN. 1905– . *North American Flora.* N.Y. Botanical Garden, New York. Volumes have been issued at irregular intervals since 1905. When complete it will comprise about 34 volumes.

NUTTALL, THOMAS. 1818. *The Genera of North American Plants,* 2 vols. Philadelphia, Penn.

PECK, M. E. 1961. *A Manual of the Higher Plants of Oregon* (2nd ed.). Oregon State University Press, Corvallis, Ore.

PIPER, C. V., and R. K. BEATTIE. 1914. *Flora of Southeastern Washington and Adjacent Idaho.* Lancaster, Penn.

PIPER, C. V., and R. K. BEATTIE. 1915. *Flora of the Northwest Coast.* Lancaster, Penn.

PORSILD, A. E. 1957. *Illustrated Flora of the Canadian Arctic Archipelago.* Ottawa, Canada.

PURSH, FREDERICK. 1814. *Flora Americae Septentrionalis,* 2 vols. London (2nd ed., 1816.)

REHDER, ALFRED. 1940. *Manual of Cultivated Trees and Shrubs Hardy in North America* (2nd ed.). Macmillan, New York.

RYDBERG, P. A. 1954. *Flora of the Rocky Mountains and Adjacent Plains* (2nd ed.). Hafner, New York.

RYDBERG, P. A. 1932. *Flora of the Prairies and Plains of Central North America.* Stechert, New York.

SCHAFFNER, J. H. 1928. *Field Manual of the Flora of Ohio and Adjacent Territory.* Columbus, Ohio.

SMALL, J. K. 1903. *Flora of the Southeastern United States.* New York.

SMALL, J. K. 1953. *Manual of the Southeastern Flora.* University of North Carolina Press, Chapel Hill, N.C.

STEVENS, O. A. 1950. *Handbook of North Dakota Plants.* North Dakota Institute for Regional Studies, Fargo, N.D.

STEYERMARK, J. A. 1963. *Flora of Missouri.* Iowa State University Press, Ames, Iowa.

TIDESTROM, IVAR. 1925. *Flora of Utah and Nevada.* Contr. U.S. Nat. Herb. 25:1–665.

TIDESTROM, IVAR, and SISTER TERESITA KITTELL. 1941. *A Flora of Arizona and New Mexico.* Washington, D.C.

TORREY, JOHN, and ASA GRAY. 1838–1840. *A Flora of North America,* 2 vols. New York.

WIGGINS, I. L., and J. H. THOMAS. 1961. *Flora of the Alaskan Slope.* University of Toronto Press, Toronto, Canada.

WOOTON, E. O., and P. C. STANDLEY. 1915. *Flora of New Mexico,* Contr. U.S. Nat. Herb. **19**:1–794.

European botanists have at their disposal hundreds of floras, covering various parts of that continent and written in various languages. Of particular interest, however, is the *Flora Europaea* (in English), now in preparation, the first of four volumes having been issued in 1964. The project is sponsored by the Linnean Society of London. It is directed by an editorial committee based in Great Britain and by a group of advisory editors and regional advisers from all over Europe. Cambridge University Press is the publisher.

Popular Treatments

These "flower books" are the delight of the novice and of persons wishing to consult nontechnical works. They also have their proper place in taxonomic literature and are frequently consulted by experts. Their chief usefulness is that they enable almost anyone to identify many of the common or more ornamental flowering plants by means of pictures alone. There are a number of such books, but perhaps the most sumptuously illustrated and beautiful volumes are *The Macmillan Wild Flower Book* by C. J. Hylander and E. F. Johnston (Macmillan, New York, 1954) and *Wild Flowers of America* by H. W. Rickett, with illustrations by Mary Vaux Walcott and Dorothy Falcon Platt. Both of these contain full-page illustrations in color, supplemented by brief text descriptions.

Treatments of Special Groups

The special interests of experts result in volumes on such subjects as grasses, trees, aquatic plants, forage plants, poisonous plants, and weeds. Most of these are well illustrated, and they are often semi-popular or nontechnical in treatment. Some, however, are aimed at serving the needs of experts and professional botanists. Many such publications are issued as bulletins by federal and state agencies.

Of particular interest to persons interested in cultivated plants is *The Standard Cyclopedia of Horticulture* by L. H. Bailey (three volumes, Macmillan, New York, 1947), in which the plants are treated in alphabetical order according to their generic names. The first volume includes a synopsis of the plant kingdom, keys to the families and genera of cultivated plants, English equivalents of Latin names, and an excellent glossary of botanical terms. This work is copiously illustrated. A smaller publication by the same author, *Manual of Cultivated Plants* (2nd ed., Macmillan, New York, 1949), is a handy reference book for the identification of cultivated plants by means of keys and descriptions, but it has few illustrations.

Those interested in trees are well served by a large array of books, from the eleven-volume, beautifully illustrated set by C. S. Sargent entitled *The Silva of North America* (reprinted in 1947 by Peter Smith) to the pocket-sized manuals such as Richard J. Preston's *North American Trees* (2nd ed., Iowa State University Press, 1960), which has good illustrations and a distribution map for each species. Many states have published useful bulletins on the local trees. The U.S. Department of Agriculture's *Yearbook of Agriculture* for 1949, called *Trees,* is another useful source. Many other references will be found in W. A. Dayton's *Bibliography of Tree Identification* (see the reference at the end of the chapter).

Persons concerned with identification of grasses may consult the *Manual of the Grasses of the United States* by A. S. Hitchcock, the second edition revised by Agnes Chase (1950) and published as Miscellaneous Publication No. 200 by the U.S. Department of Agriculture. Here will be found illustrations of most of the species and maps showing their distribution.

Aquatic plants have been treated satisfactorily on a nation-wide basis by two authors: N. C. Fassett, *A Manual of Aquatic Plants* (University of Wisconsin Press, 1940, revised by E. C. Ogden in 1957), and W. C. Muenscher, *Aquatic Plants of the United States* (Cornell University Press, 1944). Both are well illustrated and contain keys for the identification of the species included. *A Flora of the Marshes of California* by H. L. Mason (University of California Press, 1957) is another useful source book in this category.

Poisonous plants have been described and illustrated in *Poisonous Plants of the United States* by W. C. Muenscher (Macmillan, New York, 1939); and many of the states, particularly those in the West, have issued bulletins on the poisonous plants of their areas. A recent and detailed text is the third edition of *Poisonous Plants of the United States and Canada* by John M. Kingsbury (Prentice-Hall, Englewood Cliffs, 1964).

The literature dealing with weeds is very extensive, and nearly every state is served by agricultural bulletins dealing with this subject. The chief general reference is *Weeds* by W. C. Muenscher (2nd ed., Macmillan, New York, 1955).

Those interested in orchids may refer to the beautifully illustrated and very complete treatment by D. S. Correll, *Native Orchids of North America* (Ronald Press, New York, 1950), illustrated by Blanche Ames and Gordon Dillon, with cultural notes by E. T. Wherry and J. V. Watkins.

Monographic Treatments

These are detailed studies of limited scope and most often deal with single genera or groups of species. The author attempts to bring up to date all the available information concerning the group, and he usually includes detailed keys, citations to original publications, lists of synonyms, descriptions of the taxa, and lists of specimens examined. Included, also, will be attempts to interpret the phylogeny of the group and in recent years a discussion of the cytology involved. Such publications often represent years of painstaking research and constitute the chief bases for our understanding of perplexing groups of plants. It is usually this sort of research that is undertaken by graduate students in taxonomy when working toward an advanced degree. Once having developed an interest in a certain group of plants in this way, a student will often continue the study for many years and eventually become known as the expert on that group of plants. Monographs commonly appear from time to time in the recognized journals and are a valuable supplement to the more meager information contained in floras and manuals.

Notes and Observations

Throughout the literature one will find numerous short papers dealing with such matters as new or noteworthy species, minor variations, and range extensions. This type of publication, because of its varied nature and often inadequate titling, is often difficult to incorporate into a filing system, but the information contained in it may be of considerable significance.

Index, Catalogue, and Dictionary

A few such reference books are essential to the taxonomist. Apart from the usual English dictionaries, which are of general usefulness, and the foreign-language or classical dictionaries, certain more specialized botanical reference books are frequently consulted by the taxonomist. One of these is the *Index Kewensis* (1893–1895 and supplements to date), in which an attempt has been made to list all the scientific names of families, genera, and species of flowering plants, their countries of origin, their authors, and their places of publication. Synonyms, as interpreted by the staff of the Herbarium of the Royal Botanic Gardens at Kew, England, are also included, together with an indication of the valid name to which each applies. A portion of a page of this work is shown in Figure 5.

In *Genera Siphonogamarum* (Berlin, 1900–1907), edited by C. G. Dalla Torre and H. Harms (usually referred to by taxonomists merely as "Dalla Torre and Harms"), there is a listing of families, subfamilies, tribes, and genera of seed plants according to the Engler system. For each genus there is given an estimated number of species. Synonyms, author names, place of publication, and date are included. The genera are assigned numbers from 1–9629, and these numbers are frequently used in the filing system of herbarium specimens. A particularly useful part of the volume is the 284-page index at the end, which enables anyone unfamiliar with a name to place it in its family.

Of particular interest to workers in North America is the *Gray*

AMBROSIA—AMERIMNON

'6.—
'hys.
57 =
77 =
a.
16 =
'olia.
la.
156
:nui-
vii.
olia.
Afr.
nisi-
) 99
–82)
840)
misi-
seria

AMEBIA, Regel, Pl. Nov. Fedsch. 58 (1882) err. typ. = **Arnebia**, Forsk. Boragin.).

AMECARPUS, Benth. in Lindl. Veg. Kingd. 554 (1847) = **Indigofera**, Linn. (Legumin.).

AMECHANIA, DC. Prod. vii. 578 (1839) = **Agarista**, D. Don (Ericac.).
hispidula, DC. l. c. 579 (= *Leucothoë hispidula*).
subcanescens, DC. l. c. (= *Leucothoë subcanescens*).

AMELANCHIER, Medic. Phil. Bot. i. 135 (1789).
ROSACEAE, Benth. & Hook. f. i. 628.
ARONIA, Pers. Syn. ii. 39 (1807).
PERAPHYLLUM, Nutt. in Torr. & Gray, Fl. N. Am. i. 474 (1840).
XEROMALON, Rafin. New Fl. Am. iii. 11 (1836).
alnifolia, *Nutt. in Journ. Acad. Phil.* vii. (1834) 22.—Am. bor.
asiatica, Endl. in Walp. Rep. ii. 55 = canadensis.
Bartramiana, M. Roem. Syn. Rosifl. 145 = canadensis.
Botryapium, DC. Prod. ii. 632 = canadensis.
canadensis, *Medic. Gesch.* 79; *Torr. & Gray, Fl. N. Am.* i. 473.—Am. bor.; As. or.
chinensis, Hort. ex C. Koch, Dendrol. i. 186 = Sorbus arbutifolia.
cretica, DC. Prod. ii. 632 = vulgaris.
denticulata, C. Koch, Dendrol. i. 183 = Cotoneaster denticulata.
florida, Lindl. Bot. Reg. t. 1589 = alnifolia.
grandiflora, Dougl. ex M. Roem. Syn. Rosifl. 145 = canadensis.
integrifolia, Boiss. & Hohen. Diagn. Ser. I. iii. 8 = vulgaris.
intermedia, Spach, Hist. Vég. Phan. ii. 85 = canadensis ?
japonica, Hort. ex C. Koch, Dendrol. i. 179 = canadensis.
melanocarpa, Decne. in Nouv. Arch. Mus. Par. Sér. I. x. (1874) 136.—Hab. ?
oblongifolia, M. Roem. Syn. Rosifl. 147 = canadensis.
oligocarpa, M. Roem. l. c. 145 = canadensis.
orbicularis, Borck. ex Steud. Nom. ed. II. i. 76 = vulgaris.
ovalis, Medic. Gesch. 78 = canadensis.
parviflora, *Boiss. Diagn.* Ser. I. iii. 9.—As. Min.
parviflora, Hort. ex Loud. Arb. Brit. ii. 877 = alnifolia.
pisidica, Boiss. & Heldr. Diagn. Ser. I. x. 2 = parviflora.
pumila, Nutt. ex Torr. & Gray, Fl. N. Am. i. 474 = alnifolia.

AMELIA :—
media, Al
media.
minor, Al

AMELINA,
26 (1874)
Wallichii,
aequinoc

AMELLUS,
thera, l

AMELLU
POSITA
HAENEI
KRAUSS
alternifoli
anisatus, *(*
austr.
annuus, *W*
carolinian
coilopodius
diffusus,
Chiliotri
divaricatu
Belgii.
floribundu
igniariur
flosculosus
hispidus, *L*
Lychnitis,
Lychnitis,
tenuifoli
microgloss
mutabilis,
nanus, *DC*
officinalis,
pallidus, S
peduncula
916 = Tr
rosmarinij
Chiliotr
scabridus,
speciosus,
spinulosus,
strigosus,
tenuifolius
tridactylus
umbellatu
1225) =

5 Portion of a page of the *Index Kewensis*. Note the considerable amount of information given concerning genera and species.

Herbarium Card Index, which lists all new names and combinations, including infraspecific names, for the flowering plants and vascular cryptogams of the western hemisphere. Thousands of cards are found in this index, and more are being added each year.

Another very helpful index is the one, previously mentioned, published in each issue of the *Bulletin of the Torrey Botanical Club* and reprinted by the American Society of Plant Taxonomists as the *Taxonomic Index* in the journal *Brittonia.*

In *Taxon,* the official news bulletin of the International Association for Plant Taxonomy, will be found announcements of significant floras as they are published, as well as other information of general interest to taxonomists.

For a listing of recent publications in taxonomy and related fields, the taxonomist will profit by reference to *Biological Abstracts,* also mentioned earlier, in which summaries of published research will be found.

Among the commonly consulted compendia of major publications are the following: *Thesaurus Literaturae Botanicae* by G. A. Pritzel, originally published in Berlin in 1871 and republished in Italy in 1950, which lists both by author and by subject the chief botanical publications preceding it; *Guide to the Literature of Botany* by B. D. Jackson, London, 1881 (reprinted in 1964), listing some 9,000 works, and an important supplement to Pritzel; *Index Londinensis to Illustrations of Flowering Plants, Ferns, and Fern Allies* by Otto Stapf and O. C. Worsdell (1921–1935); *Catalogue of the Library of the Royal Botanic Gardens, Kew,* Bulletin of Miscellaneous Information (Kew Bulletin), Additional Series III, 1899, with supplement 1898–1915 (London, 1919); and *Catalogue of the Books, Manuscripts, and Drawings in the British Museum (Natural History),* in eight volumes (1903–1940), compiled by B. B. Woodward and A. C. Townsend (reprinted in 1964).

Specialized dictionaries and glossaries giving the meanings and often the derivations of technical terms are an essential part of any taxonomist's library. Some of these are listed here:

Gray's Lessons in Botany (New York, 1887) has long been known as one of the best-illustrated sources of information on the terminology of structural botany. It was prepared by Asa Gray, and despite its age it is still exceedingly useful at an elementary level. *A Dictionary of Flowering Plants and Ferns* by J. C. Willis (6th ed., Cambridge, England, 1931) contains alphabetically listed and described families and genera. *A Glossary of Botanic Terms* by B. D. Jackson (4th edition, London, 1949) is a standard work in this category. On the practical side we have the *Glossary of Botanical Terms Commonly Used in Range Research* by W. A. Dayton (U.S. Dept. Agr. Misc. Publ. 110, Washington, D.C., 1931). *Vocabularium Botanicum* (*Plant Terminology*) by E. F. Steinmetz (2nd ed., Amsterdam, 1953) contains about 4,000 scientific terms in English, Dutch, German, French, Latin, and Greek, and is a useful source of equivalents in various languages. For the average student with a need for a small but practical dictionary of terms there are *Taxonomic Terminology of the Higher Plants* by H. I. Featherly (Iowa State University Press, 1954), *Scientific Terminology* by J. N. Hough (Holt, Rinehart and Winston, New York, 1953), and the more complete *Composition of Scientific Words* by R. W. Brown (Baltimore, Md., 1954). Practically all regional manuals and floras also include glossaries.

Finally, since it is general practice in taxonomic publications to use abbreviations for publications cited, a very useful reference book dealing with standard abbreviations is *Abbreviations Used in the Department of Agriculture for Titles of Publications* by Carolyn Whitlock (U.S. Dept. Agr. Misc. Publ. 337, Washington, D.C., 1939). A similar list is *Abbreviations of Periodicals* by Lazella Schwarten and H. W. Rickett (*Bull. Torr. Club* 74:348–356, 1947).

Regulations

Legislation agreed upon by representatives (delegates) at International Botanical Congresses, which have met more or less regularly since 1867, is contained in the *International Code of Botanical Nomenclature.* These rules and recommendations are subject to revision at each International Congress, after which new editions are prepared. A Congress was held at Edinburgh, Scotland, in 1964, and the most recent edition of the *Code* was published in 1961 under the editorship of J. Lanjouw. Further revisions are made following each Congress, normally at four-year intervals. Such regulations apply only to scientific names and their applications.

Some time ago an attempt was made to legislate and standardize the use of common names of plants of North America. This is an almost impossible task and it has met with only partial success, though the use of standard common names is the general practice in publications of the United States government. The guide in use is the second edition of *Standardized Plant Names,* edited by H. P. Kelsey and W. A. Dayton (Harrisburg, Penn., 1942).

REFERENCES

BLAKE, S. F., and ALICE C. ATWOOD. *Geographical Guide to Floras of the World: I, Africa, Australia, North America, South America, and Islands of the Atlantic, Pacific, and Indian Oceans,* U.S. Dept. Agr. Misc. Publ. 401 (Washington, D.C., 1942). Includes an appendix of abbreviations of periodicals cited.

DAYTON, W. A. 1952. *United States Tree Books: A Bibliography of Tree Identification,* U.S. Dept. Agr. Bibliogr. Bull. No. 20.

LANGMAN, IDA K. 1964. *A Selected Guide to the Literature on the Flowering Plants of Mexico,* University of Pennsylvania Press, Philadelphia, Penn.

RENNER, F. G., *et al.* 1938. *A Selected Bibliography on Management of Western Ranges, Livestock, and Wildlife,* U.S. Dept. Agr. Miscel. Publ. No. 281 (Washington, D.C.).

SENN, H. A. 1951. *A Bibliography of Canadian Plant Geography,* Trans. Royal Canad. Instit. 1946–1947, Vol. 26, Parts 1–2; Publ. 863, Dept. Agr. Ottawa (to be continued).

CHAPTER 4

Field and Herbarium Methods

All persons actively engaged in taxonomic work have occasion to study actual plants—plants living in the field or garden, or specimens preserved in a herbarium, or both. There is no substitute for the firsthand information gained through observation of living populations in diverse situations and localities; and if some of this information is to become a part of the permanent record, for others to examine, it is essential that selected material be carefully prepared and preserved in a herbarium.

Objectives

The ultimate goal of collecting in the field and preserving in the herbarium is very simple: it is to preserve for all time a series of specimens and notes that will yield the maximum of information about the plants concerned.

The following details regarding field and herbarium techniques are applicable to a study of flowering plants, ferns, and conifers. Special methods are usually required for preserving such plants as fungi and algae, and directions applicable to those groups will be found in the references at the end of this chapter.

Field Equipment and Methods

The essential items of equipment for field collecting are few (see Fig. 6), and they are subject to modification to meet diversified field conditions and the individual requirements of the collector. The

6 Field collecting equipment. On the upper left is the vasculum, and below it are plastic bags, notebook, and pencil. Two kinds of diggers are leaning against the vasculum. On the right is a plant press containing alternating layers of corrugated cardboard ventilators and newspaper pressing sheets, the latter being folded to contain the plant specimens.

amount and exact kind of equipment will depend upon the limitations imposed on space and weight by the means of transportation, upon the nature of the terrain, upon the nature of the plant groups to be collected, and upon the individual preferences of the collector.

THE PLANT PRESS. This is a device by means of which fresh specimens are pressed flat and quickly dried.

The top and bottom frames of the press, usually 12 by 18 inches, are composed of thin, strong pieces of wood laid at right angles to one another so as to form a lattice work, the joints being securely screwed or riveted. Between the frames are driers, or sheets of moisture-absorbing material, also 12 by 18 inches, which may be in the form of blotters or smooth-faced corrugated cardboards, and between these are folded newsprint sheets, in each of which the collector places one kind of plant. The assembled press is tightly bound together by a pair of heavy straps or ropes, these being long enough to allow for expansion as more specimens are added.

Good specimens should include either the flowering or the fruiting phase of the plant, preferably both. They should be carefully displayed on the pressing sheet without unnecessary folding or hiding of parts. If a specian is too large to fit easily on the sheet, it should be bent into a V shape or a W shape, this shape being maintained, if necessary, by slotted strips of heavy paper or cards slipped over the bent portions.

The press containing the specimens is placed in the hot sun or suspended over moderate heat, and the driers (but not the papers) are changed or dried at least daily until the plants are dry. This task becomes arduous at times, but it must not be neglected, or poor and discolored specimens with poor color preservation will result.

THE FIELD NOTEBOOK. The notebook, another essential, should preferably be permanently bound, pocket-sized, and with horizontal rulings. It is used for data recorded at the time of collection: the field number of the specimen, the locality, the habitat, the date of collection, the name of the plant if that is known, the flower color, the size of the plant if it is too large to be preserved complete, its abundance, the variations observed, the associated plants, the elevation, and so on. *A number series, once started, should be continuous* throughout the life of the collector so that no confusion will arise when specimens are referred to by the collector's number. If several specimens of the same kind are taken at the same time and place, these duplicates will, of course, be given the same number; otherwise numbers should never be duplicated. Numbers corresponding to those in the field notebook should be *written on the papers containing the specimens* in the field, never on separate slips of paper, which might drop out or become detached from the specimens. Careful and farsighted workers sometimes keep duplicate records: one book for field use, the other remaining at home and containing the same data copied from the field book as soon as possible. Such a system insures against loss of older records should anything happen to the field book.

THE VASCULUM. This is a container that will somewhat preserve the freshness of specimens until it is convenient to place them in the press. No specimens are likely to be improved by postponement of pressing, but it is not always convenient to press plants as they are collected. Standard vascula are made of sheet metal, are somewhat oval on the ends, are from 18 to 24 inches long, and have on one side a hinged door that is provided with a secure fastener to keep the interior as air-tight as possible.

Recently it has been found that plastic bags make excellent containers for fresh plants; they prevent loss of moisture when closed securely, and each collection may be kept separate in the general container or vasculum. When light weight and compactness are at a premium, as in mountain collecting on foot, such a system, combined with a light pack-basket or back-pack, will be found quite satisfactory.

THE DIGGER. When one is collecting herbaceous plants, it is essential that the underground parts be made a part of the specimen. The kind of root system, and the presence or absence of rhizomes, bulbs, or other subterranean parts, constitute valuable diagnostic characters for identification. For this reason it is necessary for the collector to carry some sort of substantial digger—a heavy sheath-knife, a dandelion digger, a geologist's pick, a bricklayer's hammer with chisel blade, or any similar suitable tool. And in order to avoid loss in the field one should paint the handle a bright orange or yellow.

The Herbarium

The herbarium, which becomes the repository of the specimens and notes, permanently preserved, is a growing source of information about the vegetation of an area. It may contain millions of specimen vouchers, gradually accumulated by a large institution, and may represent the flora of a continent or more; or it may be a very modest personal and local collection.

As the herbarium grows, it is necessary that the material in it be systematically filed so as to be quickly accessible to all who use it. It is equally important that the specimens be adequately prepared for handling and filing. All of this has led to some standardization of herbarium equipment and methods, but much individuality still remains in the various large collections.

MOUNTING. This is the process by which specimens are prepared for the permanent files. The pressed and dried plants or plant parts are securely attached to one side of a sheet of mounting paper, and a label is attached at the lower right-hand corner.

The mounting paper used in most American herbaria measures 11½ by 16½ inches (European institutions often use a longer sheet); it should be fairly stiff and of an all-rag content for permanence. Various supply houses and paper companies can supply suitable mounting sheets.

Specimens are attached to the sheets in various ways, but the commonest method involves the use of strips of white gummed cloth (known as Holland cloth), supplemented by a non-staining glue, which one applies to the flat surfaces by means of a spatula or by laying the specimen on a glue-coated glass plate and then transferring it to the mounting sheet. The strips should not obscure critical parts, which may need to be studied. Cellulose tape is to be avoided because it is not permanent. One should place thick parts of specimens in different regions of the sheets in order to avoid bulkiness at one spot or at one end. Loose valuable parts, such as seeds, fruits, or dissected parts, are placed in paper packets or envelopes, which are glued to the mounting paper.

In recent years many herbaria have adopted the plastic method of mounting specimens, described by W. Andrew Archer in 1950 and elaborated on by R. C. Rollins in 1955 (see the references and Fig. 7). This method is based on the use of ethyl cellulose and a resin dissolved in a mixture of toluene and methyl alcohol to give a thick, syrupy, adhesive liquid, which is applied by means of a pressure oil gun or a slim-nosed squeeze bottle of the type used to dispense catsup or mustard. One disadvantage of this method was that it was necessary to purchase large quantities of the ingredients, but lately small quantities have been offered for sale, mixed and ready, by the Carolina Biological Supply Company, Elon College, North Carolina. The big advantage of the method lies in the rapidity with which specimens can be mounted and the permanence and pleasing appearance of the mounted plants.

THE LABEL. The label, usually glued by one edge at the lower right-hand corner of the sheet, supplies information taken from the field book. It varies in size and format, but is usually about 3 by 5 inches or a little less. It should include at least the following: (1) a heading that indicates the state, province, or country of the collection, and usually the name of the person or institution with which the specimen originated; (2) the genus and species, with authority (some include the family as well); (3) the locality of collection (some workers even include a small printed map with a spot on it to indicate the locality); (4) the habitat; (5) the date of collection; (6) the name of the collector; (7) the collector's field number; (8) the name of the person who identified the specimen, if not the collector. Additional information might include: the names of associated plants found growing in the immediate vicinity, the color of the flowers or other information about flower parts not readily seen in pressed specimens, the height of the plant if the specimen is not complete, the abundance of the plant in the area, the altitude of the locality, and, if the plant is a large shrub or a tree, the nature of the bark.

7 Mounting herbarium specimens by the plastic method, using a pressure oil gun to dispense the plastic. The sheets at the left are stacked up to dry.

Labels, above all, should be legible, neat, and permanent. A specimen without an adequate label is practically worthless, however well it was pressed and mounted. The best labels are printed, but clear typewritten labels or even carefully handwritten ones are quite satisfactory. Large printed forms to be filled out are seldom satisfactory, but this kind is used by several governmental agencies. Examples of several sorts of labels are shown in Figure 8.

FILING. This is the process of placing mounted material in a systematic arrangement for storage until needed. In major herbaria the specimens are housed in steel filing cases, compartmented into pigeonholes, and provided with tight-fitting doors. A card on the outside designates the nature of the contents. In small collections the filing system may be entirely alphabetical by genus and species, but in most herbaria the filing system follows a phylogenetic sequence so as to bring closely related things together, in which case the sequence employed is often that of C. G. Dalla Torre and H. Harms in their *Genera Siphonogamarum* (1900–1907). According to this system, each family and each genus has a place in a numbered sequence based on supposed relationships. At the end of this reference is an index of generic names and the families to which they are assigned.

In all major herbaria the specimen sheets are filed inside manila folders slightly larger than the sheets. Each of these folders has written on it, on the front edge, the scientific name of the included specimens. There may be a single folder for an entire genus of rare plants, or there may be several folders for a single commoner species. In some herbaria the folders, and the specimens with them, are segregated geographically, and various color schemes for folders are employed by some to indicate political or physiographic areas.

CARE OF SPECIMENS. A properly cared-for specimen should last indefinitely (Fig. 9), but carelessness or abuse can ruin valuable collections very quickly. This does not mean that the resources of a herbarium should not be used, for that is its purpose.

In most areas some precautions must be taken to safeguard against insect damage; these usually consist of fumigation with chemicals such as paradichlorobenzene or of poisoning the specimens by brushing them with a solution of bichloride of mercury. Cabinet doors should be kept tightly closed when specimens are not in use, and a fumigant should be kept inside each cabinet. Care and judgment should be exercised in the examination and handling of specimens: they should not be bent, turned over like leaves in a book, or subjected to injury

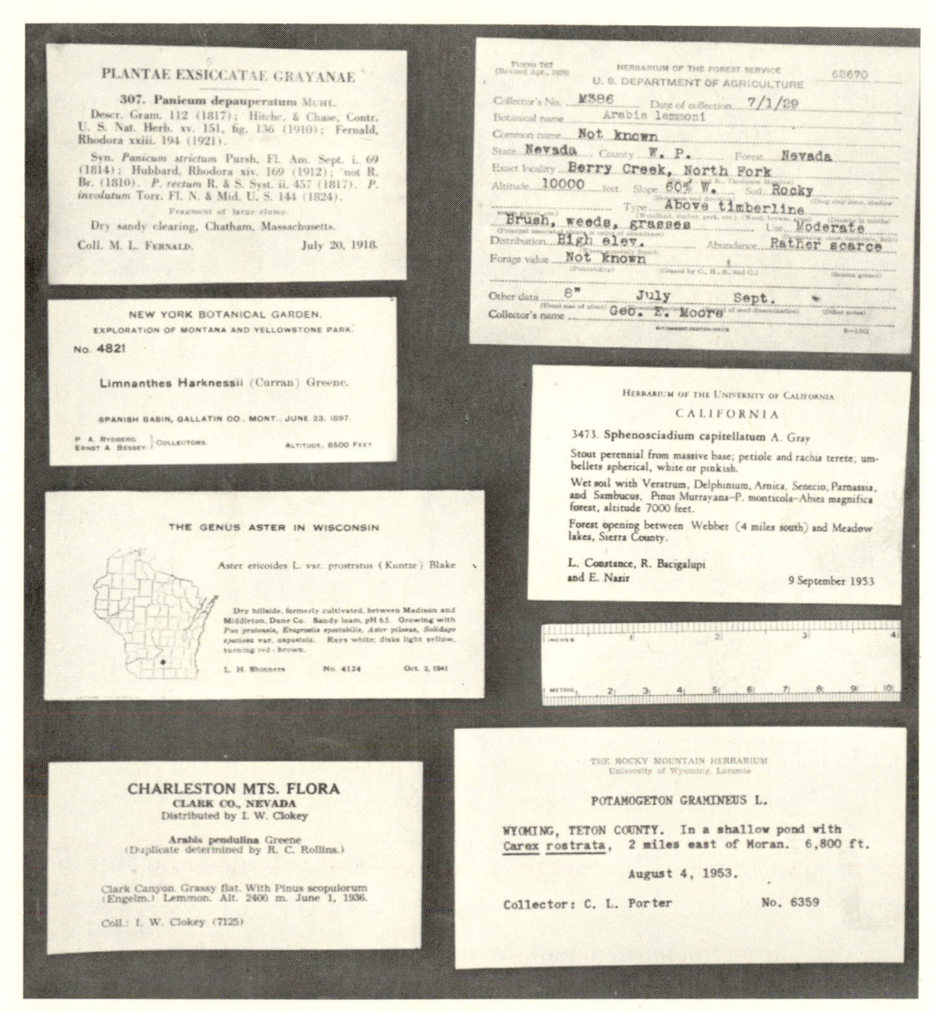

8 Some examples of herbarium labels. The more modern labels generally give more detailed information than the older labels.

by abrasion or pressure. Any dissected fragments or parts that have become detached should be placed in small packets.

If corrections in name become necessary, these notations should be written or typed on separate slips of paper, known as *annotation labels,* which are then affixed to the sheets above the regular labels or at some other nearby place. An annotation should include the name of the person making it (not merely his initials) and the date.

In order to prevent undue wear and possible abuse to type specimens, many herbaria have them segregated from the general collection. This is a very desirable practice.

Finally, every precaution should be taken to prevent loss or damage

9 Two herbarium specimens, the left-hand one collected in 1855, that on the right collected in 1956. Specimens should last indefinitely if not abused.

due to fire. This entails storage in fire-resistant cabinets and in a fireproof building.

LOANS. Specimens are lent by the curators of major herbaria to responsible persons requesting them. This service to research workers also benefits the lending institution, for those workers are often able to make corrections and revisions in nomenclature and identification. Loans are generally made for a specified period of time, and the borrower usually pays the costs of transportation both ways. Borrowers should be meticulously careful to adhere to all the provisions of loans and to supply adequate and carefully prepared annotations.

EXCHANGES. Exchanges of specimens among scientific institutions and collectors are a standard means of augmenting a collection at little cost. As a rule, the exchanges are conducted on a specimen-for-specimen basis, the shipper paying transportation costs. Specimens for exchange are unmounted, but they should be supplied with adequate labels.

REFERENCES

ALLARD, H. A. 1951. "Drying herbarium specimens slowly or rapidly," *Castanea* **16**:129–134.

ANDREWS, F. M. 1932. "Preservation of dry plant material," *Ind. Acad. Sci. Proc.* **41**:80–81.

ARCHER, W. A. 1950. "New plastic aid in mounting herbarium specimens," *Rhodora* **52**:298–299.

BAILEY, L. H. 1946. "The palm herbarium," *Gentes Herb.* **7**:153–180.

BAILEY, W.W. 1881. *The Botanical Collector's Handbook* (Salem, Mass.).

BAILEY, W. W. 1904. "Collecting plants for identification," *Am. Bot.* **7**:9–11.

BEAMAN, J. M., R. C. ROLLINS, and A. H. SMITH. 1965. "The herbarium in the modern university," *Taxon* **14**:113–133.

BENSON, LYMAN. 1939. "Notes on taxonomic techniques," *Torreya* **39**:73–75.

BLAKE, S. F. 1919. *Directions for the Preparation of Plant Specimens for Identification,* U.S. Bur. Plant Indus., Office Econ. and System. Bot. (Circ. 1), 2 pp.

BLAKE, S. F. 1920. *Directions for Collecting Flowering Plants and Ferns,* U.S.D.A. Dept. Circular 76, pp. 1–7.

BLUMER, J. C. 1907. "A simple plan for collectors of ecological sets of plants," *Plant World* **10**:40–42.

BUCHHOLZ, J. T. 1931. "A practical drier for botanical specimens," *Ill. Acad. Sci. Trans.* **24**:103–107.

CAMP, W. H. 1946. "On the use of artificial heat in the preparation of herbarium specimens," *Bull. Torr. Club* **73**:235–243.

CHAMBERLAIN, E. B. 1903. "Mounting moss specimens," *Bryologist* **6**:75–76.

CLARK, C. S. 1903. "Mounting mosses," *Bryologist* **6**:102–103.

CLEMENTS, F. E. 1904. "Formation and succession herbaria," *Nebr. Univ. Studies* **4**:329–355.

COLLINS, J. F. 1906. "Mounting mosses—some hints," *Bryologist* **9**:60–63.

COLLINS, J. F. 1910. "The use of corrugated paper boards in drying plants," *Rhodora* **12**:221–224.

COLLINS, J. F. 1932. "Better herbarium specimens," *Rhodora* **34**:247–249.

COVILLE, F. V. 1895. *Directions for Collecting Specimens and Information Illustrating the Aboriginal Uses of Plants,* U.S. Nat. Mus. Bull. 39, 8 pp.

CURTIS, A. H. 1901. "Hints on herborizing," *Plant World* **4**:61–66, 81–87.

DERR, H. B., and C. H. LANE. 1914. *Collection and Preservation of Plant Material for Use in the Study of Agriculture,* U.S.D.A. Farmers' Bull. 586, 24 pp.

DIETERLE, JENNIE V. A. 1960. "Sandbags as a technical aid in mounting plants," *Rhodora* **62**:322–324.

ENGLEMANN, BEBB, BAILEY, SCRIBNER, MORONG, *et al.* 1886. "How to collect certain plants," *Bot. Gaz.* **11**:135–150.

FERNALD, M. L. 1945. "Injury to herbarium specimens by extreme heat," *Rhodora* **47**:258–260.

FOGG, J. M., JR. 1940. "Suggestions for collectors," *Rhodora* **42**:145–157.

FOSBERG, F. R. 1939. *Plant Collecting Manual for Field Anthropologists* (Philadelphia, Penn.).

FOSBERG, F. R. 1947. "Formaldehyde in plant collecting," *Science* **106**:250–251.

FOSBERG, F. R., and M. H. SACHET. 1965. *Manual for Tropical Herbaria* (Utrecht, Netherlands).

GATES, B. N. 1950. "An electrical drier for herbarium specimens," *Rhodora* **52**: 129–134.

GATES, F. C. 1935. "Paradichlorobenzene, an effective herbarium insecticide," *Science* **81**:438–439.

GENTRY, H. S. 1952. "The belt plant press," *Bull. Torr. Club* **79**:84–86.

GILBERT, B. D. 1904. "Mounting mosses," *Bryologist* **7**:61–62.

GLEASON, H. A. 1933. "Annotations on herbarium sheets," *Rhodora* **35**:41–43.

GLEASON, H. A., and A. C. SMITH. 1930. "Methods of preserving and arranging herbarium specimens," *J. N.Y. Bot. Gard.* **31**:112–125.

GRIFFITHS, D. 1907. "Preparation of specimens of *Opuntia*," *Plant World* **9**:278–284.

HEDGECOCK, G. G., and P. SPAULDING. 1906. "A new method of mounting fungi grown in cultures for the herbarium," *J. Myc.* **12**:147.

HITCHCOCK, A. S. 1900. "Collecting sets of plants for exchange," *Plant World* **3**: 148–151.

HITCHCOCK, A. S. 1925. *Methods of Descriptive Systematic Botany* (New York).

HITCHCOCK, A. S. 1931. *Field Work for the Local Botanist* (Washington, D.C.).

HITCHCOCK, A. S., and AGNES CHASE. 1909. *Directions for Preparing Herbarium Specimens of Grasses,* U.S.D.A., Bur. Plant Ind. Doc. 442, 4 pp.

HODGE, W. H. 1947. "The use of alcohol in plant collecting," *Rhodora* **49**:207–210.

HOLZINGER, J. M. 1903. "Some notes on collecting," *Bryologist* **6**:37–38.

HOWARD, R. A. 1947. "The use of DDT in the preparation of botanical specimens," *Rhodora* **49**:286–288.

JOHNSTON, I. M. 1939. *The Preparation of Botanical Specimens for the Herbarium,* 33 pp. (The Arnold Arboretum, Jamaica Plain, Mass.)

KOBUSKI, C. E., C. V. MORTON, MARION OWNBEY, and R. M. TRYON. 1958. "Report of the Committee for Recommendations on Desirable Procedures in Herbarium Practice and Ethics," *Brittonia* **10**:93–95.

LITTLE, E. L. 1952. "Preparing specimens of Picea and Tsuga," *Rhodora* **54**:232–234.

LUNDELL, C. L. 1946. "A useful method for drying plant specimens in the field," *Wrightia* **1**:145, 161–162.

LUNELL, J. 1918. "Collecting, drying and mounting of plant specimens," *Am. Midl. Nat.* **5**:191–195.

MARTIN, G. W. 1925. "Paradichlorobenzene in the herbarium," *Bot. Gaz.* **79**:450.

MERRILL, E. D. 1926. "An economical herbarium case," *Torreya* **26**:50–54.

MERRILL, E. D. 1926. "An efficient and economical herbarium paste," *Torreya* **26**:63–65.

MERRILL, E. D. 1934. "The significance of the compiler's data in field work," *Bull. Torr. Club* **61**:71–74.

MERRILL, E. D. 1937. "On the technique of inserting published data in the herbarium," *J. Arnold Arbor.* **18**:173–182.

MILLISPAUGH, C. F. 1925. "Herbarium organization," *Field Mus. Nat. Hist. Publ., Mus. Tech. Ser.* **1**:1–18.

MOORE, H. E., JR. 1950. "A substitute for formaldehyde and alcohol in plant collecting," *Rhodora* **52**:123–124.

NICHOLS, G. E., and H. ST. JOHN. 1918. "Pressing plants with double-faced corrugated paper boards," *Rhodora* **20**:153–160.

NIEUWLAND, J. A. 1909. "The mounting of algae," *Bot. Gaz.* **47**:237–238.

OGDEN, E. C. 1945. "Display pocket for cryptogams," *Bryologist* **48**:194.

PEEBLES, R. H. 1942. "Preservation of cactus material," *Cact. Succ. J.* (January).

RICHARDS, H. M. 1901. "New methods of drying plants," *Torreya* **1**:145–146.

RICKER, P. L. 1913. "Directions for collecting plants," *U.S.D.A., Bur. Pl. Indus. Circ.* **126**:27–35.

ROBINSON, B. L. 1903. "Insecticides used at the Gray Herbarium," *Rhodora* **5**:237–247.

ROLLINS, R. C. 1955. "The Archer Method for mounting herbarium specimens," *Rhodora* **57**:294–299.

SANFORD, S. N. F. 1936. *The Collection and Preservation of Flowering Plants,* Boston Soc. Nat. Hist. Bull. 79, 23 pp.

SAVILE, D. B. O. 1962. *Collection and Care of Botanical Specimens* (Ottawa, Canada).

SCHRENK, J. 1888. "Schweinfurth's Method of preserving plants for herbaria," *Bull. Torr. Club* **15**:292–293.

SCHULTES, R. E. 1947. "The use of formaldehyde in plant collecting," *Rhodora* **49**:54–60.

SCULLY, F. J. 1937. "Preservation of plant material in natural colors," *Rhodora* **39**:16–19.

SHARP, A. J. 1935. "An improvement in the method of preparing certain gymnosperms for the herbarium," *Rhodora* **37**:267–268.

SHOPE, P. F. 1936. "Paradichlorobenzene as a herbarium insecticide," *Science* **83**:19.

SMITH, MARTINDALE, CHICKERING, BESSEY, *et al.* 1886. "Specimens and specimen making," *Bot. Gaz.* **11**:129–134.

SPAULDING, V. M. 1907. "Suggestions to plant collectors," *Plant World* **10**:40.

STANDLEY, P. C. 1909. "Herbarium notes," *Torreya* **9**:74–77.

STEVENS, O. A. 1949. "A simple plastic collecting bag," *Rhodora* **51**:393.

STEYERMARK, J. A. 1947. "Notes on drying plants," *Rhodora* **49**:220–227.

STONE, G. E. 1900. "Formalin as a preservative for botanical specimens," *J. Appl. Microscopy* **2**:537–540.

SWINGLE, C. F. 1930. "Oxyquinoline sulphate as a preservative for plant tissue," *Bot. Gaz.* **90**:333–334.

TEMPLETON, B. C. 1932. "Methods of preserving cacti for herbarium use," *Desert* **3**:127.

THIERET, J. W., and R. J. REICH. 1960. "The formaldehyde method of collecting plant specimens," *Turtox News* **38**:114–115.

TRAUB, H. P. 1954. "PDB in plastic envelopes for pest control in the small herbarium," *Taxon* **3**:84–88.

U.S. DEPARTMENT OF AGRICULTURE, FOREST SERVICE. 1915. *Suggestions for the Collection of Range Plant Specimens on National Forests,* U.S. Forest Service, unnumbered leaflet, 4 pp.

U.S. DEPARTMENT OF AGRICULTURE, FOREST SERVICE. 1925. *Instructions for National Forest Range Plant* Work, U.S. Forest Service, unnumbered circular, 4 pp.

WALKER, E. 1942. "Recording localities on specimen labels," *Chron. Bot.* **7**:70–71.

WHERRY, E. T. 1949. "A plastic spray for coating herbarium specimens," *Bartonia* **25**:86.

WILLIAMS, R. S. 1910. "On collecting mosses," *Bryologist* **13**:56–57.

WOLLEY-DOD, A. H. 1920. "On collecting roses," *J. Bot.* **58**:23–24.

CHAPTER 5

Nomenclature

As the number of plants known to man increased, it became apparent that some uniform and generally acceptable set of principles had to be adopted if confusion in names was to be avoided. Professional botanists have therefore gradually adopted a system of naming plants and plant groups according to international agreements reached at meetings. Such meetings, as already noted, are known as International Congresses, and the rules adopted and published by them are known as the *International Code of Botanical Nomenclature*. These meetings and rules deal only with the use and application of scientific names.

The same degree of accord, unfortunately, has not been attained for the use of common names of plants, among which there is great confusion. Many a plant has several common names, used in various parts of the country—there are some 140 common names for *Verbascum thapsus,* the common Mullein; and often the same common name is used for quite different plants. Examples of this are the use of the name "Cedar" for quite different groups of trees and the use of the name "Foxtail" for various different grasses. Some plants, furthermore, have no common names, and that is confusing to the layman. An attempt to reduce this confusion was made by a Joint American Committee on Horticultural Nomenclature, which published a list of *Standardized Plant Names* (1923), but few persons outside governmental agencies have adopted this list. It is practically impossible to legislate on common names as on scientific names.

Nomenclature may be defined as the *system of naming* plants, animals, or other objects, or groups of plants, animals, or objects. In botanical nomenclature the names given to plants either are Latin names or are names taken from some other language and Latinized.

The Binomial System

The scientific name of any plant is in two parts: (1) the name of the genus, or *generic name;* (2) the *specific epithet.* For example, the scientific name of the White Oak is *Quercus alba,* and the scientific name of the White Poplar is *Populus alba.* (Scientific names should always be given in italics when printed, or underlined when typed or written by hand.) The scientific name of any species of plant consists, then, of two Latin or Latinized words. The first, or generic name, is sometimes given merely as an abbreviation consisting of the capital initial (followed by a period) if the context makes its meaning clear.

According to the International Code, there can be only one group of plants (one genus) with the name *Quercus,* reserved for the Oak group, and only one with the name *Populus,* reserved for the Poplar group, and within each genus there can be only one valid specific epithet *alba;* but the same specific epithet may be applied to plants that are not members of the same genus.

Should duplicate names arise, one of them (the more recent usually) must be discarded. Should additional names for a validly named plant arise through misconceptions, by errors of one kind or another, or for whatever reason, they must be discarded in favor of the one *legitimate name,* the others becoming *synonyms.*

THE GENERIC NAME. This is always a noun, in the singular, nominative case, and it is always written with a capital initial letter. It may be (1) descriptive, with reference to some characteristic prominent in the included species, such as *Xanthoxylum* (yellow wood), *Liriodendron* (lily or tulip tree), or *Cercocarpus* (coiled fruit); (2) the aboriginal name of the plant, such as *Quercus, Fagus,* and *Betula,* which were old Greek names for Oak, Beech, and Birch; (3) a name in honor of a person, such as *Jeffersonia* for Thomas Jefferson, *Linnaea* for Linnaeus, *Grayia* for Asa Gray, one of the fathers of American botany, and *Lewisia* for Captain Meriwether Lewis, leader of the Lewis and Clark expedition.

THE SPECIFIC EPITHET. This may be any of the following:

1. An adjective, agreeing with the generic name in gender, and usually indicating a distinguishing characteristic of the species, or

sometimes referring to a locality where the species was first discovered: *Rosa alba,* the White Rose; *Ulmus americana,* the American Elm; *Erigeron peregrinus,* the Wandering Daisy; *Ranunculus jovis,* a Buttercup first collected on the Thunderer, in Yellowstone National Park, the name referring to the God of Thunder. Sometimes, when botanists find that a certain species is particularly difficult to define or distinguish, the epithet *perplexus* is used.

2. A noun in apposition, hence always in the nominative case to agree with the generic name, but not necessarily agreeing with it in gender, as in *Pyrus malus,* the Apple, in which the epithet is another generic name used as a specific epithet and, according to present rules, either capitalized or not at the discretion of the person using it.
3. A noun in the genitive singular, such as occurs when the species is named in honor of a person, and thus of either masculine or feminine gender: *Carex davisii,* named for a Mr. Davis; *Gilia piersonae,* named for a Miss Pierson. The specific epithet taken from the name of a person may be capitalized or not at the discretion of the user.
4. A common name in the genitive plural, usually indicating something about the habitat of the species: *Polygonum dumetorum,* meaning "of the thickets"; *Convolvulus sepium,* meaning "of the hedges"; *Carex paludosum,* meaning "of the swamps."

Names of taxa superior to the genus, such as orders, families, and subdivisions of such groups, are also formed in accordance with generally accepted principles. Some of these are given below.

The Order

This is the major taxon immediately superior to the family and is often referred to in taxonomic literature. There is still some disagreement among botanists, unfortunately, as to its interpretation and application. We form the name of the order, or the ordinal name, by adding *ales* to the stem of an included generic name: *Poales,* from *Poa,* for the order including grasses and sedges; *Liliales,* from *Lilium,* for the order that includes the lilies. A few ordinal names, still in use by some botanists, are exceptions to this general rule, the exceptions being of the sort that will be explained below in the discussion of family names.

The Family

A family consists of a group of related genera, rarely of a single genus, and it is usually a fairly clear-cut taxon agreed upon by most botanists. (Even here, however, we may find some disagreement as to just what should be included in a single family.) We form its name, except for a few that antedate the standardized system, by adding *aceae* to the stem of an included generic name: *Rosaceae,* from *Rosa,* for the Rose Family; *Ranunculaceae,* from *Ranunculus,* for the Buttercup Family.

A few families have long been designated by names that antedate this system, and such names are often merely the common names used by the ancients and are not based upon generic names. Since a uniform system has certain advantages, however, it has been agreed that substitute names, formed in accordance with the method explained above, may be used in place of the older names at the discretion of the users. Some of these are the following:

	OLD NAME	NEW NAME
Grass Family	*Gramineae*	*Poaceae*
Mustard Family	*Cruciferae*	*Brassicaceae*
Pea Family	*Leguminosae*	*Fabaceae*
Parsnip Family	*Umbelliferae*	*Apiaceae*
Mint Family	*Labiatae*	*Lamiaceae*
Sunflower Family	*Compositae*	*Asteraceae*

THE SUBFAMILY. This is a major subdivision of a family and is sometimes used when the size of the family justifies it and when the included genera may be naturally so grouped. We form the name by adding *oideae* to the stem of an included generic name, such as *Festucoideae,* from *Festuca,* and *Panicoideae,* from *Panicum,* for two subfamilies of the Grass Family.

THE TRIBE. A tribe is a subdivision of a family, subordinate to the subfamily when that taxon is employed. We form the name by adding *eae* to the stem of an included generic name, such as *Festuceae,* from *Festuca,* for the Fescue Tribe of the Grass Family.

The Authority

Since some person, or occasionally two or more persons, originally published an account of and described each taxon, whether family, genus, or species, and gave it a name, the name of the person or persons written after the scientific name is known as the authority of the name. The author's name may be written out, but more commonly it is indicated by a standardized abbreviation. Since, for example, *Poa pratensis* (Kentucky Bluegrass) was first named and described by Linnaeus, he becomes the authority for that name, and it is written as *Poa pratensis* L. Similarly, *Erythronium grandiflorum*

Pursh, for the Glacier Lily, shows that Pursh first named the species. *Lomatium montanum* C. & R., a member of the Parsnip Family, was first named and described by two men, Coulter and Rose, working together on a revision of that group of plants. Supplying the authority for a scientific name is a standard procedure in all serious work; it pins down the name with greater certainty and avoids the misunderstanding that might arise if two or more persons, unknown to each other, publish the same name.

THE PARENTHETICAL AUTHORITY. When the rank of a plant or plant group is changed, or when a species is transferred from one genus to another, or whenever similar changes in nomenclature are made, the name of the original author is placed in parentheses and is followed by the name of the person making the change. Thus *Medicago polymorpha* L., variety *orbicularis* L., originally so designated by Linnaeus, is raised to specific rank by Allioni, and then becomes *Medicago orbicularis* (L.) All. And when *Ferula foeniculacea* Nutt., first described by Nuttall, is transferred to the genus *Lomatium* by Coulter and Rose, it becomes *Lomatium foeniculaceum* (Nutt.) C. & R.

The Principle of Priority

In order to avoid confusion in the application of scientific names and to eliminate duplication, botanists have agreed that there will be a system of priority: This means, in general, that names published earlier take precedence over names of the same rank published later; the first validly published name of a species or other taxon becomes its valid name. If other names are subsequently published for the same taxon, they shall become *synonyms,* or invalid names. In order to have a definite starting point for such a system of priority, botanists have agreed that for the flowering plants the starting point shall be Linnaeus' *Species Plantarum,* published in 1753. Exceptions to this rule, applying only to generic names at present, are agreed upon from time to time and are listed in the *International Code of Botanical Nomenclature* as *nomina generica conservanda et rejicienda,* a list of conserved and rejected names. The starting point for accepted family names is A. L. de Jussieu's *Genera Plantarum* (1789), and a list of conserved family names may be found in the *International Code.*

Thus, as stated earlier, there can be only one legitimate name for any plant, but it may have many synonyms in the literature. For this reason it is also common practice, at least in detailed treatments, to cite the publication and date for each name. This is known as the *citation.* For example, "*Cypripedium knightae* A. Nels., Bot. Gaz. **42:**

48 (1906)" means that the species was proposed by Aven Nelson in the *Botanical Gazette,* volume 42, page 48, in 1906.

Summary of Taxa Used in Classification
(Applied to *Poa pratensis,* Kentucky Bluegrass)

- KINGDOM—*Plantae,* the plant kingdom.
  - DIVISION—*Embryophyta,* the embryo plants.
    - SUBDIVISION—*Phanaerogama,* the seed plants.
      - BRANCH—*Angiospermae,* the angiosperms, seeds enclosed in an ovary.
        - CLASS—*Monocotyledoneae,* the monocotyledons.
          - SUBCLASS—*Glumiflorae,* those with chaffy flowers.
            - ORDER—*Poales,* the grasses and sedges.
              - FAMILY—*Poaceae,* the Grass Family.
                - SUBFAMILY—*Festucoideae,* those resembling *Festuca.*
                  - TRIBE—*Festuceae,* the Fescue Tribe.
                    - GENUS—*Poa,* the Bluegrasses.
                      - SECTION—*Pratenses,* those with rhizomes.
                        - SPECIES—*Poa pratensis,* Kentucky Bluegrass.

Infrageneric Taxa

Sometimes the taxon *subgenus* is used between the genus and the section, particularly if the genus is naturally divisible into only a few major groups. The name of a subgenus is a substantive resembling the generic name, such as *Trifoliastrum* in the genus *Trifolium,* or *Drabella* in the genus *Draba.* But the rules also provide that the subgenus containing the type of a generic name (see below for the meaning of types) must bear that name unaltered; the subgenus of *Astragalus* that includes the type of that genus (*Astragalus christanus* L.) is therefore named *Astragalus,* not, as formerly, *Euastragalus.*

If there are several or more fairly equal subdivisions of a genus, the taxon *section* is applied to each. Or the genus may be first divided into subgenera and each of these divided into sections. Names of sections are formed in the same manner as names of subgenera.

If a section is large, containing many species, it may be found desirable to divide it into *subsections,* in which case the names are preferably plural adjectives agreeing with the generic name in gender, and written with a capital initial letter.

Infraspecific Taxa

When it is desired to show subdivisions of a species, several designations are possible. These are, in descending order of magnitude, the *subspecies,* the *variety,* and the *forma,* or form. Unfortunately, clear distinctions are not always drawn between subspecies and varieties, and the two categories are used more or less interchangeably. Subspecies and varieties are usually associated with inheritable differences, races, etc., and their scientific names are formed in the same way as those of species, being trinomials with the appropriate designation interposed between the last two epithets. The forma is usually associated with environmentally caused differences of a minor nature, and its name also is formed in the same way as the specific epithet.

The typical phase of the species is now designated by a trinomial that repeats the specific epithet exactly. The typical phase of *Sidalcea candida* Gray is *Sidalcea candida* Gray var. (or subsp.) *candida,* without the addition of any author's name for the second epithet. The typical phase of the species, of course, is the one that includes the type specimen of the species, whether or not this happens to be the commonest or the best-known phase. Trinomials of other phases than the typical one should be followed by an author's name, as in *Sidalcea candida* var. *glabrata* Hitchcock.

The Type Method

In order to stabilize the concepts of taxa from species and their subdivisions upward through orders, botanists have generally adopted the type method. Above the rank of order this method, it is felt, cannot be applied with profit at the present time.

The method requires, briefly, that the author of a species must designate a certain specimen (or an acceptable substitute for a specimen) as the type of that species. This thereafter becomes the nomenclatural type of that species, permanently associated with it in fixing the application of the name given it. The nomenclatural type of a genus then becomes the species on which the generic name was based; the nomenclatural type of a family is the genus on which the family name was based; and the nomenclatural type of an order is the family on which the ordinal name was based. It follows that, if the type of a name is excluded from a taxon for any reason, the name of the taxon must be changed. It also follows that the type of a taxon is not necessarily its most characteristic or representative phase; it is merely that element on which the name was originally based. For example, a certain specimen carefully preserved in a herbarium is the type of the species *Poa pratensis* L. This specimen fixes definitely the strict concept of that species and the application of its name; the genus *Poa* is typified by the species *Poa pratensis* L.; the family *Poaceae* is typified by the genus *Poa;* and the order *Poales* is typified by the family *Poaceae.*

Since catastrophes such as bombings, or fires or other accidents, may damage or destroy preserved botanical specimens (including types, which are the most valuable of all specimens), and in order to put the system into effect in groups not recently revised, botanists have devised a system for the designation of acceptable substitutes. The following terminology of type materials will indicate these and other procedures:

holotype (type): the particular specimen or element designated by the author, which automatically fixes the application of the name.

lectotype: a specimen or element selected by a competent worker from the original material studied by the author to serve as a substitute for the holotype if the latter was not designated in the original publication or is missing. A lectotype takes precedence over the following type.

neotype: a specimen selected to serve as a substitute for the holotype when all material on which the name was based is missing.

isotype: any specimen, other than the holotype, that duplicates the holotype (from the same collection, with the same locality, date, and number as the holotype).

paratype: any specimen, other than the holotype, referred to in the original publication of the taxon. Earlier workers often referred to these as "co-types."

syntype: one of two or more specimens or elements used by the author of a taxon if no holotype was designated; or one of two or more specimens designated as types simultaneously in the original publication. The designation "co-type" has also been used here.

topotype: a specimen collected at the same locality as the holotype and therefore probably representing the same population.

REFERENCES

BAILEY, L. H. 1933. *How Plants Get Their Names* (New York). Reprinted by Peter Smith, Gloucester, Mass.

BAILEY, L. H. 1949. "Problems in taxonomy" (Symposium on Botanical Nomenclature, VII), *Am. J. Bot.* **36**:22–24.

BLAKE, S. F. 1943. "Cotype, Syntype, and other terms referring to type material," *Rhodora* **45**:481–485.

BLAKE, S. F. 1949. "Byways of nomenclature" (Symposium on Bontanical Nomenclature, III), *Am. J. Bot.* **36**:8–10.

BLAKE, S. F. 1956. "Terms used to designate type material," *Madroño* **13**:207.

BULLOCK, A. A. 1959. 'Nomina familiarum conservanda proposita," *Taxon* **8**:154–181, 189–205.

CROIZAT, LEON. 1945. "History and nomenclature of the higher units of classification," *Bull. Torr. Club* **72**:52–75.

CROIZAT, LEON. 1953. On nomenclature: The type method,'" *Taxon* **2**:105–107, 124–130.

CRONQUIST, ARTHUR. 1960. "The divisions and classes of plants," *Bot. Rev.* **26**:425–482.

DRESS, W. J. 1955. "On the gender of scientific plant names," *Baileya* **3**:59–63.

EPLING, CARL. 1939. "An approach to classification," *Sci. Monthly* **49**:360–367.

FRIZZELL, D. L. 1933. "Terminology of types," *Am. Midl. Nat.* **14**:637–668.

HALL, H. M. 1926. The taxonomic treatment of units smalled than species," *Proc. Int. Bot. Congr.* **2**:1461–1468.

HITCHCOCK, A. S. 1921. The type concept in systematic botany," *Am. J. Bot.* **8**:251–255.

HITCHCOCK, A. S. 1926. The relation of nomenclature to taxonomy," *Proc. Int. Bot. Congr.* **2**:1434–1439.

LANJOUW, J. (*editor*). 1956. *International Code of Botanical Nomenclature* (Utrecht, Netherlands).

MERRILL, E. D. 1949. "Adventures in locating validly published but unlisted binomials" (Symposium on Botanical Nomenclature, V), *Am. J. Bot.* **36**:14–19.

PENNELL, F. W. 1919. "Concerning duplicate types," *Torreya* **19**:14.

PENNELL, F. W. 1949. "Toward a simple and clear nomenclature" (Symposium on Botanical Nomenclature, VI), *Am. J. Bot.* **36**:19–22.

PESANTE, ALDO. 1961. "About the use of personal names in taxonomic literature," *Taxon* **10**:214–221.

RICKETT, H. W. 1948. "Orthography in botanical nomenclature," *Brittonia* **6**:365–368.

RICKETT, H. W. 1953. "Expediency vs. priority in nomenclature," *Taxon* **2**:117–124.

RICKETT, H. W., and W. H. CAMP. 1948. "The nomenclature of hybrids," *Bull. Torr. Club* **75**:496–501.

RICKETT, H. W., and W. H. CAMP. 1950. "The application and use of botanical names," *Bull. Torr. Club* **77**:245–261.

ROSENDAHL, C. O. 1949. "The problem of subspecific categories" (Symposium on Botanical Nomenclature, VIII), *Am. J. Bot.* **36**:24–28.

SAVORY, THEODORE. 1963. *Naming the Living World* (Wiley, New York).

SMITH, A. C. 1945. "The principle of priority in biological nomenclature," *Chron. Bot.* **9**:114–119.

SMITH, A. C. 1957. "Fifty years of botanical nomenclature," *Brittonia* **9**:2–8.

TIPPO, O. 1942. "A modern classification of the plant kingdom," *Chron. Bot.* **7**: 203–206.

WEATHERBY, C. A. 1949. "Botanical nomenclature since 1867" (Symposium on Botanical Nomenclature, II), *Am. J. Bot.* **36**:5–8.

CHAPTER 6

Concepts of Taxa

If our systems of classification and nomenclature are to be meaningful and stable, we must arrive at a reasonable interpretation of the various taxa, or units of classification, and these concepts must be widely acceptable. Much work has been done in recent years to crystallize our thinking along these lines, most of it aimed at the interpretation of the species and its subdivisions. Far less thought has gone into the understanding of the larger taxa, such as the order.

The Concept of the Species

Before the doctrine of evolution was accepted, it was generally believed that the different kinds of organisms owed their origin to *special creation,* and that these discrete and immutable entities, or species, were incapable of change or intergradation by hybridization or other means. This early concept of the species, a simple, arbitrary, and comforting one, was summarized by the youthful Linnaeus in his *Classes Plantarum* (1738) in the following words: "There are as many species as there were originally created diverse forms."

In his more mature years, however, Linnaeus radically revised his earlier concept in accord with his discovery that distinct species of plants could be hybridized, and in his *Systema Vegetabilium* (1774) he stated:

Let us suppose that the Divine Being in the beginning progressed from the simpler to the complex; from few to many; similarly that He in the beginning of the plant kingdom created as many plants as there were natural orders. These plant orders He himself, therefrom producing, mixed among themselves until from them originated those plants which today exist as genera.

Nature then mixed up these plant genera among themselves through generations of double origin and multiplied them into the existing species, as many as possible (whereby the flower structures were not changed), excluding from the number of species the almost sterile hybrids, which are produced by the same mode of origin.

This theory of evolution, proposed by Linnaeus almost a hundred years before that of Darwin, was overlooked by most investigators, who were preoccupied with naming and describing all the new plants being discovered at that time. The significance of Linnaeus' hybridization experiments, as well as those of J. G. Koelreuter in 1761–1766 and C. F. Gaertner in 1849, lies in the establishment of the fact that hybridization can occur, that species hybrids are mostly sterile (but not always completely so), and that varieties of a single species may be crossed to produce fertile offspring.

The next step in the evolution of the concept of species was the observation by Alexis Jordan, a French botanist who published his findings in 1846, that races of one Linnaean species of *Viola* (Violets and Pansies) remained distinct and recognizable when grown under standard conditions in a garden. He interpreted these local races as species, however, and named them accordingly; but what he had really discovered was the fact that *many species consist of local populations* whose members are interfertile yet maintain themselves as recognizable units, often occupying separate ecological niches.

Charles Darwin, in *The Origin of Species* (1859), also called attention to the considerable degree of variation that exists in living things, and brought out *the significance of natural selection and its effect upon survival.*

A major contribution to our understanding of the problem of species was the research by Gregor Mendel, an Austrian monk, who in 1865 discovered the *basic principles governing inheritance,* although his studies did not come to the attention of biologists until 1900. Mendel's work provided a working basis for understanding the mechanism of inheritance as well as a partial explanation of the great variation that was known to occur in various organisms.

Wilhelm Johannsen, experimenting with cultivated beans (1903–1911), showed that *two kinds of variation occurred within a species* and jointly determined its outward appearance (phenotype): (1) variation due to inheritance, which could be transmitted to the offspring; (2) variation caused by the environment, which was not inheritable. It naturally followed that the forces of natural selection could act effectively only on inheritable variations.

The studies of W. A. Cannon, W. S. Sutton, and M. F. Guyer in 1902, and those of T. H. Morgan in 1911, gave us our knowledge of the behavior of chromosomes at the time of formation of gametes, of *how chromosomes pair* (enter into synapsis) during reduction division and are then redistributed to the sex cells. This knowledge was most significant, for inability to pair (asynapsis), due to the coming together of different numbers of chromosomes, to differences in the size and shape of chromosomes, or to any genetic dissimilarity, might account, in part, for the sterility barrier that usually exists between species. And a little later, in 1917, Öjvind Winge showed that related plant species might differ cytologically by having various multiples of a basic chromosome number, thus *establishing the fact of polyploidy,* now known to occur very commonly in plants. Polyploid populations are ranked as species in some instances, as varieties in others, and sometimes they are not even assigned any taxonomic rank.

It has been estimated by Stebbins (see References) that about half the species of angiosperms and three-fourths of the grasses are polyploids or of polyploid origin. Furthermore, it appears that polyploids occur more frequently in severe climates of the mountains and the northern latitudes than in moderate climates.

Polyploids differ in kind: *autoploids,* in which a single kind of genome is multiplied; and *alloploids,* in which different genomes are combined. Autoploids, therefore, are less likely to be of taxonomic significance than are alloploids.

Sometimes polyploids are visibly different from diploids. They often have larger flowers and fruits, and their cells, including pollen grains and epidermal cells, are often larger. They often have a darker green color of foliage, they may contain more vitamins, and they may have a stockier habit. Polyploids are likely to be more variable and to show an ability to tolerate a wider range of habitats. Hybrids of diploids are often sterile, but tetraploid hybrids are often fertile.

The *discovery of mutations* about 1900 opened up a new avenue of attack on the problem of speciation. These spontaneous and unpredictable changes sometimes resulted in striking inheritable differences in the outward appearance of plants and animals. We now know that mutations may involve invisible characteristics as well. Examples of mutations in flowering plants are double-flowered and cut-leaved individuals, and white-flowered plants of species normally having colored flowers. These mutations would, of course, come under the selective influence of the environment; so they might or might not have survival value. In more recent years we have learned that gene mutations can sometimes be induced or accelerated by artificial means, such as exposure of the organism to X-radiation, ultraviolet rays, variations in temperature, and mustard gas. At present we are all concerned about the possible effects of radioactive fallout.

We also know, through cytological examination of thousands of plants and animals in various species, genera, and families, that *the basic chromosome number and morphology frequently vary* considerably among the species as well as among the larger taxa. In some genera the evolutionary tendency has been toward a reduction in chromosome number, in others toward an increase. One genus in which the species have different somatic numbers is *Nymphaea* (Pond Lily), some of whose species and chromosome numbers are *stellata* 28, *lotus* 56, *odorata* 84, *candida* 112, and *gigantea* about 224. Such studies enable the taxonomist to work out a system of classification that reflects the true relationships of the plants more closely than gross morphological criteria alone do. They may also validate systematic arrangements and relationships that had been inferred by other means. It is important, of course, that cytological investigations be backed up by specimen vouchers of the plants investigated, and that these specimens be deposited in permanent collections where anyone may consult them if any doubt should arise as to their true identity.

Göte Turesson, in 1921–1931, pointed out that *the survival of a plant depends on its physiological fitness to its environment* rather than on its morphological characteristics. He showed that *species occupying large geographic areas are composed of ecological races, or what he called ecotypes.* Many taxonomists had recognized the existence of such races but had often interpreted them as species.

In recent years it has been found that a considerable number of plants are capable of reproducing not only sexually, in the normal manner, but also by the development of an embryo without fertilization. Such reproduction is termed *apomixis,* and the resulting plants are called *apomicts.* This method of reproduction is common in such genera as *Festuca* (Fescue grasses), *Poa* (Bluegrasses), *Crepis* (Hawksbeard), *Hieracium* (Hawkweeds), and *Taraxacum* (Dandelions), to name only a few. It is also known that there are many degrees of apomixis, some plants being obligate apomicts and incapable of sexual reproduction, others being facultative apomicts and capable of sexual reproduction as well. It will be readily seen that in facultative apomicts the sterility barrier that usually accompanies hybridization between sexual species could be completely removed by apomixis, and that the apomicts resulting from any one of these hybrids would be genetically uniform. An example is Wheeler Bluegrass (*Poa nervosa*), a common species of the western United States, with uniform apomictic plants occurring east of the Cascade Range and variable sexual plants occurring west of the Cascades. The loss of variability in a species, brought about by apomixis, could well be a detriment to survival because of a loss of adaptability.

It seems logical, therefore, that apomicts, reproducing asexually, as plants do from cuttings, can hardly be classified as species in the same sense as plants of normal sexual origin. Some of the "species" we know are definitely of this type. The recommendations of the *International Code* for the designation of apomicts are as follows: (1) if the group is considered to be a species, the abbreviation "ap." is placed between the generic name and the specific epithet; (2) if the group is below the rank of species, the abbreviation "ap." is placed before the final infraspecific epithet.

Much experimentation on several groups of plants has been carried on in California by Jens Clausen, D. D. Keck, and W. M. Hiesey (1936–1952), who have gone into the problems of speciation from the standpoints of the morphologist, the ecologist, the cytologist, the geneticist, and the physiologist. Their findings they have summarized as follows:[1]

> . . . plants are organized into groups, the members of each of which are able to interchange their genes freely in all proportions without detriment to the offspring. Such groups are separated from one another by internal barriers that are of a genetic-physiologic nature (including chromosomal barriers) [and] that prevent such free interchange. These natural groups correspond fairly closely to the species of the moderately conservative taxonomists working with plants that reproduce sexually.
>
> This criterion for species, now substantiated by experiment, is the same that Turesson (1922) previously applied to the ecospecies. Consequently, we use his terminology to distinguish species whose status has been determined by experiment. The ecospecies becomes the experimental homologue of the taxonomic species. Also Dobzhansky (1937) has recently called attention to the importance of the internal ("physiologic") barriers separating species, noting that commonly they coincide with the delimitations of the species as accepted by systematists. [*Am. J. Bot.* **26**:104 (1939.]

The following tabular form presents the concept given above in condensed form, as subsequently published by Clausen.

[1] The quotation is used with the kind permission of the Editor-in-Chief of the *American Journal of Botany* and of Dr. Jens Clausen, the senior author.

CHART SHOWING THE EVOLUTIONARY ENTITIES WITHIN THE GENUS OR COMPARIUM

<table>
<tr><th rowspan="2">MORPHOLOGY</th><th rowspan="2">ECOLOGY</th><th colspan="3">GENETIC RELATIONSHIPS</th></tr>
<tr><th>HYBRIDS FERTILE (2nd generation vigorous)</th><th>HYBRIDS PARTIALLY STERILE (2nd generation weak)</th><th>HYBRIDS STERILE (Or none)</th></tr>
<tr><td rowspan="2">Morphologically distinct</td><td>In distinct environments</td><td>Distinct subspecies (ECOTYPES) of one species</td><td>Distinct species (ECOSPECIES)</td><td rowspan="2">Distinct species complexes (CENOSPECIES)</td></tr>
<tr><td>In the same environment</td><td>Local variations of one species, or varieties (BIOTYPES)</td><td>Species overlapping in common territory (hybrid swarms)</td></tr>
<tr><td rowspan="2">Morphologically similar</td><td>In distinct environments</td><td>Distinct ecotypes of one species, subspecies</td><td colspan="2" rowspan="2">Genetic species only (autoploidy or chromosome repatterning)</td></tr>
<tr><td>In the same environment</td><td>Taxonomically the same entity</td></tr>
</table>

Source: Slightly modified from Jens Clausen, *Stages in the Evolution of Plant Species.* Used by permission of the author.

In the foregoing tabular form, the systematic units based on experimental evidence are in capitals, their homologues based on external characteristics in italics. It should be emphasized that the terms BIOTYPES, ECOTYPES, ECOSPECIES, and CENOSPECIES should be used only when experimental evidence is at hand to validate their use; otherwise the commonly used terms *variety*, *subspecies*, *species*, and *species complex* should be employed. Two kinds of barriers to interbreeding are recognized: (1) internal barriers, which are of a genetic-physiologic nature expressed through incompatibility and intersterility or through weakness of the hybrid offspring; (2) external barriers, which are environmental and ecologic-geographic. The internal barriers are the more enduring, for the external barriers may be broken down by changes in the earth's surface.

It must be apparent that no universal definition of species is likely to be forthcoming even though a definite concept may be formulated for any group of plants. When plants reproduce by purely sexual means, in the usual fashion, the problem is more easily resolved on the basis of sterility barriers and over-all morphology and geography. But when plants reproduce asexually, by apomixis or other means, only experience and judgment can bring about a reasonable working system of classification. Perhaps it is reasonable, for practical purposes, to interpret a species as a recognizable and self-perpetuating population that is more or less isolated genetically as well as by its geographic distribution and its environment.

The Concept of the Genus

Perhaps the oldest of all concepts of taxa developed by man is the generic one. Even a cursory examination of the literature of all languages shows that all have words that express well-known generic concepts: in English such words as Oak, Pine, Buttercup, Violet, Maple, and Clover. Even primitive races have fairly sound concepts of natural groups of plants, which are roughly (and sometimes exactly) equivalent to genera, and words to identify them. In fact, whatever the culture of the people, and whatever their language—Chinese, Malay, Egyptian, Greek, Eskimo, or Aztec—all types of ancient and modern civilizations were aware of the group, or generic, idea, with subdivisions to indicate kinds, or species. In a sense, this idea has brought about the development of our present system of nomenclature. When we name people, the surname takes on the generic sense and indicates the group, while the given name indicates the individual. (In our civilization, however, the sequence of names is the reverse of that used in science, wherein the generic name precedes the specific epithet.)

The modern, Latinized, single words for genera are usually thought of as dating back to Tournefort (about 1700), who is also credited with giving the first consistent characterization of genera, although Brunfels' herbal (1532) had previously contained many such generic names. Linnaeus merely adopted Tournefort's generic concepts and enlarged upon them. He also adopted the generic names proposed by Charles Plumier (1703), a contemporary of Tournefort, who named more than 900 American plants and assigned each to a definite genus. Linnaeus did believe, however, that a genus should be a *natural* entity whose species show close genetic affinities.

The criteria for valid genera, then, should be morphological similarity and genetic affinity of the included species. No rule of thumb that will always apply can be set down; but, as suggested by Tournefort, a *similarity of flowers and fruits* often makes the best criterion.

Similarities of other organs, such as roots, stems, and leaves, may also be used; and the characteristics of seeds, seedlings, and embryos are often generically diagnostic, as in the *Chenopodiaceae* and *Brassicaceae.* The form and arrangement of the leaves of various conifers, for example, yield generic distinctions into Pine, Spruce, Fir, and so on, whereas the reproductive structures would unite these into one genus. Linnaeus held much the same view, believing that *morphological combinations* furnished the best clues to generic segregation. Anatomy, as pointed out by I. W. Bailey (1953) and others, may also yield clues to the proper disposition of misfit genera in certain families; it has been used quite effectively, in fact, in the disposition of recently discovered relic genera whose affinities are in doubt. The presence, for example, of vesselless xylem in certain primitive angiosperms found in the Southwest Pacific region seems to indicate that these plants are what is left of an ancient, diversified, woody, dicotyledonous flora. Similar studies have been made of the anatomy of other parts of plants, including flowers. Much more work of this nature needs to be done in order to clear up some of the confusion in various parts of our classification system. Just as cytological investigations have become an almost routine part of monographic studies, so should anatomical studies be made with the same ultimate goal of establishing a sound basis for delimiting families, genera, and species.

There has been a tendency in modern times to narrow down the generic concepts of Linnaeus when research has shown that certain groups were not natural entities. Some workers even take a numerical approach and split genera that contain unusually large numbers of species, forgetting that *large genera may be just as natural as small genera.*

In order to comprehend a genus fully, one should study it throughout its entire range; otherwise misconceptions are likely to arise through provinciality of knowledge. Persons familiar with *Senecio* (Old Man) in the northern part of its range, where the plants are herbaceous and sometimes only a few inches high, might form a quite different concept of that genus than those knowing it in tropical regions, where the plants may become tree-like; and the arborescent *Cornus florida* (Flowering Dogwood), which extends into the southern United States, is quite different in habit from the diminutive and herbaceous *Cornus canadensis* (Bunchberry), which extends into the Arctic. The tropical and northern counterparts of many genera are often, in fact, very unlike in growth form; but transitional forms often connect them.

Convenience is also to be reckoned with when we delimit genera;

it is desirable to be able to assign at least a generic name to a plant that is unknown. If, for example, generic distinctions hinge largely or entirely on fruit characters, and if a plant does not produce flowers and fruits at about the same time, specimens in flower alone could not be placed in a genus with certainty, and we should have no name at all to give the plant. A situation such as this has arisen in the large and diversified genus *Astragalus* (Vetch), the American species of which were once divided among twenty-eight genera. Other genera that were once dismembered but have been reunited are *Oenothera* (*Anogra, Onagra, Lavauxia, Pachylophis, Galpinsia, Meriolix, Taraxia, Sphaerostigma, Chylismia,* etc. for various Evening Primroses) and *Haplopappus* (*Oonopsis, Pyrrocoma, Stenotus, Macronema, Isocoma, Oreochrysum, Isopappus, Ericameria,* etc. for a large group of *Asteraceae*).

Cytological investigations have yielded some useful information concerning relationships at the generic level as well as at the level of the species. In the *Ranunculaceae* (Buttercup Family), for instance, the following basic numbers and relative sizes of chromosomes have been determined, showing that a combination of chromosome number and size provides evidence in support of the integrity of the genera listed:

Ranunculus: 7 or 8, medium
Paeonia: 5, large
Anemone: 8, large
Hepatica: 7, large
Caltha: 16, medium
Coptis: 9, small
Thalictrum: 14, small
Aquilegia: 7, small
Nigella: 6, large
Aconitum: 8, medium

Since the modern tendency, and a good one, is to take a rather conservative view of generic concepts, it should not be difficult to give any plant at least a generic name, a convenient handle for the nonspecialist, and at least a temporary home among its relatives in the herbarium. The desirability of a reasonably stable generic concept is further emphasized by the fact that every change from established usage involves creating a new set of binomials for the species.

The Concept of the Family

What has been said about the concept of the genus also applies in large part to that of the family. The common human habit of grouping things, involving categories larger than the genus, may embrace the family as well, as is illustrated by such well-known concepts and terms as *grasses, legumes,* and *orchids.*

The delimitation of families of flowering plants has been fairly well worked out by taxonomists, leaving only a few disagreements here and there—whether, for example, to unite or keep as discrete families the various elements of the *Liliaceae* (Lily Family), the *Fabaceae* (Pea Family), the *Ericaceae* (Heath Family), and the *Asteraceae* (Aster Family).

The characteristics forming the basis of family distinctions are as diverse as those used to distinguish genera, and in general are of the same sort—namely, *combinations of morphological features*, particularly those of flowers and fruits. Because of the frequent diversity of things included within a single family, and since hard-and-fast definitions frequently necessitate an allowance for exceptions, it is often profitable to think of family characteristics as *tendencies*. As examples of combinations of morphological features we might consider the following: *Brassicaceae* (Mustard Family), mostly herbs with pungent watery juice and alternate exstipulate leaves, the flowers often produced in ebracteate racemes, hypogynous, polypetalous, 4-merous, regular, the pistil of 2 united carpels separated by a persistent septum, the stamens usually 2 short and 4 long, the fruit a silique or silicle; or the *Lamiaceae* (Mint Family), plants aromatic, with square stems and opposite leaves, the flowers hypogynous, sympetalous, with irregular corolla, the stamens 2 long and 2 shorter, or only 2, the ovary deeply 4-lobed, and the fruit of 4 seed-like nutlets.

Again it should be emphasized that the ultimate goal in developing concepts of families is to recognize *natural groups*, or combinations of genera that express separate evolutionary trends and relationships. Some families may be very small; others may be huge but may nevertheless be natural groups.

The Concept of the Order

Just as the genus is a natural group of species, and the family a natural group of genera (occasionally these taxa may consist of single components), so the order, ideally, includes one or more families that show definite affinities and similar evolutionary trends. At this point, however, we find the greatest degree of disagreement among the classifiers. The generalization that an order should be a natural group of families is agreeable to all, but many different ideas have been expressed about the setting of ordinal limits. Should we have a few large orders, as suggested by Bessey, or many smaller orders, each containing only one or a few families, as proposed by Hutchinson?

In either case, it is again combinations of morphological characters that determine the limits. The *Rosales* (Rose Order), for example, might be thought of as including a large segment of the perigynous *Polypetalae*, the *Rosaceae, Saxifragaceae, Fabaceae*, etc.; or the *Fabaceae* (Pea Family) may be excluded from this group to make a separate order, based on the leguminous fruit character and a tendency toward irregular corollas and diadelphous stamens.

As the amplitude of a taxon increases beyond that of the genus, the scope of the problem often becomes too great for solution by any one person. A lifetime of research may result in a working knowledge of a genus, if it is not excessively large, or one may become well versed in the intricacies of a family, again if it is not too large; but the chances are slight that one person could become an expert on a large order. Through cooperation between experts, however, and by use of all the information available, a reasonably good understanding of the larger taxa may be achieved. One effort of this sort is Arthur Cronquist's "Outline of a new system of families and orders of dicotyledons," which presents some realignments, circumscribes the orders the author recognizes, gives the evolutionary trends of each, and offers a distinguishing key to the orders.

Recent Developments

Science rarely advances in a steady, orderly manner. New techniques, new tools, and new discoveries often furnish a sudden spurt of thinking and research along a particular line, especially when this is abetted by grants or other financial inducements. Unfortunately, some of these developments take on a fashionable aspect and become particularly attractive to the younger worker. Taxonomy is no exception to this.

There are, however, two promising developments that are now influencing taxonomic thinking and interpretation. Time and much research will determine their worth, but both show much promise of helping to solve problems relating to evolution, classification, and speciation, and both may help to clear up matters pertaining to the validity of certain plant groupings. These are *chemical taxonomy*, an offshoot of biochemistry, and *numerical taxonomy*, an offshoot of statistics and computer technology.

Chemical taxonomy is not new. It has been employed since ancient times by medical practitioners interested in drugs, by the spice industry, by the manufacturers of perfumes and dyes, and by many others. Plant poisons have long been known: we are told of the death of Socrates brought about by an alkaloid from a plant known as Poison Hemlock (*Conium maculatum*) of the Parsnip Family. Insecticides, fish poisons, tranquilizers, and many other products derived from plants have long been used. The serum diagnostic work of Karl Mez, discussed later, and the methods of bacteriologists involve chemical taxonomy and are of long standing.

With the advent of refined biochemical methods and the use of such modern laboratory tools as apparatus for chromatography and electrophoresis, we possess the means of extending our knowledge of relationships based on plant biochemistry. The presence or absence of certain compounds and the particular compounds produced in a plant group may be determined, or at least it can readily be shown that a compound or group of compounds, known or unknown, appears in a given sample. Some of the compounds being investigated include alkaloids, terpenes, pigments, flavones, fatty acids, and amino acids (the last probably of least systematic value).

Chemical taxonomy, then, is another weapon we may employ in our struggle to arrive at valid concepts of various taxa, large or small. There are, however, certain pitfalls to guard against: the same convergent and reticulate evolutionary patterns occur here as elsewhere. Students wishing further information on this subject may consult the publications of Alston and Turner, including their fine text, the book edited by Swain, and the book by Trevor Robinson.

Numerical taxonomy or "taximetrics," as it has been called, is not entirely new either. It is based on a principle set down about two hundred years ago by the French botanist Michel Adanson, who opposed weighting of characters possessed by plants in determining a phylogenetic system.

The development of sophisticated computers and the newer methods of the science of statistics have given an impetus to this kind of research, and there has emerged a science of "computer taxonomy" that is ridiculed by some and highly recommended by others. All known, inherent characters of each taxonomic unit (at least fifty) are recorded and coded, and comparisons are made by machine, by the use of a formula for coefficients of association, and by the use of a matrix that shows the similarity between the units being tested. The main objective is to eliminate personal bias in arriving at results. Many taxonomists, however, are unwilling to believe that all the characters recorded are equally important, and therein lies the difference of opinion among taxonomists as to the validity of the method. Just how successful this approach will be remains to be seen. One thing that should not be overlooked is the efficiency of that original model of all computers, the human brain! Students seeking further information on numerical taxonomy may consult the following references: Kendrick, Michener and Sokal; Sokal and Sneath; and Rogers.

REFERENCES

ALLEN, J. A. 1908. "Another aspect of the species question," *Am. Nat.* **42**:592–600.

ALSTON, R. E., T. J. MABRY, and B. L. TURNER. 1963. "Perspectives in chemotaxonomy," *Science* **142**:545–552.

ALSTON, R. E., and B. L. TURNER. 1963. *Biochemical Systematics,* Prentice-Hall, Englewood Cliffs.

ANDERSON, EDGAR. 1937. "Supra-specific variation in nature and in classification—from the viewpoint of botany," *Am. Nat.* **71**:223–235.

ANDERSON, EDGAR. 1941. "The technique and use of mass collections in plant taxonomy," *Am. Mo. Bot. Gard.* **28**:287–292.

ANDERSON, EDGAR, and E. C. ABBE. 1934. "A quantitative comparison of specific and generic differences in the *Betulaceae,*" *J. Arn. Arbor.* **15**:43–49.

ANDERSON, EDGAR, and T. W. WHITAKER. 1934. "Speciation in *Uvularia,*" *J. Arn. Arbor.* **15**:28–42.

ANDERSON, EDGAR, and G. L. STEBBINS. 1954. "Hybridization as an evolutionary mechanism," *Evolution* **8**:378–388.

BABCOCK, E. B. 1930. "Cytogenetics and the species concept," *Proc. 5th Int. Bot. Congr.,* pp. 216–218.

BABCOCK, E. B. 1934. "Basic chromosome numbers in plants with special reference to the *Compositae,*" *New Phytol.* **33**:386–388.

BAILEY, I. W. 1953. "The anatomical approach to the study of genera," *Chron. Bot.* **14**:121–125.

BAILEY, L. H. 1930. "Statements on the systematic study of variables," *Proc. 5th Int. Bot. Congr.,* pp. 1427–1433.

BARTLETT, H. H., *et al.* 1940. "The concept of the genus," *Bull. Torr. Club* **67**:349–389.

BEAUDRY, J. R. 1960. "The species concept: Its evolution and present status," *Rev. Canad. Biol.* **19**:219–240.

BESSEY, C. E., *et al.,* 1908. "Aspects of the species question," *Am. Nat.* **42**:218–281.

CAMP, W. H. 1951. "Biosystematy," *Brittonia* **7**:113–127.

CAMP, W. H., and C. L. GILLY. 1943. "The structure and origin of species," *Brittonia* **4**:323–385.

CASPARI, ERNST. 1948. "Cytoplasmic inheritance," *Adv. Genet.* **2**:1–66.

CLAUSEN, JENS. 1951. *Stages in the Evolution of Plant Species.* Hafner, New York.

CLAUSEN, JENS. 1959. "Gene systems regulating characters of ecological races and subspecies," *Proc. X Int. Congr. Genet.* **1**:434–443.

CLAUSEN, JENS, D. D. KECK, and W. M. HIESEY. 1936–1941. *Experimental Taxonomy,* Carnegie Inst. Wash. Yearbooks 35–40.

CLAUSEN, JENS, D. D. KECK, and W. M. HIESEY. 1939. "The concept of the species based on experiment," *Am. J. Bot.* **26**:103–106.

CLAUSEN, JENS, D. D. KECK, and W. M. HIESEY. 1941. "Regional differentiation in plant species," *Am. Nat.* **75**:231–250.

CLAUSEN, JENS, D. D. KECK, and W. M. HIESEY. 1947. "Heredity of geographically and ecologically isolated races," *Am. Nat.* **81**:114–133.

CLAUSEN, JENS, D. D. KECK, and W. M. HIESEY. 1950. *Experimental Studies on the Nature of Species:* I, *Effect of Varied Environments on Western North American Plants,* Carnegie Inst. Wash. Publ. 520 (second printing).

CRONQUIST, ARTHUR. 1957. "Outline of a new system of families and orders of dicotyledons," *Bull. Jard. Bot.* (*Bruxelles*) **27**:13–40.

DARLINGTON, C. D., and A. P. WYLIE. 1956. *Chromosome Atlas of Flowering Plants* (2nd ed.). Macmillan, New York.

DAVIDSON, J. F. 1947. "The polygonal graph for simultaneous portrayal of several variables in population analysis," *Madroño* **9**:105–110.

DOBZHANSKY, T. 1951. *Genetics and the Origin of Species* (3rd ed.). Columbia University Press, New York.

DU RIETZ, G. E. 1930. "The fundamental units of biological taxonomy," *Svensk. Bot. Tidskr.* **24**:333–428.

EAMES, A. J. 1953. "Floral anatomy as an aid in generic limitation," *Chron. Bot.* **14**:126–132.

EARNSHAW, F. 1953. "The nature of ecotypes," *Proc. Int. Congr. Plant Sci.* 1950, pp. 269–271.

ERDTMAN, G. 1952. *Pollen Morphology and Plant Taxonomy.* Ronald Press, New York.

ERDTMAN, G. 1958. *Pollen and Spore Morphology and Plant Taxonomy.* Ronald Press, New York.

FOSBERG, F. R. 1941. "For an open-minded taxonomy," *Chron. Bot.* **6**:368–370.

GATES, R. R. 1951. "The taxonomic units in relation to cytogenetics and gene ecology," *Am. Nat.* **85**:31–50 (1951).

GRANT, VERNE. 1957. "The plant species in theory and practice," in *The Species Problem,* pp. 39–80. A.A.A.S., Washington, D.C.

GRANT, VERNE. 1963. *The Origin of Adaptations.* Columbia University Press, New York.

GREGOR, J. W. 1942. "The units of experimental taxonomy," *Chron. Bot.* **7**:193–196.

GUSTAFSON, A. 1946–1947. "Apomixis in the higher plants," I–III, Lunds Univ. Arsskr. N.F. 42, n. 3; 43, n. 3 & 12; 370 pp.

HALL, H. M., and F. E. CLEMENTS. 1923. *The Phylogenetic Method in Taxonomy.* Carnegie Inst. Wash. Publ. 326.

HARPER, R. A. 1929. "Significance of taxonomic units and their natural basis," *Proc. Int. Congr. Plant Sci. 1926* **2**:1588–1589.

HEILBORN, O. 1929. "Significance of taxonomic units and their natural basis from the point of view of cytology," *Proc. Inst. Congr. Plant Sci. 1926,* **2**:1576–1577.

HESLOP-HARRISON, J. 1956. *New Concepts in Flowering Plant Taxonomy.* Harvard University Press, Cambridge, Mass.

HESLOP-HARRISON, J. 1929. "The species concept," *Proc. 5th Int. Cong. Plant Sci. 1926,* **2**:1576–1577.

HEYWOOD, V. H., and J. MCNEILL (editors). 1964. *Phenetic and Phylogenetic Classification* (London).

HURST, C. C. 1930. "The new species concept," Proc. 5th Int. Bot. Congr., pp. 222–223.

HUXLEY, JULIAN (editor). 1940. *The New Systematics.* Oxford University Press, Oxford.

JOHNSON, A. W., and J. G. PACKER. 1965. "Polyploidy and environment in Arctic Alaska," *Science* **148**:237–239.

JORDAN, ALEXIS. 1846. "Observations sur plusieurs plantes nouvelles, rares ou critiques de la France," II, *Ann. Soc. Linnéenne de Lyon.*

KENDRICK, W. B. 1965. "Complexity and dependence in computer taxonomy," *Taxon* **14**:141–154.

KNOBLOCH, I. W. 1959. "A preliminary estimate of the importance of hybridization in speciation," *Bull. Torr. Club* **86**:296–299.

LAWRENCE, G. H. M., *et al.* 1953. "Plant genera," *Chron. Bot.* **14**:89–160. (Some of the parts of this symposium are also listed separately in this bibliography.)

LÖVE, Á. 1951. "Taxonomical evaluation of polyploids," *Caryologia* **3**:263–284.

LÖVE, Á. 1962. "The biosystematic species concept," *Preslia* **34**:127–139.

LÖVE, Á. 1964. "The biological species concept and its evolutionary structure," *Taxon* **13**:33–45.

MAYR, ERNST. 1942. *Systematics and the Origin of Species.* Dover, New York.

MCNAIR, J. B. 1934. "The evolutionary status of plant families in relation to some chemical properties," *Am. J. Bot.* **21**:427–452.

MICHENER, C. D., and R. S. SOKAL. 1957. "A quantitative approach to a problem in classification," *Evolution* **11**:130–162.

NEWMAN, D. W. (editor). 1964. *Instrumental Methods of Experimental Biology.* Macmillan, New York.

NYGREN, AXEL. 1954. "Apomixis in the angiosperms," II, *Bot. Rev.* **20**:577–649.

ROBINSON, B. L. 1906. "The generic concept in the classification of the flowering plants," *Science* n.s. **23**:81–92.

ROBINSON, TREVOR. 1963. *The Organic Constituents of Higher Plants—Their Chemistry and Interrelationships.* Burgess, Minneapolis, Minn.

ROGERS, D. J. 1963. "Taximetrics—New name, old concept," *Brittonia* **15**:285–290.

ROGERS, D. J., and T. T. TANIMOTO. 1960. "A Computer Program for Classifying Plants," Part I, *Science* **132**:1115–1118.

ROGERS, D. J., and HENRY FLEMING. 1964. "A Computer Program for Classifying Plants," Part II, "A Numerical Handling of Non-numerical Data," *Bioscience* **14**:15–28.

ROLLINS, R. C. 1952. "Taxonomy Today and Tomorrow," *Rhodora* **54**:1–19.

ROLLINS, R. C. 1953. "Cytogenetic Approaches to the Study of Genera," *Chron. Bot.* **14**:133–139.

ROLLINS, R. C. 1965. "On the Bases of Biological Classification," *Taxon* **14**:1–6.

SENN, H. A. 1938. "Chromosome Number Relationships in the Leguminosae," *Bibliog. Genet.* **12**:175–336.

SHARP, A. J., *et al.* 1963. "Modern Species Concepts: a Symposium," *Bryologist* **66**:93–124.

SHARP, L. W. *Fundamentals of Cytology* (New York, 1943), especially Chap. XVII, "Cytology and Taxonomy," pp. 234–250.

SHULL, G. H. 1929. "Significance of Taxonomic Units and Their Natural Basis from the Point of View of Genetics," *Proc. Int. Cong. Plant Sci. 1926*, **2**:1578–1586.

SIMPSON, G. G. 1951. "The Species Concept," *Evolution* **5**:285–298.

SIMPSON, G. G. 1961. *Principles of Animal Taxonomy* (New York).

SINNOTT, E. W., L. C. DUNN, and T. DOBZHANSKY. 1950. *Principles of Genetics*, 4th edition (New York).

SOKAL, R. R., and P. H. A. SNEATH. 1963. *Principles of Numerical Taxonomy* (Freeman, San Francisco).

STEBBINS, G. L., JR. 1942. "Polyploid Complexes in Relation to Ecology and the History of Floras," *Am. Nat.* **76**:36–45.

STEBBINS, G. L., JR. 1942. "The Role of Isolation in the Differentiation of Plant Species," Biol. Symposia 1.

STEBBINS, G. L., JR. 1947. "Types of Polyploids: Their Classification and Significance" in C. Demerec, *Adv. Genet.* **1**:403–429.

STEBBINS, G. L., JR. 1950. *Variation and Evolution in Plants* (New York).

STEBBINS, G. L., JR., and E. B. BABCOCK. 1939. "The Effect of Polyploidy and Apomixis on the Evolution of Species in Crepis," *J. Heredity* **30**:519–530.

SWAIN, T. (editor). 1963. *Chemical Plant Taxonomy.* Academic Press, New York.

SWANSON, C. P. 1957. *Cytology and Cytogenetics.* Prentice-Hall, Englewood Cliffs, N.J. (The student will be particularly interested in chapters 6, 16, and 17.)

TISCHLER, G. 1937. "On some problems of cytotaxonomy and cytoecology," *J. Indian Bot. Soc.* **16**:165–169.

TURESSON, G. 1922. "The genotypical response of the plant species to the habitat," *Hereditas* **3**:341–347.

TURESSON, G. 1925. "The plant species in relation to habitat and climate," *Hereditas* **6**:147–236.

TURRILL, W. B. 1936. "Contacts between plant classification and experimental botany," *Nature* **137**:563–566.

VALENTINE, D. H. 1949. "The units of experimental taxonomy," *Acta Biotheoretica* **9**:75–88.

VAVILOV, N. I. 1922. "Law of homologous series in variation," *J. Genetics* **12**:47.

WANSCHER, J. H. 1934. "The basic chromosome number of the higher plants," *New Phytol.* **33**:101–126.

WHITE, O. E. 1940. "Temperature reaction, mutation, and geographical distribution in plant groups," *Proc. 8th Amer. Sci. Congr.* **3**:287–294.

CHAPTER 7

The Construction and Use of Keys

A key is a device for easily and quickly identifying an unknown object. The user is presented with a sequence of choices, usually between two statements but occasionally among more, and by always taking the correct choice he arrives at the name of his unknown object. The statements in the choices are based on the characteristics of the unknown object and, if the object is a flowering plant, are usually concerned with flowers, fruits, seeds, stems, leaves, and roots.

Keys based on successive choices between only two statements, known as *dichotomous* (forking) keys, are preferred to those that offer several equal statements to choose from. In using keys, however, one should be on the lookout for occasional places where three or even four equal choices occur and not overlook some of them.

Using such a key may be likened to traveling a well-marked road that forks repeatedly, each fork bearing directions; if the traveler is correctly informed and follows directions carefully, he will always arrive at his destination. If either his information or the directions along the way are inaccurate, he will, naturally, become lost.

Two general types of construction are used in the preparation of botanical keys, the *indented* type and the *bracket* type. Most botanists prefer the indented type, and most zoologists prefer the bracket type. The plants used in the following examples are common genera of the

Ranunculaceae (Buttercup Family): *Anemone* (Anemone or Windflower), *Aquilegia* (Columbine), *Clematis* (Clematis), *Delphinium* (Larkspur), and *Ranunculus* (Buttercup).

Indented Key

1. Fruit a group of akenes; flower not spurred
    2. Petals none
        3. Sepals usually 4; involucre none . . . . CLEMATIS
        3. Sepals usually 5; involucre present . . . . ANEMONE
    2. Petals present . . . . RANUNCULUS
1. Fruit a group of follicles; flowers spurred
    4. Flowers regular; spurs 5 . . . . AQUILEGIA
    4. Flowers irregular; spur 1 . . . . DELPHINIUM

The first choice, if we are concerned only with the genera above, is between "Fruit a group of akenes; flowers not spurred" and "Fruit a group of follicles; flowers spurred," these paired statements being given the same indention and the same number. If the latter choice is taken, the next choice, as shown by the indention and the number, is between "Flowers regular; spurs 5" and "Flowers irregular; spur 1." Thus, if the plant in question has follicles and irregular flowers with a single spur, it must be a *Delphinium.*

The bracket type of key is based on the same principle of contrasting choices, but these are always placed in adjacent lines of the key, not separated by intervening lines. Such a key requires less room on a page and has the advantage of keeping coordinate choices together, but it does not show relationships as well and backtracking is more difficult if an error has been made. An example follows:

Bracket Key

1. Fruit a group of akenes; flowers not spurred 2
1. Fruit a group of follicles; flowers spurred 4
    2. Petals none 3
    2. Petals present . . . . RANUNCULUS
3. Sepals usually 4; involucre none . . . . CLEMATIS
3. Sepals usually 5; involucre present . . . . ANEMONE
    4. Flowers regular; spurs 5 . . . . AQUILEGIA
    4. Flowers irregular; spur 1 . . . . DELPHINIUM

The number at the right end of a line in the bracket key indicates the next numbered pair of choices to be considered. A similar numbering system may be employed in indented keys if they are long enough to warrant it, particularly when they extend over more than a single page. But in a short indented key the indention, and usually the first words of the lines, will indicate which are equivalent lines of the key.

General Suggestions

1. Always read both choices even if the first seems to be the logical one to take. The second may be even better.
2. Be sure you understand the meaning of the terms involved. Do not guess.
3. When measurements are given, use a calibrated scale. Do not guess.
4. When minute objects are concerned, use a lens of sufficient magnifying power to show clearly the feature you need to see.
5. Since living things are always somewhat variable, do not base your conclusion on a single observation, but arrive at an average by studying several parts or specimens. It is surprising how often students will find the one unusual or aberrant sample in a large assortment of normal things!
6. As in traveling a forking road, if the choice of division is not clear, or if you have no way of making a choice because you do not have sufficient information, try both divisions, arrive at two possible answers by doing so, and then read descriptions of each in order to make a choice. A key is only a shortcut to identification; it is not essential if descriptions are available.
7. In constructing keys, keep the following in mind:
   a. Use constant characteristics rather than variable ones.
   b. Use measurements rather than terms such as "large" and "small."
   c. Use characteristics that are generally available to the user of the key rather than seasonal characteristics or those seen only in the field.
   d. When possible, group to show relationships rather than construct an entirely artificial key.
   e. If possible, start both choices of a pair with the same word, and always capitalize the first word. And, if possible, start different pairs of choices with different words.
   f. Precede the descriptive terms by the name of the part to which they apply. For example:

GOOD:	POOR:
Flowers red or purple	Red or purple flowers
Leaves toothed	Toothed leaves

   g. Construct a comparison chart of the objects to be keyed before proceeding with the key itself. This forces one to make complete comparisons and to avoid mentioning a character in one statement and not in the corresponding one. The heavy vertical lines in the following comparison chart indicate separations between possible key characters or important differences. Reading vertically we get a description of each group; reading horizontally we see similarities and differences.

COMPARISON CHART

	Clematis	*Anemone*	*Ranunculus*	*Aquilegia*	*Delphinium*
Fruits	akenes	akenes	akenes	follicles	follicles
Flowers regular or irregular	regular	regular	regular	regular	irregular
Number of spurs	none	none	none	5	1
Flowers with or without petals	without	without	with	with	with
Usual number of sepals	4	5	5	5	5
Involucre present	no	yes	no	no	no

8. Finally, having arrived at an answer in a key, do not accept this as absolutely reliable, but check a description of the plant to see if it agrees with the unknown specimen. If not, an error has been made somewhere, either in the key or in its use. The ultimate check on identifications is a comparison of the unknown with an authentically named specimen in a herbarium.

CHAPTER 8

Phytography and the Terminology of Plant Description

Phytography is that part of taxonomy which deals with descriptions of plants and their various organs. Its two chief objectives are accuracy and completeness of description without undue wordiness.

Because the space available in taxonomic publications is necessarily limited, it is common practice to limit descriptions to the characters deemed necessary for recognition. Complete descriptions, consequently, are rarely found. Only within the last generation of botanists, in fact, were detailed descriptions felt to be necessary, the earlier workers often having been content to give only a few words of description when publishing accounts of species and genera. The early accounts of vegetation were often profusely illustrated, however, by highly accurate drawings, which indicated that the taxonomists of years gone by were keen observers.

Some of the basic features of the organs of flowering plants—those that are important in the formulation of descriptions—are described below, and the terms commonly used are introduced. The *vegetative organs* (roots, stems, and leaves) are treated first, and then the *reproductive organs* (inflorescences, flowers, fruits, and seeds).

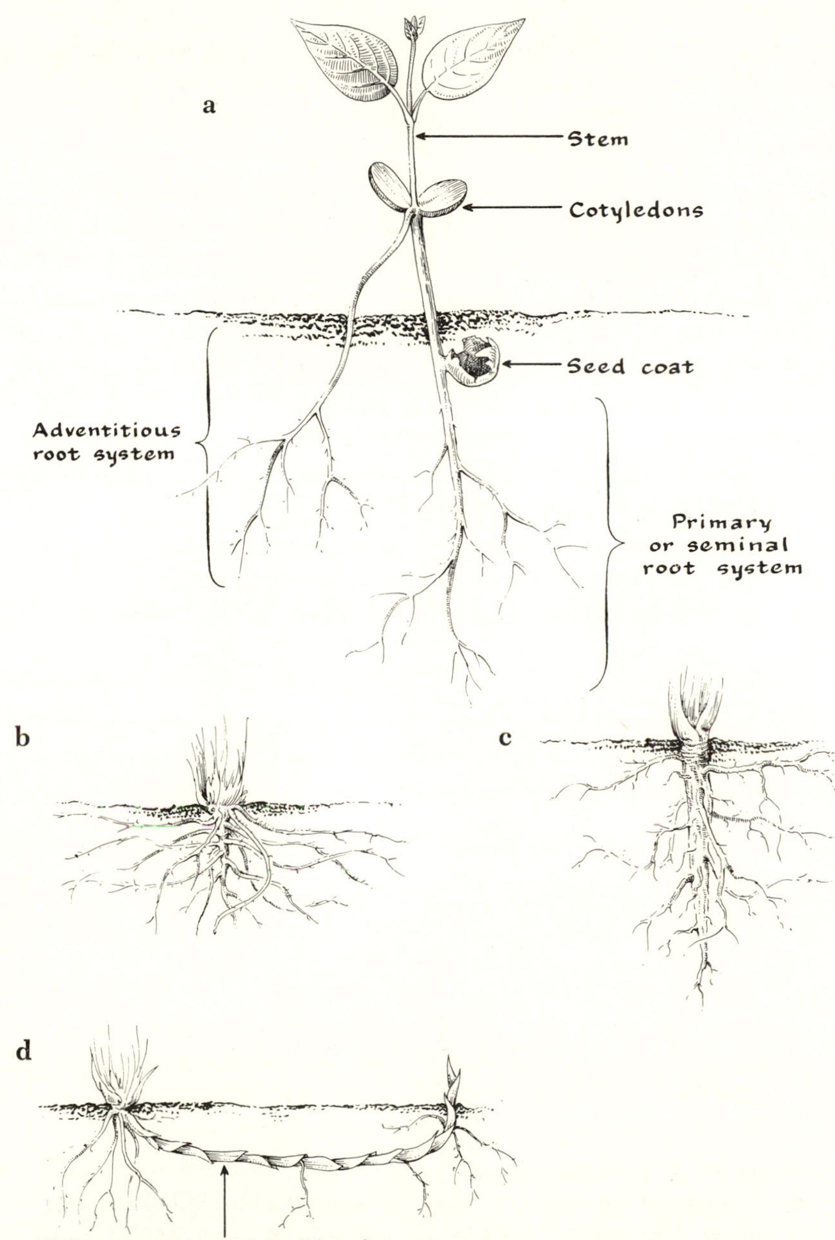

10 Some characteristics of root systems. (**a**) Diagram of a hypothetical plant to illustrate the two principal types of root systems. (**b**) Adventitious root system originating from a vertical, unmodified stem. (Annual Bluegrass, *Poa annua.*) (**c**) Root system in which the primary root is maintained. (Dandelion, *Taraxacum.*) (**d**) Adventitious root system involving a modified stem. (Kentucky Bluegrass, *Poa pratensis.*)

Roots

Root systems are seldom used extensively in the classification of flowering plants. Too often, indeed, they are ignored completely, perhaps because adequate and generally accepted methods of classification of root systems are lacking.

Roots may be thought of as belonging to two general types: (1) those derived from the primary, or seminal (seed), root; (2) adventitious roots, which are derived in some other way. Root systems then fall into three broad classifications: (1) those in which the primary, or seminal, root is maintained; (2) those in which the adventitious root system originates from the lower portion of a vertical unmodified stem, this root system early replacing the primary root; (3) those in which adventitious roots involve various kinds of modified stems, such as rhizomes, stolons, tubers, corms, and bulbs. Figure 10 illustrates some of these characteristics of roots.

In addition to these types, which are characteristically subterranean, there are various types of aerial roots, such as the climbing, or holdfast, roots of certain vines, and the pendent aerial roots of many epiphytes, or "air plants."

Certain plants, notably some forest trees and members of the Heath Family, have root systems that are characteristically associated with fungi in a more or less symbiotic relationship. This fungus-root association is known as a *mycorhiza.*

Stems

The stems of flowering plants may be quite varied in size, duration, position above or below the ground level, and direction in which they grow. They are sometimes modified for special functions.

Differences in size and duration are indicated by the general terms *herb, shrub,* and *tree,* following a distinction that was made in the earliest writings. More precisely, however, we recognize the following terms:

herbaceous: dying down to the ground every year, the stems containing very little woody tissue, and the duration *annual* if for one year only; *biennial* if the plant blooms the second year, after preliminary vegetative growth the first year, and is short-lived; *perennial* if the plant continues to live for an indefinite period of years and blooms ordinarily every year after the first.

suffrutescent: semishrubby, the lowest parts of the stems becoming woody and remaining alive over the winter when the higher parts die back (such plants are also referred to as semishrubs).

shrubby, or **fruticose:** woody more or less throughout, and large, commonly with several main stems but no main trunk.

arborescent: becoming tree-like in size, and woody, usually with a single main trunk.

Most plants produce ordinary leafy stems, with appendages such as leaves, buds, and flowers from their nodes. Such plants with leafy

11 Basal portion of plant of the grass *Calamovilfa longifolia* (Hook.) Scribn., showing strong creeping rhizomes. This makes the plant an excellent sand-binder.

stems are termed *caulescent* (having a *caulis,* or true stem); when there is no evident aerial stem, the leaves being all basal (actually on a shortened stem at the base), and the flower stalk, or *scape,* is leafless, the plants are termed *acaulescent* (stemless) or *scapose* (with a scape), as Dandelions and some kinds of Violets. These forms are shown in Figure 12.

There are many modifications of stem structure, among them the following, which are quite common:

stolons, or **runners:** stems trailing above ground, which often root at their nodes and thus tend to produce new plants when the connection with the parent plant is broken, as in the Strawberry.

rhizomes, or **rootstocks:** underground stems, often creeping horizontally and producing new shoots at their tips, as in Kentucky Bluegrass.

tubers: thickened and fleshy subterranean stems that serve as organs for food storage and reproduction, as in the Irish Potato.

corms: fleshy, upright, subterranean stems that bear papery modified leaves or scales, as in the Crocus.

bulbs: upright, subterranean stems, the stem part of which is much smaller than in the corm and is surrounded by thickened, fleshy leaves or scales, as in the Onion.

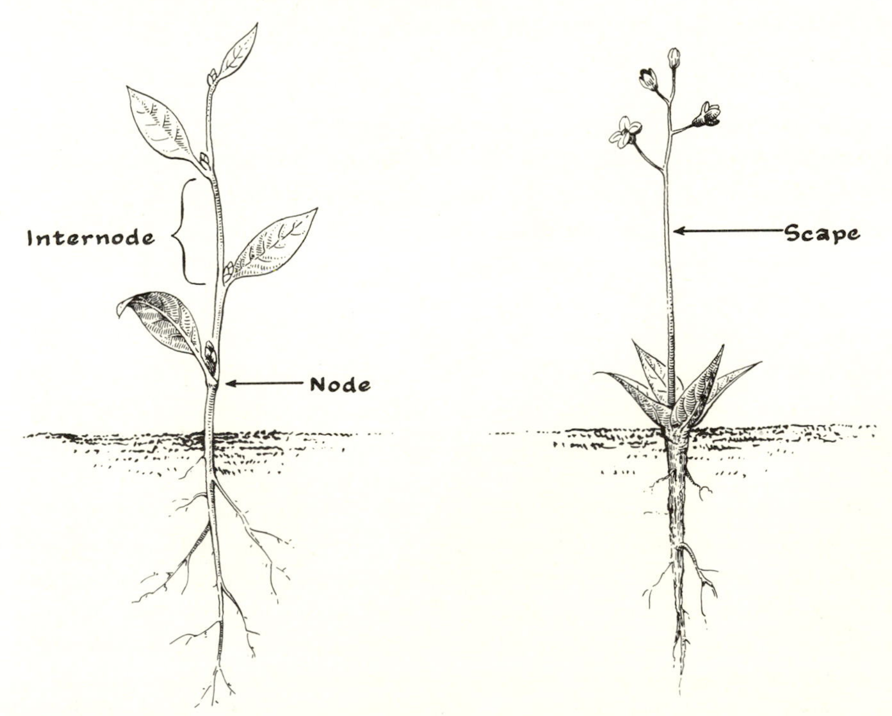

12 (Left) Caulescent plant. (Right) Acaulescent or scapose plant.

tendrils: may be slender, twining branches used for support by climbing plants such as Grapes, or, in other plants, may be of leaf origin.
spines, or **thorns:** are often sharp and stunted branches, but may be modified leaves or parts of leaves.

Leaves

Most stems of flowering plants produce leaves from their nodes, usually of a form and in an arrangement peculiar to the species, which one can often recognize by the leaves alone. A few species, such as many cactus plants, are leafless, and a few, such as Sassafras and Mulberry trees and many aquatic plants, show considerable variation in the shape of the leaves.

LEAF ARRANGEMENT. If the leaves are arranged singly on the stem, one leaf at a node, they are *alternate* in arrangement, but the angle between the leaf and the one directly above it may vary considerably and may be expressed by a numerical fraction indicating the degree of rotation of the internode before the next leaf is reached. If the leaves are paired on the stem, two at each node, they are *opposite.* If three or more leaves occur at a single node, they are *whorled.* Most plants maintain a single kind of arrangement, or *phyllotaxy,* of their leaves throughout; but occasionally a plant may be variable in this characteristic, having perhaps alternate leaves below and opposite leaves higher on the stem.

LEAF PARTS. A leaf may consist of three main parts: the *blade,* or expanded portion, the *petiole,* or leaf stalk, and the *stipules,* which are a pair of appendages at the base of the petiole. Any of these parts may be lacking: infrequently a blade is reduced to only a midrib or is completely lacking, and the leaf is then *bladeless;* if the blade is present but much reduced, the leaves may be called *phyllodia;* the petiole may be lacking and the blade attached directly to the branch, such leaves being called *sessile;* often stipules are not produced, and such leaves are *exstipulate.* In some plants the stipules may be modified into thorn-like structures, which are then called *stipular spines.*

SIMPLE AND COMPOUND LEAVES. A leaf with a single blade (which may be variously indented or deeply cut) is a *simple leaf.* A leaf with more than one blade is a *compound leaf,* and its blades are called *leaflets.* The leaflets of a compound leaf may be arranged like the spokes of a wheel or the fingers of a hand when the leaf is *palmately* or *digitately compound;* or they may be arranged on either side of an elongated axis, or *rachis,* when the leaf is *pinnately compound,* resembling a feather. When the leaflets are in threes, as in most Clover leaves, the leaf is usually termed a *trifoliolate* leaf; if the leaf blade is divided more than once into threes, as in some members of the Parsnip Family, it is called a *ternate* leaf. Compound leaves are sometimes twice divided or even thrice divided, in which case they are *twice-compound* or *decompound*, the latter term generally designating very numerous and fine divisions. Some of these terms are illustrated by Figure 13, a–f.

LEAF MARGINS. The edge of the leaf blade is its *margin*, and there is a great diversity in the margins of the leaves of different kinds of plants. The illustrations (Fig. 13, g) show some of the common types of margins and their terminology.

LEAF VENATION. The system of principal veins in the leaf blade constitutes its *venation*. The illustration (Fig. 13, h) shows the three chief types, which are *parallel* (sometimes called *nerved*), *pinnate*, and *palmate*, the second and third being *reticulate*, or *net-veined*, types. Occasionally we find combinations of these types, as in the Buckthorn (*Rhamnus*), in which some species have three principal veins from the base but each is pinnate above.

LEAF SHAPES. The general outline of the blade, or of all the leaflets of a compound leaf, constitutes the *shape* of the leaf. Usually the shape is described as including only the blade, omitting the petiole. We approximate the general shape of lobed leaves and compound leaves by drawing an imaginary line round all projecting parts, ignoring the indentions and the areas between leaflets. *The student should be careful to discriminate between the shape of a compound leaf and the shape of its leaflets.*

The illustrations (Fig. 14) show some of the common terms describing leaf shapes. It is common practice to use two or more terms to describe intermediate shapes, hyphenating the terms used; thus a leaf that is intermediate between ovate and lanceolate in outline may be termed ovate-lanceolate. The prefix *ob* means that the shape is inverted: a heart-shaped leaf upside down is thus *obcordate*.

LEAF SURFACES. The presence or absence of hairs, the kind of hairs, and the presence or absence of other surface features, such as glands, all offer the taxonomist useful recognition characters. These are best examined through a good hand lens or microscope. The following

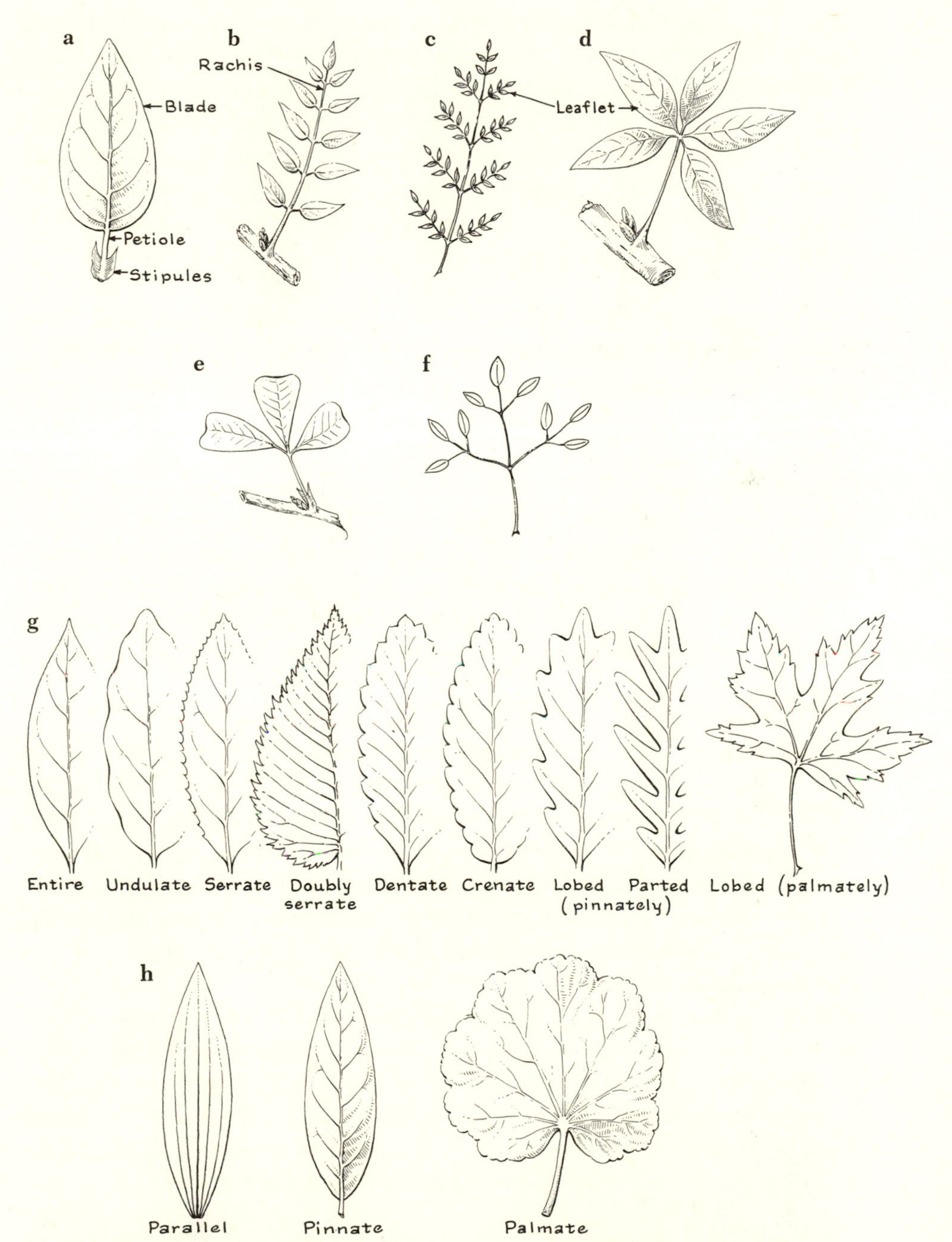

13 Some terms used in descriptions of leaves. (**a**) Simple leaf and its parts. (**b, c**) Pinnately compound leaves, once pinnate (**b**) and twice pinnate (**c**). (**d**) Palmately or digitately compound leaf. (**e**) Trifoliolate leaf. (**f**) Ternate leaf. (**g**) Leaf margins. (**h**) Leaf venation.

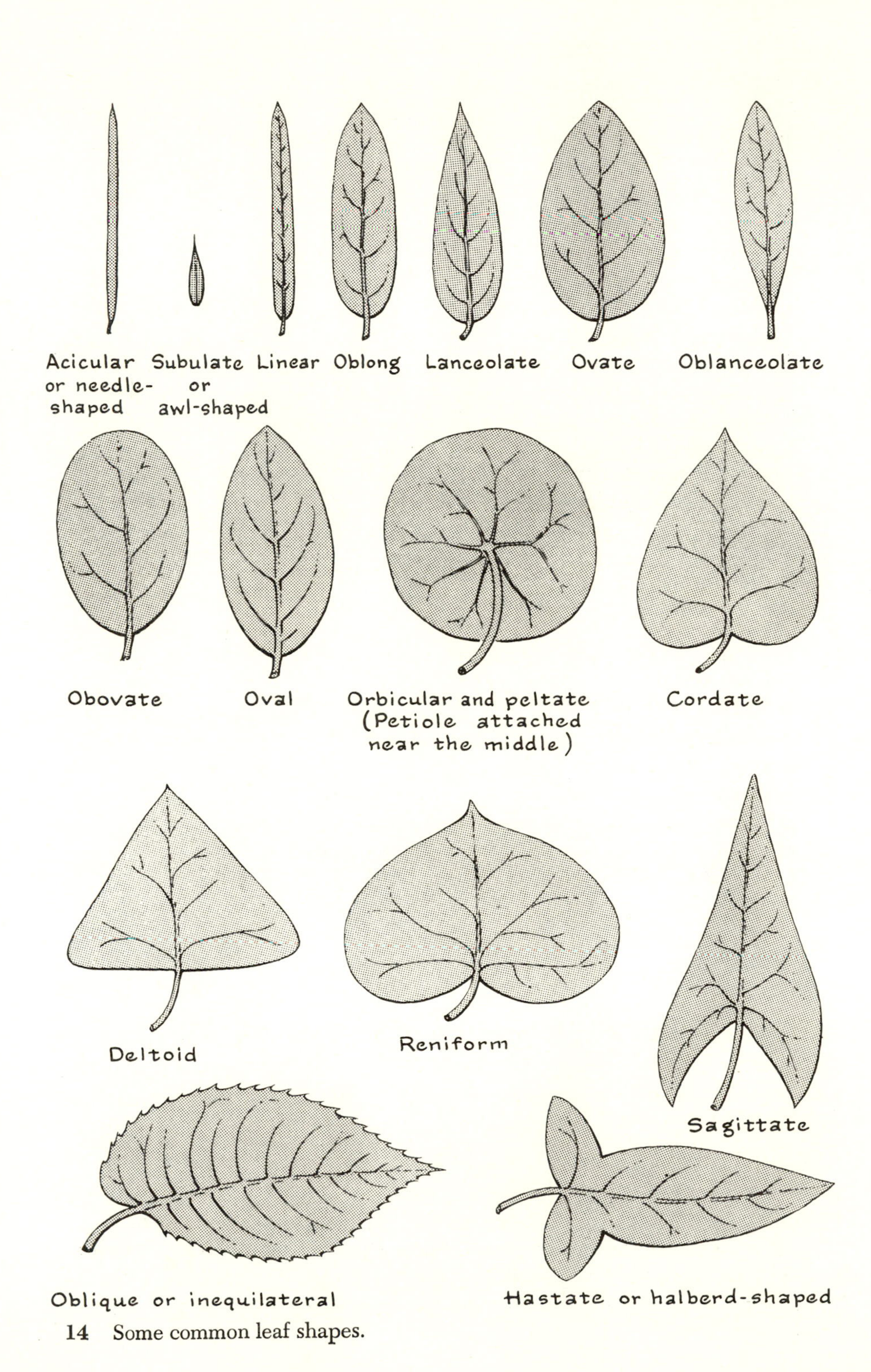

14 Some common leaf shapes.

terms are commonly used to describe the surfaces of leaves and other parts of plants:

barbellate: barbed down the sides (applied to hairs).
canescent: covered or becoming covered with grayish hairs.
ciliate: with soft hairs on the margin forming a fringe.
comose: having a tuft of hair, as many seeds.
fimbriate: with a fringe.
floccose: with tufts of woolly hairs that rub off easily.
glabrescent or **glabrate**: becoming glabrous (hairless) in age.
glabrous: without hairs.
glandular: with glands (usually hairs having enlarged cells at the tip).
glaucous: covered with a bloom, a whitish substance that rubs off.
glochidiate: having hairs that are barbed at the tip.
hirsute: having moderately stiff separate hairs.
hispid: with stiff or bristly hairs.
lanate: covered with woolly, tangled hairs.
pilose: covered with soft, rather long, shaggy hairs.
puberulent: with very soft, minute, downy hairs.
pubescent: covered with short, soft hairs (the term *pubescence* refers to any kind of hairiness).
punctate: having dots or pits, these often waxy or glandular.
rugose: wrinkled.
scabrous: with very short, stiff hairs or projections that one can feel by lightly passing the fingers over the surface.
scurfy: covered with minute scales.
sericeus: silky.
stellate: like a star (hairs having radiating branches).
strigose: with sharp-pointed, straight, appressed hairs.
tomentose: densely woolly, with matted hairs.
uncinate: with a hook at the tip, as some hairs or spines.
villous: with long, soft, shaggy hairs that are not matted.
viscid: sticky (usually from glandular hairs exuding a sticky liquid).

The student will find it profitable to get good, typical, representative parts of various plants, showing these and other surface features, and to make small, permanent mounts of them so that he will have examples to refer to when in doubt about the application of terms. Fragments of leaves or other parts may be mounted on cards and protected by an overlay of clear plastic, the edges of the mount being bound with tape.

MODIFIED LEAVES. Most plants produce modified leaves that are somewhat different from the ordinary foliage leaves. Buds are covered by *bud scales;* bulbs are surrounded by *bulb scales;* some *tendrils* are modified leaves or parts of leaves serving as thread-like attachment organs; *bracts* are more or less modified leaves in an inflorescence or flower cluster; scale-like leaves occur on rhizomes; and sometimes leaves are modified into contrivances for capturing insects, as in the Pitcher Plant (*Sarracenia*), whose leaves are tubular and hold water. Some plants have their leaves reduced to spines. And, as will be pointed out later, the parts of a flower—sepals, petals, stamens, and carpels—are generally believed to have evolved from leaves.

Inflorescences

An inflorescence is the arrangement of flowers on a plant, or the mode of flowering. It may be very simple and readily distinguishable, or it may be a highly complex structure whose precise nature is not evident at a glance. We may try to force all flower arrangements into a set of rigidly defined types, but it should be remembered that it is not always possible to make variable plants fit such a system.

An old concept of inflorescences was based on the sequence of blooming and the position occupied by the oldest flower. Inflorescences were classified by this method into *determinate* types, in which the oldest flower terminated the main axis and the general progression of blooming was downward or outward, and *indeterminate* types, in which the youngest flower was terminal or central and the progression of blooming was upward or inward. Careful observation has shown, however, that, though the sequence of blooming may be significant in some instances, it may be quite irregular in others. The flowers of some inflorescences appear to open almost simultaneously; and sometimes the terminal flower aborts.

The main supporting stalk of the whole inflorescence is called a *peduncle.* The stalks supporting single flowers are called *pedicels.* These parts are indicated in Figure 17.

An inflorescence may have *bracts* (Fig. 17), which are modified leaves or scales from the axils of which flowers or flowering branches are produced. If there are no bracts, the inflorescence is said to be *ebracteate.* Bracts are generally unlike the foliage leaves, often being smaller or of a different shape or texture. Clusters or whorls of bracts make up an *involucre* (Figs. 22 and 25); if secondary involucres occur (as in compound umbels), these are known as *involucels of bractlets* (Fig. 25). A single, conspicuous bract that subtends a flower cluster (usually a fleshy spike) is called a *spathe* (Fig. 27); spathes are often ornamental.

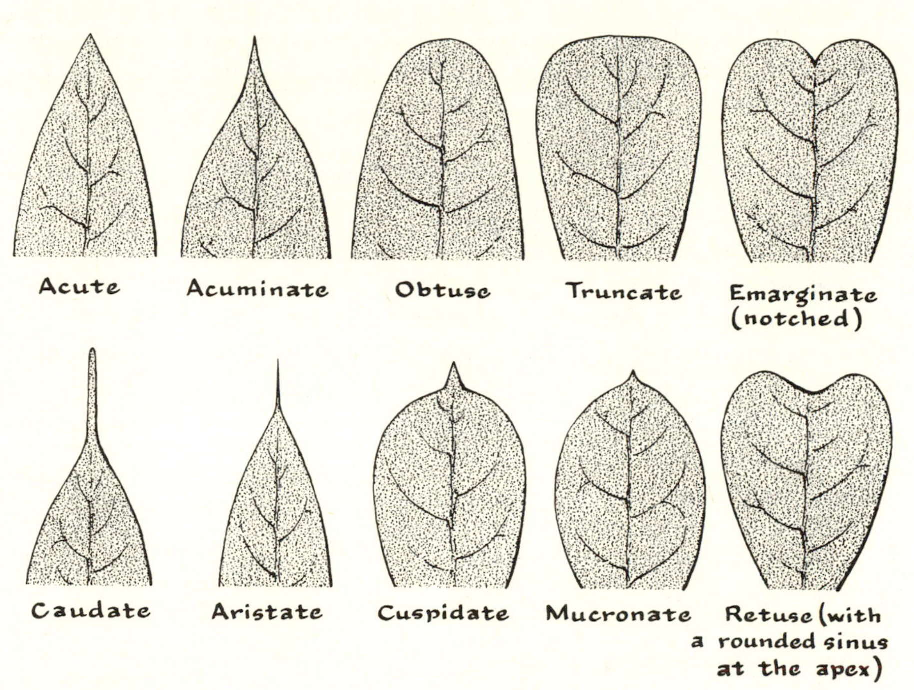

15 Leaf apices.

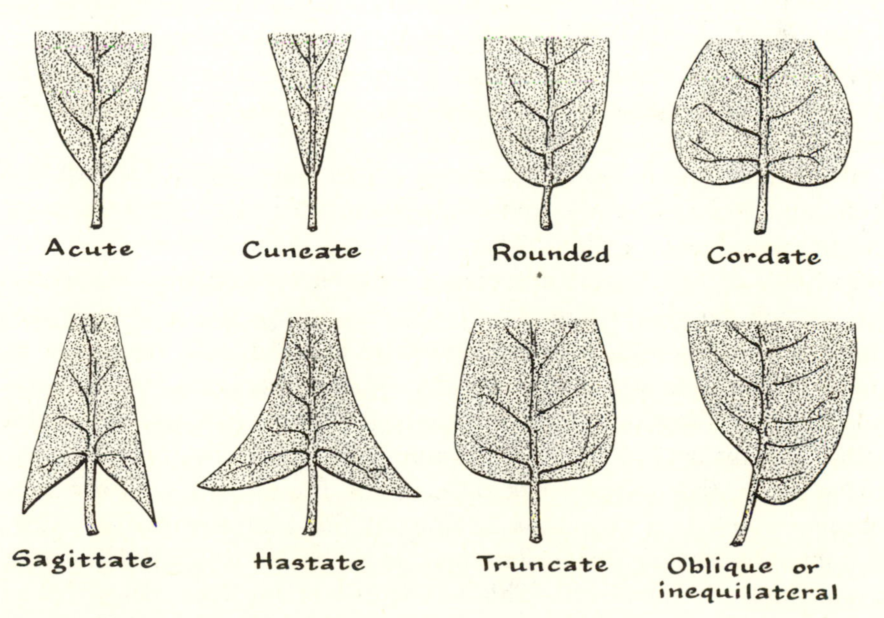

16 Leaf bases.

Following are descriptions of the common kinds of inflorescence. Each is illustrated by a diagram in which small circles represent flowers; if there is a significant order of blooming, the larger circles represent older flowers.

DICHASIUM. A dichasium is a peduncle bearing a terminal flower and a pair of branches that produce lateral flowers. The oldest flower is the central one. This *simple dichasium* is a common unit making up parts of many more complex inflorescences. A repetition of this on a lateral pair of branches produces a *compound dichasium.* A dichasium may be called a *cyme,* but that term has been loosely applied to any compound, more or less flat-topped inflorescence, particularly if the oldest flowers terminate the main axes. Dichasia are shown in Figure 17.

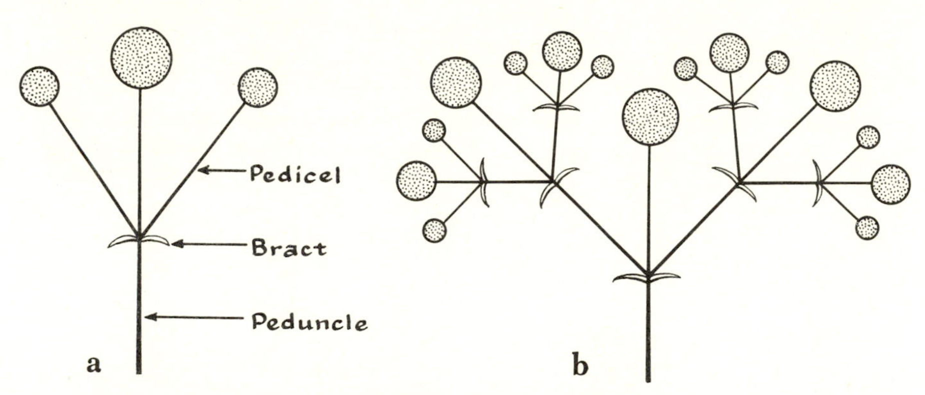

17 (**a**) Simple dichasium. (**b**) Compound dichasium (cyme).

MONOCHASIUM. A monochasium is a peduncle bearing a terminal flower and, below it, one branch that produces a single lateral flower. The terminal flower is older. This is a *simple monochasium.* A repetition of this on the lateral branches produces a *compound monochasium,* which may be of four types: (1) the *bostryx,* or helicoid cyme, which is spirally coiled round the vertical axis; (2) the *cincinnus,* or scorpioid cyme, in which the flowers appear alternately to one side and the other along one side of the axis, the whole inflorescence often coiling downward; (3) the *rhipidium,* which is a bostryx flattened in one plane, the inflorescence often being fan-shaped; (4) the *drepanium,* which is a cincinnus having all the branches on one side, the inflorescence thus being flattened in one plane and usually coiled to one side and downward. This last is one of the commonest forms of the monochasium and is also a kind of scorpioid cyme. Monochasia are illustrated by Figure 18.

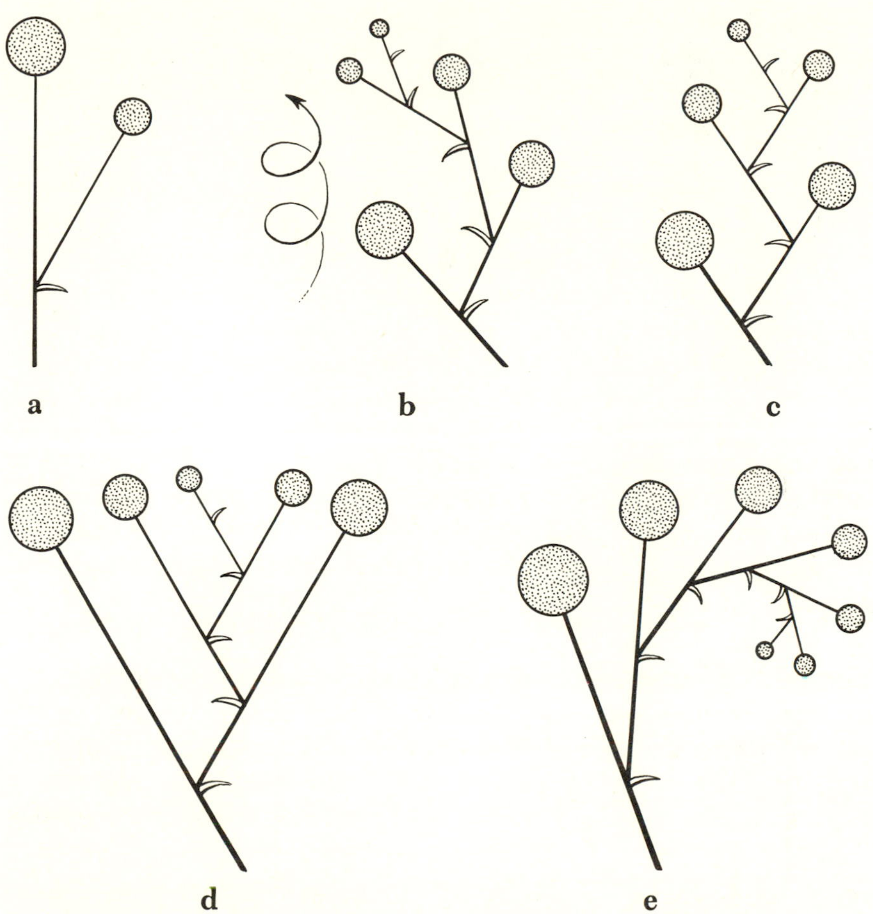

18 Monochasia. (**a**) Simple monochasium. (**b**) Bostryx (the direction of coiling indicated by the arrow). (**c**) Cincinnus (often coiled downward). (**d**) Rhipidium (flattened in the plane of the paper). (**e**) Drepanium (flattened in the plane of the paper and coiled to one side).

PANICLE. A panicle is a more or less elongated inflorescence with a central axis along which there are branches that are themselves branched. There may be a sequence of blooming from the base upward, but some panicles are made up wholly of small dichasia. A congested, more or less cylindrical panicle is sometimes called a *thyrse*. A typical panicle is shown in Figure 19.

RACEME. A raceme is an elongated inflorescence with a central axis along which are simple pedicels of more or less equal length. There is usually an order of blooming from the base upward, but some racemes have flowers opening almost simultaneously or irregularly. A raceme is shown in Figure 20.

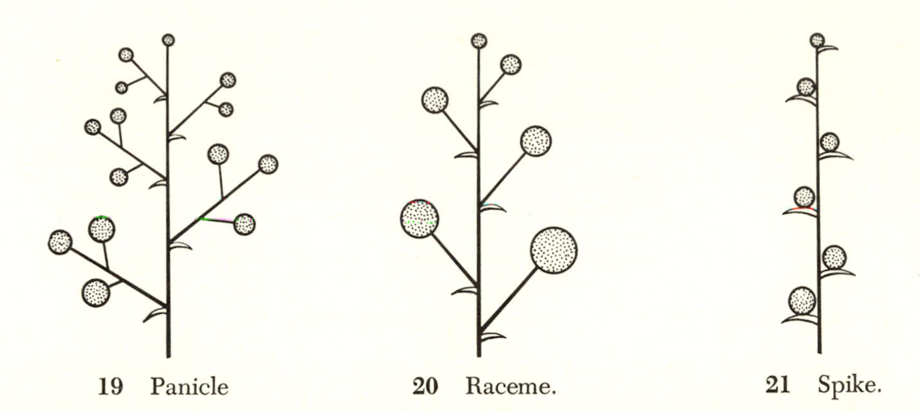

19 Panicle **20** Raceme. **21** Spike.

SPIKE. A spike is an elongated inflorescence with a central axis along which are sessile or subsessile flowers. The usual order of blooming is from the base upward. Very small spikes, particularly in grasses and sedges, are known as *spikelets*. These may be grouped into various arrangements such as panicles, racemes, or spikes. See Figure 21.

CORYMB. A corymb is a more or less flat-topped inflorescence having a main vertical axis and pedicels or branches of unequal length produced along it. The side branches may branch, or they may be simple pedicels as shown in Figure 22. The blooming sequence is usually from the outside toward the center, but it may be irregular.

HEAD. A head, or *capitulum*, is a rounded or flat-topped cluster of sessile flowers. Many heads show a progression of blooming from the outside toward the center. Heads may be solitary, or they may be aggregated into various arrangements, such as panicles, racemes, spikes, or corymbs. Heads such as those found in *Aster* and other members of the Aster Family may resemble flowers, for they often consist of two kinds of flowers, the outer more ornamental and resembling petals. Figure 23 shows such a head in longitudinal section.

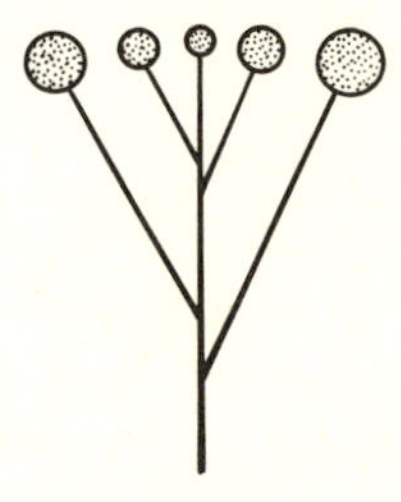

22 Corymb.

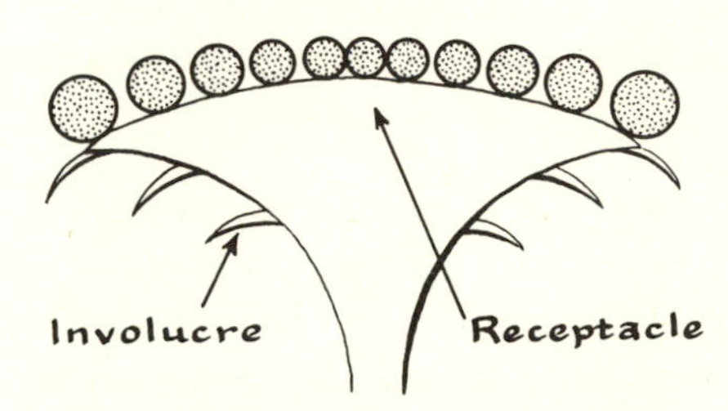

23 Head in longitudinal section.

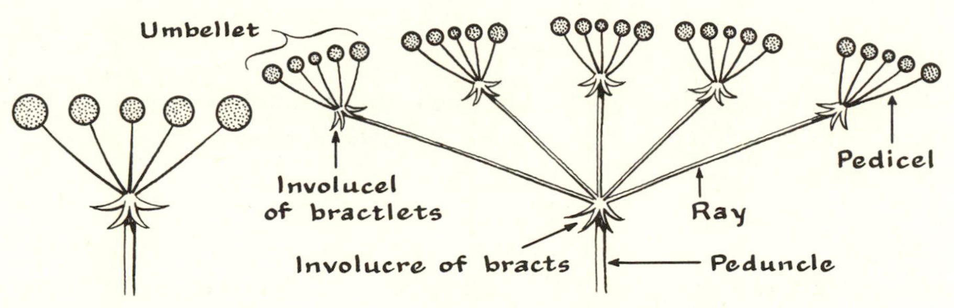

24 Simple umbel. **25** Compound umbel.

UMBEL. An umbel is an inflorescence having several branches arising from a common point at the summit of the peduncle. If these branches end in flowers, we have a *simple umbel* (Fig. 24); if they end in secondary *umbellets,* we have a *compound umbel* (Fig. 25). The blooming sequence is generally from the outside toward the center. An umbel may have a group or whorl of bracts, collectively called an *involucre,* at the summit of the peduncle. If there are similar bracts in groups at the base of an umbellet, they are called an *involucel* and the individual members are called *bractlets.* The main branches of a compound umbel are called *rays,* and the corresponding members of the umbellet are called *pedicels.*

CATKIN. A catkin, or *ament,* is a spike, raceme, or dichasium composed of unisexual flowers without petals and falling as a whole. It may be erect or pendent, and it may be long or short. The flowers are usually very small, and usually each is subtended by a little bract, or *scale.* Sometimes one or two leaf-like bracts (perhaps true leaves) occur at the base of the catkin and fall with it. This type of inflorescence is shown in Figure 26.

SPADIX. A spadix is a spike with a thickened, fleshy axis, usually enveloped by a conspicuous or colored bract called a *spathe.* The flowers are often very minute and frequently unisexual. See Figure 27.

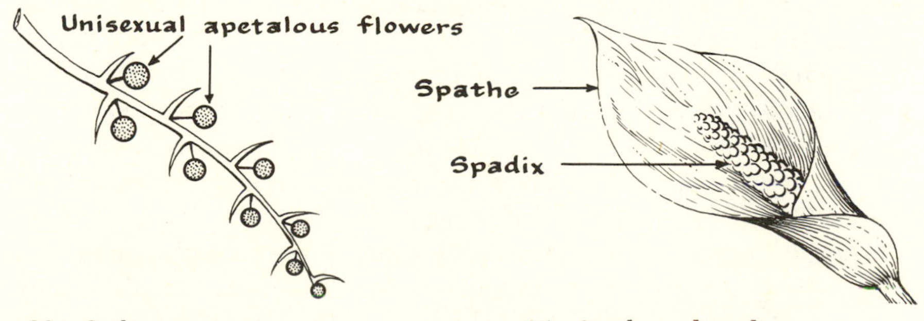

26 Catkin, or ament. **27** Spathe and spadix.

Flowers

What we might think of as a typical, unmodified flower, such as that of a Lily or Flax, is made up of four sets of flower parts: sepals, petals, stamens, and pistil or pistils.

SEPALS. Sepals, collectively called the *calyx,* are the outermost parts and are commonly leaf-like and green, but they may be colored like the petals and have a thinner texture, in which case they are described as being *petaloid.* In the unmodified flower they are all alike in size and shape. The sepals enclose the flower in bud and may or may not persist for the life of the flower.

PETALS. Petals, collectively called the *corolla,* normally occupy a position in the flower between the sepals and the stamens. Petals are commonly delicate in texture and are often colored. They are often larger than the sepals, and they may be shed soon after the flower opens. Their color or fragrance may attract insects. The sepals and petals together make up the *perianth,* or floral envelope, a term that is especially useful when the distinction between sepals and petals is not obvious.

STAMENS. Stamens, collectively called the *androecium,* are the male reproductive parts of the flower and occupy a position inward from the perianth. They vary widely in size and number. Each stamen usually consists of two parts: the *anther,* the sac-like part, which contains the pollen; and the *filament,* or stalk, which connects the anther to the floral axis or some other part.

PISTILS. Pistils, collectively called the *gynoecium,* are the female reproductive parts of the flower and occupy a central position. The gynoecium may consist of a single pistil, as in the Lily, of several or many pistils, as in the Buttercup. Each pistil usually consists of three parts: the *stigma,* the pollen-receptive part at the summit, which may be single or variously lobed or branched; the *style,* the stalk-like portion below the stigma; and the *ovary,* the enlarged portion at the base, which contains one or more ovules or immature seeds. In some pistils the style may be lacking, but the stigma and the ovary are essential to the functioning of the organ.

The basic unit of construction of a pistil is the *carpel,* which is a single megasporophyll, or modified seed-bearing leaf. A pistil may consist of a single carpel, as in the Sweet Pea, or of two or more carpels partly or completely joined together, as in the Lily (three carpels) or the Mustard (two carpels). One can usually determine the number of carpels in a pistil by sectioning the ovary and counting the number of partitions or rows of ovules or seeds (placentae), or by counting the number of styles or stigmas.

If a flower contains two or more separate carpels, it is called an *apocarpous* flower; if it contains a single pistil that consists of two or more united carpels, it is called a *syncarpous* flower.

A more detailed account of the androecium and the gynoecium will be found further along in this chapter.

FLORAL DIAGRAMS. A floral diagram represents a cross-section of a flower as it would appear if all parts were at the same level. It might also be thought of as a sort of aerial view of a flower in diagrammatic form. For uniformity and convenience, the various parts are represented in diagrams by standardized symbols, those used in this text being shown in Figure 28.

In general, lines connecting parts of a flower in the diagram indicate that those parts are connected. This and other features of floral diagrams are shown in Figure 29.

NUMERICAL PLAN OF THE FLOWER. Most flowers are constructed upon a definite numerical plan. In the monocotyledons the flowers usually have a numerical plan of three: 3 sepals, 3 petals, and usually 3 or a multiple of 3 stamens. These are called *3-merous* flowers. Dicotyledon flowers are usually constructed on a numerical plan of four or five and are then referred to as *4-merous* or *5-merous* flowers. The numerical plan of the flower is most evident in the sepals and petals, and to some extent it is carried through to the stamens; but this feature of the construction very often does not apply to carpels or pistils.

ALTERNATION OF PARTS. The parts of a flower are usually arranged in such a manner that the petals alternate with the sepals, and the stamens alternate with the petals (inner sets of stamens may alternate with outer sets). The pistils or carpels, however, are often opposite the sepals. This feature of alternation of parts gives us a valuable clue to the location of vestigial or missing flower parts in certain groups. If, for example, a single set of stamens is placed so that each stamen comes opposite a petal, as in the Primrose Family, we assume that an outer set of stamens has been lost in the process of evolution.

COMPLETE AND INCOMPLETE FLOWERS. When all the characteristic parts—sepals, petals, stamens, and pistils—are present, the flower is

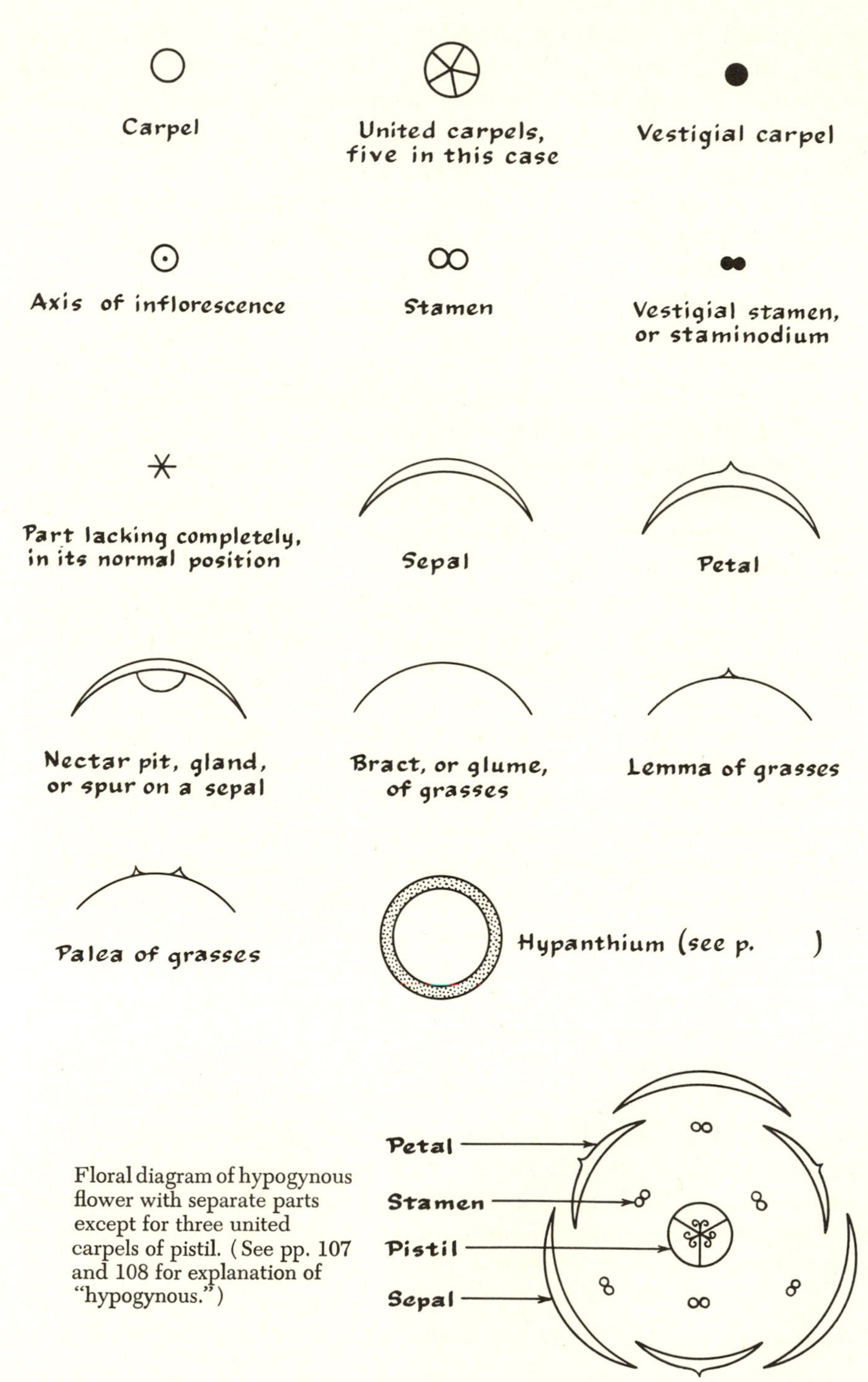

Floral diagram of hypogynous flower with separate parts except for three united carpels of pistil. (See pp. 107 and 108 for explanation of "hypogynous.")

28 Standardized symbols used in floral diagrams.

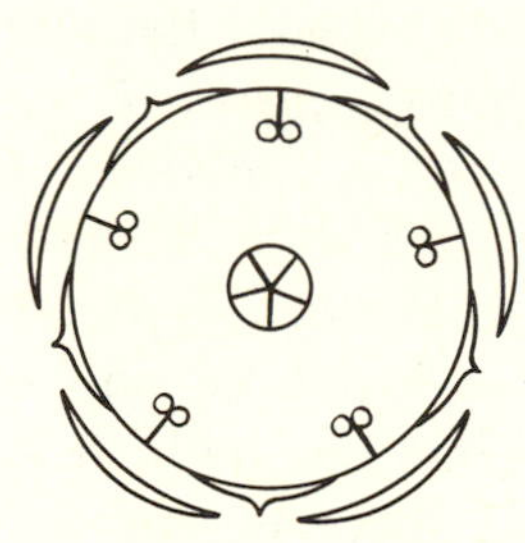

29 Corolla of united petals; stamens inserted on the corolla tube.

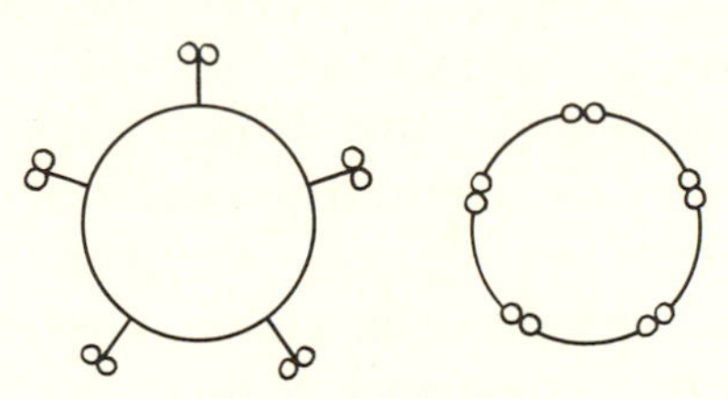

30 Stamens united (left) by their filaments (monadelphous) and (right) by their anthers (syngenesious).

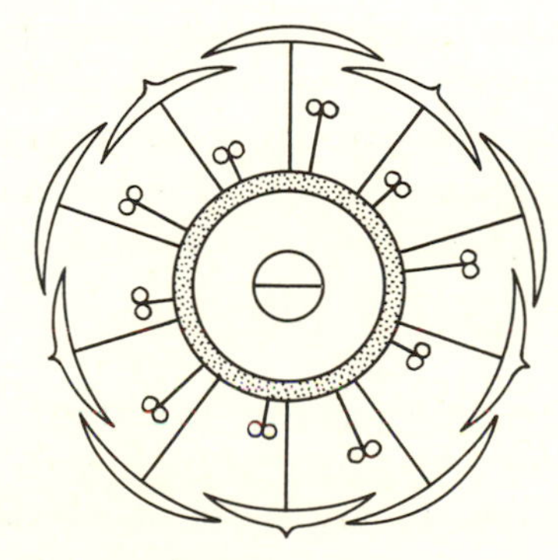

31 Floral diagram of a perigynous flower, the stamens, petals, and sepals inserted on the hypanthium.

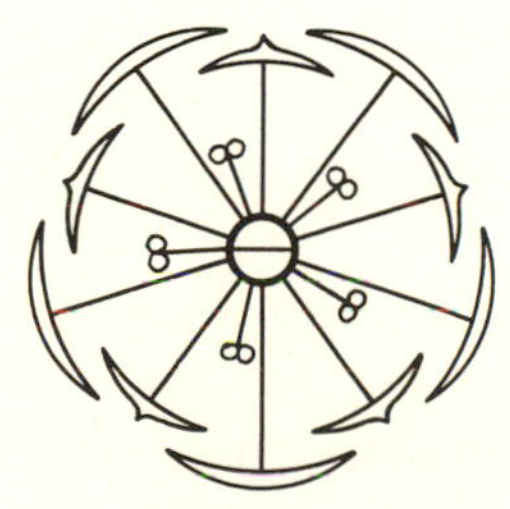

32 Floral diagram of an epigynous flower without a hypanthium, the floral parts inserted directly on the ovary.

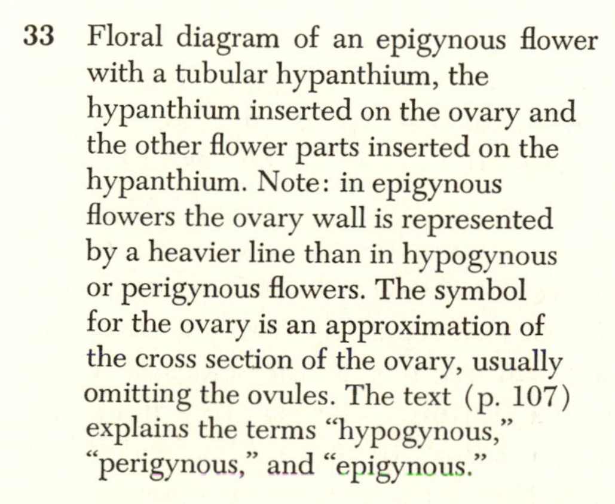

33 Floral diagram of an epigynous flower with a tubular hypanthium, the hypanthium inserted on the ovary and the other flower parts inserted on the hypanthium. Note: in epigynous flowers the ovary wall is represented by a heavier line than in hypogynous or perigynous flowers. The symbol for the ovary is an approximation of the cross section of the ovary, usually omitting the ovules. The text (p. 107) explains the terms "hypogynous," "perigynous," and "epigynous."

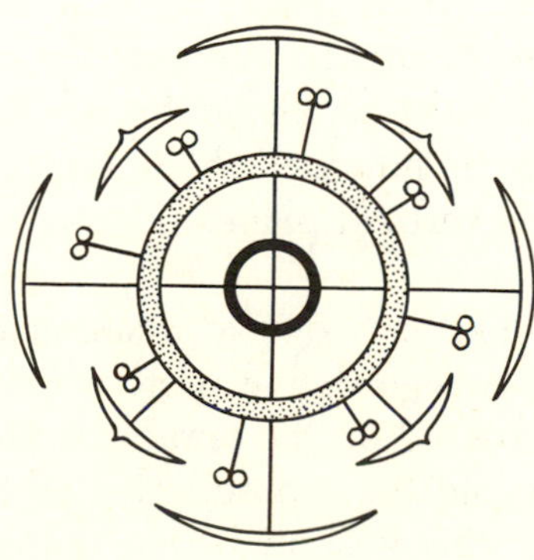

complete. But we frequently find that one or more of these parts may be lacking, in which case the flower is *incomplete.* Should the whole perianth of sepals and petals be lacking, the flower is *naked,* as in Willows. If only the petals are lacking, the flower is *apetalous,* and then the sepals frequently take on the aspect of petals, as in Anemones.

Flowers having both stamens and pistils, regardless of the presence or absence of other parts, are *perfect* flowers. If a flower has only one kind of sexual organ, either stamens or pistils but not both, it is a *unisexual* flower; male flowers, or those having stamens, are called *staminate* flowers, and female flowers, or those having pistils, are called *pistillate* flowers. When the flowers are unisexual and both sexes occur on the same plant, as in Oaks and Birches, the plants are known as *monoecious* plants; when unisexual flowers occur on separate plants, as in Willows and Poplars, the plants are *dioecious.* Intermediate conditions of sexuality also occur in some plants, such as species of the Rhubarb Family, some flowers being perfect while others are unisexual, and such a condition is known as *polygamous.* Carrying this a little further, we may designate the sexual condition as either *polygamo-monoecious* or *polygamo-dioecious,* depending on the placement of the different kinds of flowers.

REGULAR AND IRREGULAR FLOWERS. These terms apply to the perianth, as a rule, and rarely involve the reproductive parts. *Regular flowers* (sometimes called *actinomorphic* flowers) are those in which the perianth parts of each kind are similar in size and shape, so that the flower may be divided into equal halves by a vertical plane in various directions, the flower being radially symmetrical. Lilies and Buttercups, for example, have regular flowers. *Irregular flowers* (sometimes called *zygomorphic* flowers) are those in which the perianth parts of each kind are dissimilar in size and shape, some petals being unlike other petals, or some of the sepals being dissimilar in size or shape to others. Irregularity usually involves the petals, but it may involve the sepals or the whole perianth. Sweetpeas and Pansies have irregular flowers, which may be divided into equal halves only by a single vertical plane.

SPIRAL AND CYCLIC ARRANGEMENTS OF FLOWER PARTS. In what are usually regarded as more primitive flower types, the various parts may be inserted on the floral axis in a spiral manner, as the scales of a pine cone and the reproductive parts of a Buttercup flower. In a floral diagram this is shown by a dotted spiral line. In more advanced flower types, such as found in the Lily, the various parts are inserted in whorls, each whorl at a slightly different level, and this is known as a cyclic arrangement.

HYPOGYNOUS, PERIGYNOUS, AND EPIGYNOUS FLOWERS. Flowers may be hypogynous, perigynous, or epigynous, depending on the way the flower parts are inserted in relation to one another.

Hypogynous flowers have the sepals, petals, and stamens inserted round the base of the gynoecium and free from it; or the stamens may be inserted on the petals, which are inserted below the base of the gynoecium but free from it. Hypogynous flowers never have a hypanthium.

Perigynous flowers have the sepals, petals, and stamens inserted on the rim of a shallow or deep saucer-like or cup-like structure called the hypanthium, which arises at the base of the gynoecium and may be free from it or adnate to it. The gynoecium, then, is situated inside the hypanthium and above its base, never beneath it. Perigynous flowers always have a hypanthium, whether it is large and conspicuous or abbreviated and inconspicuous. If the hypanthium becomes adnate to the ovary, as in the Apple, the flower takes on the appearance of being epigynous, but a section of the flower across the ovary and hypanthium will always show a band of tissue (the hypanthium) surrounding the ovary. The term "hypanthium" is a recent one, older treatments having often confused the structure with a calyx-tube. As must be evident, the hypanthium is more than a calyx-tube, for it also gives rise to petals and stamens. In apetalous and unisexual flowers, however, it is often difficult or impossible to determine the true nature of this kind of structure, whether calyx-tube or hypanthium, and the best clues lie in anatomical features such as vasculature.

Epigynous flowers have a single ovary, which is beneath the flower parts, and these arise directly from its summit. According to some workers, epigynous flowers must have an external tissue or sleeve connecting the sepals, petals, and stamens to the base of the ovary; but in the writer's experience this cannot be demonstrated by a section across the ovary, and we must assume that those flower parts are attached directly to the ovary. It seems likely, however, that a perigynous flower with an adnate hypanthium might have given rise to an epigynous flower by losing the hypanthium.

Hypogynous flowers are believed to have originated first and to have been followed in evolution by perigynous and epigynous types. In a few families, however, such as the Heath Family, both hypogynous and epigynous flower types occur; this feature, therefore, cannot always be used as a criterion of evolutionary position.

These flower types are discussed in greater detail below under the title of Six Basic Morphological Flower Types. The parts of an ordinary, hypogynous, regular, complete flower, their arrangement and insertion, are shown in Figures 34–36.

SIX BASIC MORPHOLOGICAL FLOWER TYPES. The flowers of angiosperms, while showing great diversity of form and structure, may be considered as fitting into six basic morphological types. These represent increasing degrees of specialization, and within each type additional specialization may occur.

1. *The hypogynous flower.* Although this is the basic type, without much specialization, it includes advanced groups in which some parts, such as the corolla and stamens, are greatly modified. Hypogynous flowers of a simple type occur in Buttercups and Lilies, of an advanced type in Snapdragons and Mints. The sepals, petals, and stamens are inserted at the base of the gynoecium axis (or the stamens may be on the petals), and this renders the gynoecium superior in position. In primitive types the carpels and stamens are indefinite in number, often numerous, and often spirally arranged, becoming fewer and cyclic in higher types. In some plants, such as the Lotus (*Nelumbo*), the axis may be expanded. The parts of the perianth may be separate (*distinct*) or united, and they may be regular or irregular. There is no hypanthium. Figure 37 illustrates this type.

2. *The perigynous flower with free hypogynous hypanthium.* Here there has been an outgrowth of tissue, apparently from a primordium surrounding the base of the gynoecium axis. This is undifferentiated at first, but it eventually gives rise to stamens, petals, and sepals. This outgrowth, called the *hypanthium,* is free from the gynoecium and forms a cup-like or saucer-shaped structure. The gynoecium may consist of a single pistil, as in the Cherry (*Prunus*); or it may consist of several or numerous pistils (carpels), as in the Cinquefoil (*Potentilla*) or in the Strawberry (*Fragaria*), the latter having a fleshy gynoecium axis in fruit. The number of stamens may also vary from few to many. In most flowers of this type the corolla is regular, but in some groups, as in many members of the Pea Family, it is irregular. Perigynous flowers are believed to have been derived from hypogynous types and to manifest a higher degree of specialization. They are especially well represented in the *Rosales.* Figure 38 illustrates this type.

3. *The perigynous flower with the floral axis expanded continuously.* The floral organs are produced on a cup-like structure, and often there is a transition zone, near the middle of the cup, at which the female

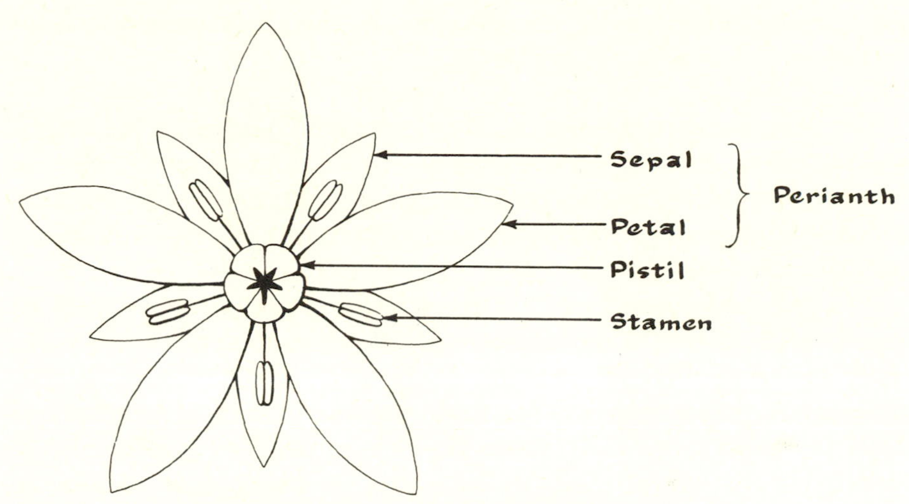

34 Viewed from above, a 5-merous, regular, complete flower.

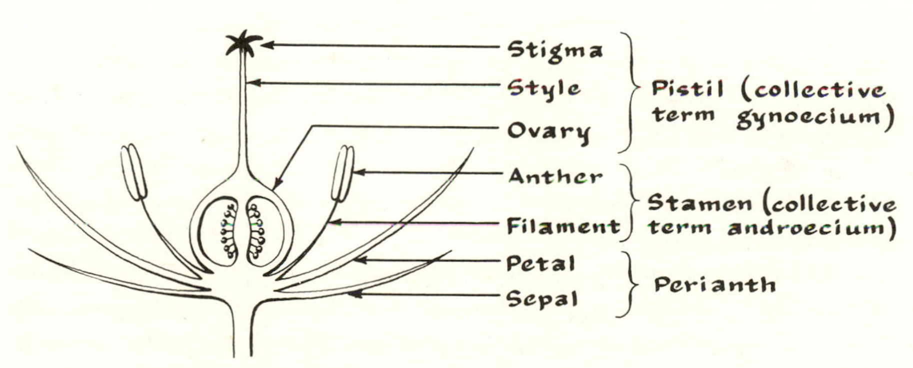

35 Longitudinal section, showing the flower to be hypogynous.

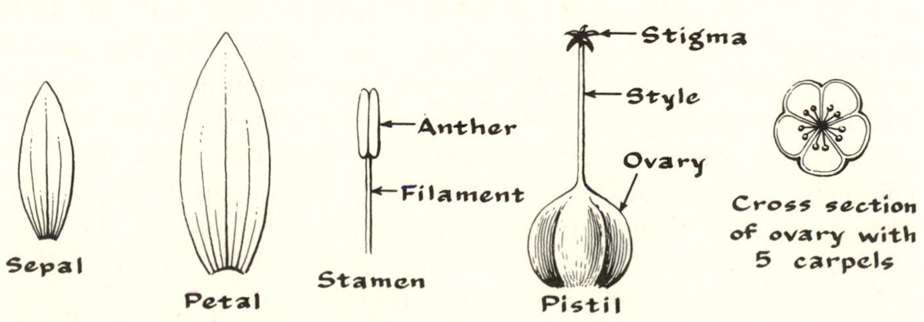

36 Flower parts represented individually.

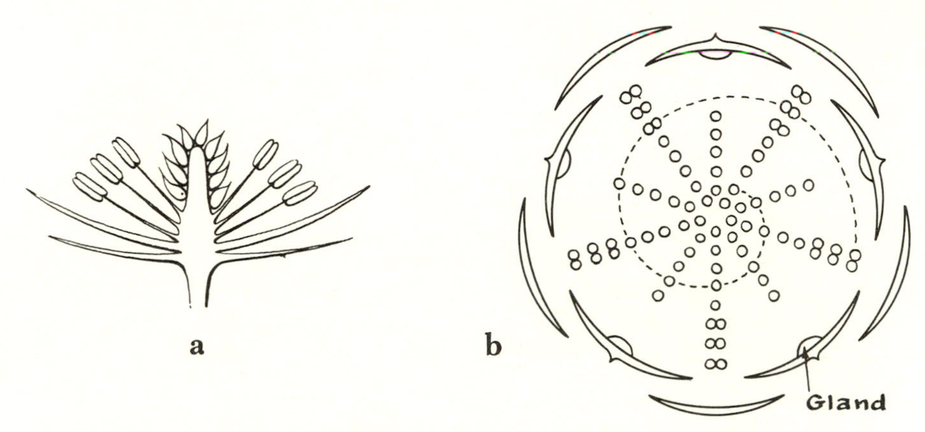

37 Hypogynous flower of a Buttercup (*Ranunculus*): (**a**) longitudinal section; (**b**) floral diagram. The dotted spiral line indicates that the stamens and carpels are spirally arranged.

portion gives way to the male portion. The transition zone often produces sterile carpels and stamens. This uncommon floral type is best illustrated by the Calycanthus Family (*Calycanthaceae*) and, in more advanced condition, by the Rose (*Rosa*) in the *Rosaceae.* In the latter the transition zone is completely sterile and produces no organs. Figure 39 illustrates this type.

4. *The perigynous flower with adnate hypogynous hypanthium.* This is similar to the preceding type, but the hypanthium grows up round the ovary and is adnate to it, often becoming fleshy in fruit. This gives the flower the appearance of epigyny, the ovary appearing to be inferior, but the appearance is due to the adnation of the hypanthium, inside which the ovary is situated. This occurs in Apple and Pear (*Pyrus*) flowers and in the Walnut (*Juglans*). The flesh of the apple fruit is hypanthium, and the husk round the nut of walnuts is also hypanthium. An interesting indication of the manner in which this type may have originated is shown by the Russian Olive (*Elaeagnus*): the flowering stage has a free hypanthium, but with the enlargement of the ovary in fruit the hypanthium-cup becomes completely filled and adnate to the ovary. Figure 40 illustrates this morphological type.

5. *The epigynous flower without a hypanthium.* Here the ovary is inferior in position to the rest of the flower parts, the sepals, petals,

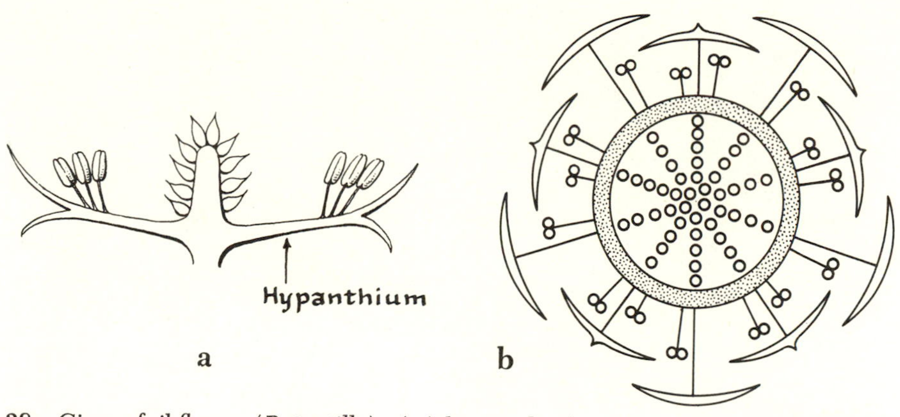

38 Cinquefoil flower (*Potentilla*): (**a**) longitudinal section; (**b**) floral diagram.

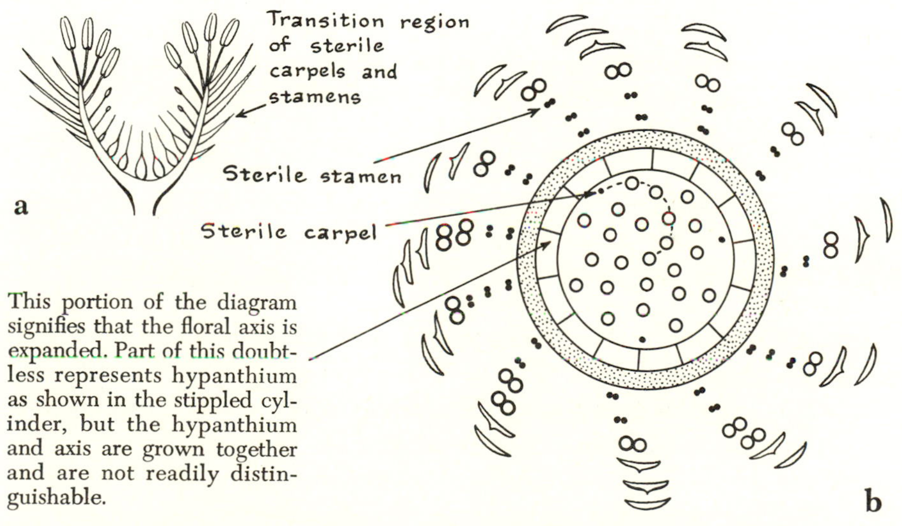

39 Calycanthus flower (*Calycanthus*): (**a**) longitudinal section; (**b**) floral diagram. When floral parts are numerous, it is better to omit connecting lines to the hypanthium to avoid confusion. The sepals, petals, and stamens must always be on the hypanthium if it is present; there can be no other place of insertion.

and stamens being inserted near the summit of the ovary. There is no hypanthium. This rather common type is illustrated by the Parsnip Family (*Apiaceae*), the Dogwood Family (*Cornaceae*), and the Orchid Family (*Orchidaceae*). It represents a somewhat higher development than the hypogynous and perigynous types. Figure 41 illustrates this type.

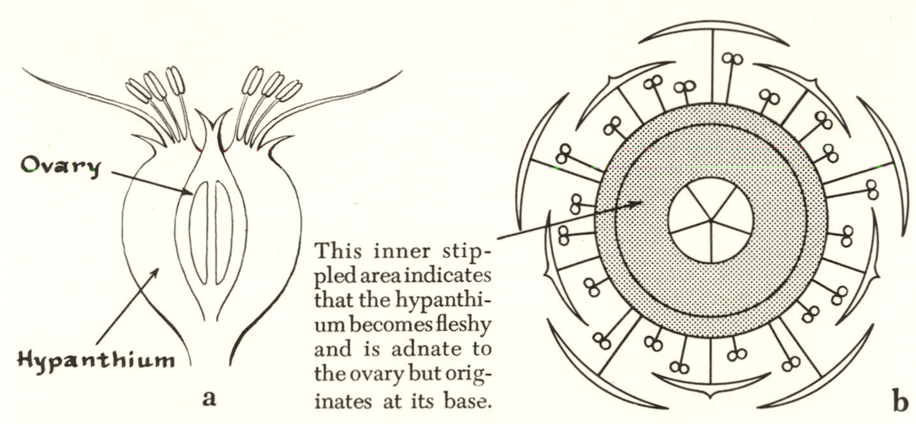

40 Apple flower (*Pyrus*): (**a**) longitudinal section; (**b**) floral diagram.

6. *The epigynous flower with an epigynous hypanthium.* This is similar to the preceding type but differs in having a hypanthium which originates from near the summit of the ovary and extends above it as a funnel, tube, or rod. This hypanthium may be very short or may be several inches long. In many members of the Evening Primrose Family (*Onagraceae*) the hypanthium is tubular or funnelform; in the Iris (*Iris*) it is solid and rod-shaped. This condition of epigyny with a hypanthium somewhat parallels the condition in No. 5 in evolution, some families having both types. Figure 42 illustrates this type.

THE ANDROECIUM. The androecium of a flower is made up of the male reproductive parts—that is, the microsporophylls, or stamens. It occupies a position outward from the gynoecium and inward from the corolla, but the stamens are sometimes inserted on the corolla

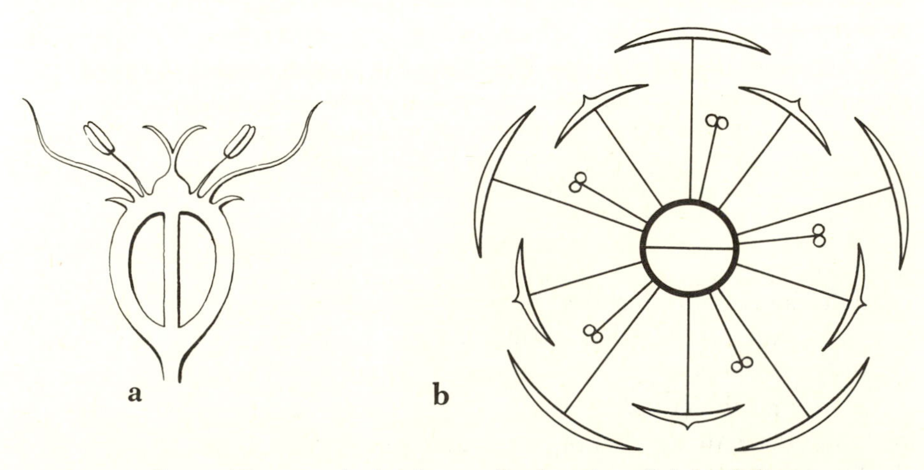

41 Parsnip flower (*Pastinaca*): (**a**) longitudinal section; (**b**) floral diagram.

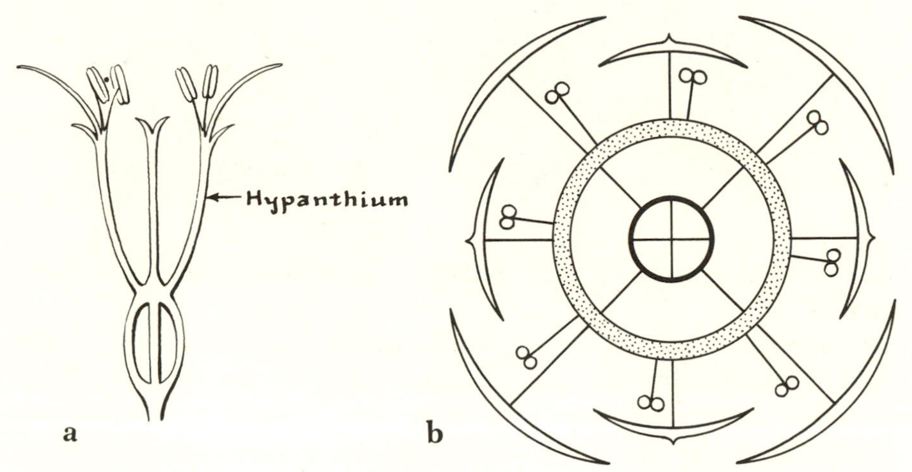

42 Evening Primrose flower (*Oenothera*):
(**a**) longitudinal section; (**b**) floral diagram.

(*epipetalous*), and infrequently, as in Orchids, they are attached to the pistil (*gynandrous*). In all perigynous flowers the stamens are inserted on a hypanthium.

Stamens are attached to one another in two general ways: by union of their filaments into one group (*monadelphous*), two groups (*diadelphous*), etc., and by union of their anthers into a ring (*syngenesious*).

When the stamens are of unequal lengths, two long and two shorter, they are termed *didynamous*. When there are four long and two shorter stamens, they are termed *tetradynamous*.

Attachment of the anther to the filament is of three general sorts: *innate* when the anther sacs are chiefly terminal and do not face inward or outward in the flower; *adnate* when the anther sacs are elongated and attached lengthwise to the filament, facing either inward (*introrse*) or outward (*extrorse*) in the flower; and *versatile* when the anther is attached by its middle with both ends free. The portion of the stamen that represents a continuation of the filament between the anther sacs is termed the *connective*.

All parts of the stamen are subject to modification. The filament is variously elongated, often dilated and petaloid, or, infrequently, lacking. The connective sometimes becomes greatly broadened transversely so that the anther sacs are widely separated (see *Salvia* in the Mint Family). Either or both of the anther sacs may be reduced and sterile; if both, we have sterile stamens, or *staminodia*, which are frequently very elaborate and much unlike the normal stamens. It is

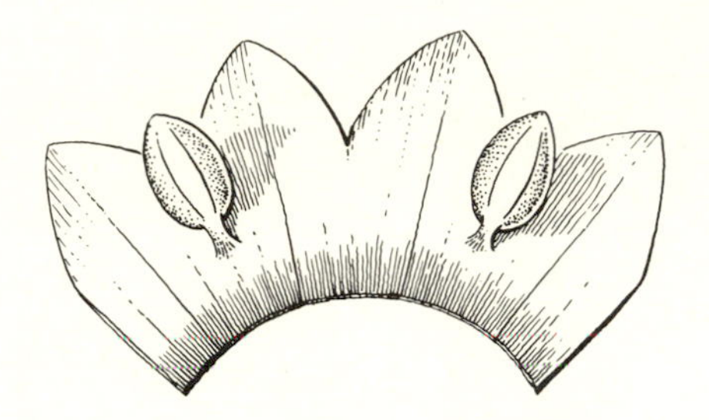

43 Epipetalous stamens of the Olive flower.

believed that glands in some flowers may represent much reduced stamens.

Primitive flowers, in general, are believed to have had numerous and more or less spirally arranged stamens, and from this condition flowers with progressively fewer and often more elaborate stamens have evolved. Irregular flowers usually have their stamens modified or reduced from the normal number (less than the number of sepals or petals). Figures 43–52 illustrate these terms and features of the androecium.

THE GYNOECIUM. The gynoecium of a flower is made up of the female reproductive parts—that is, the carpels or pistils. It occupies a central position in the flower.

When the pistil is composed of a single megasporophyll, or *carpel,* it is termed a *simple pistil.* When it is composed of two or more carpels that are more or less united, it is a *compound pistil.*

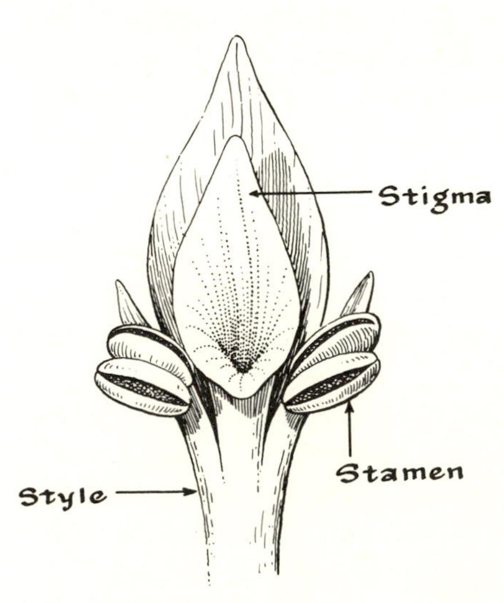

44 Gynandrous stamens of an Orchid.

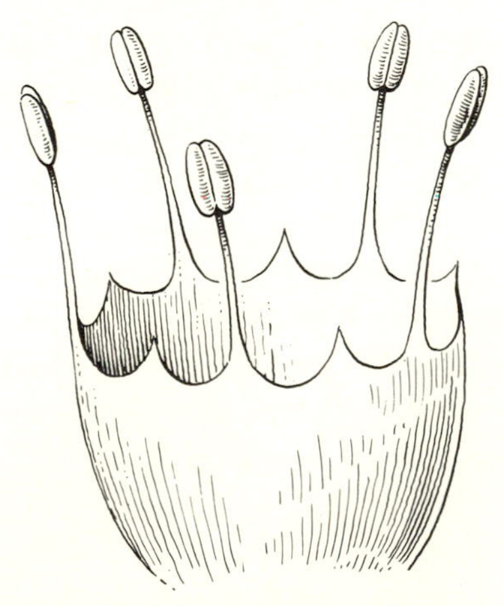

45 Monadelphous stamens of a Flax flower. The alternating points on the tube are probably the remnants of what were formerly stamens.

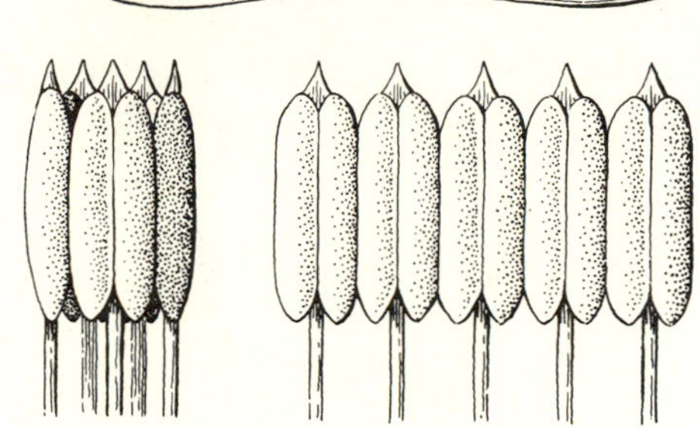

46 Diadelphous stamens of a Sweet Pea flower, nine in one group and one in the other.

47 Syngenesious stamens of a member of the Sunflower Family. At the left they are shown as they would be in the flower; at the right they are shown spread out flat.

Each carpel has two ribs, seams, or *sutures,* one representing the midrib of the leaf and called the *dorsal suture* (not really a seam), the other representing the joined edges of the leaf and called the *ventral suture.* The ovules are attached to the inside of the ventral suture along a line or at a point called the *placenta.*

Figure 53 illustrates some of the types of simple and compound pistils and shows one way in which they may have originated.

Detailed studies of the morphology of the carpels of certain primitive dicotyledons have been made recently by I. W. Bailey and his associates (1951). These give us a more modern view of the nature

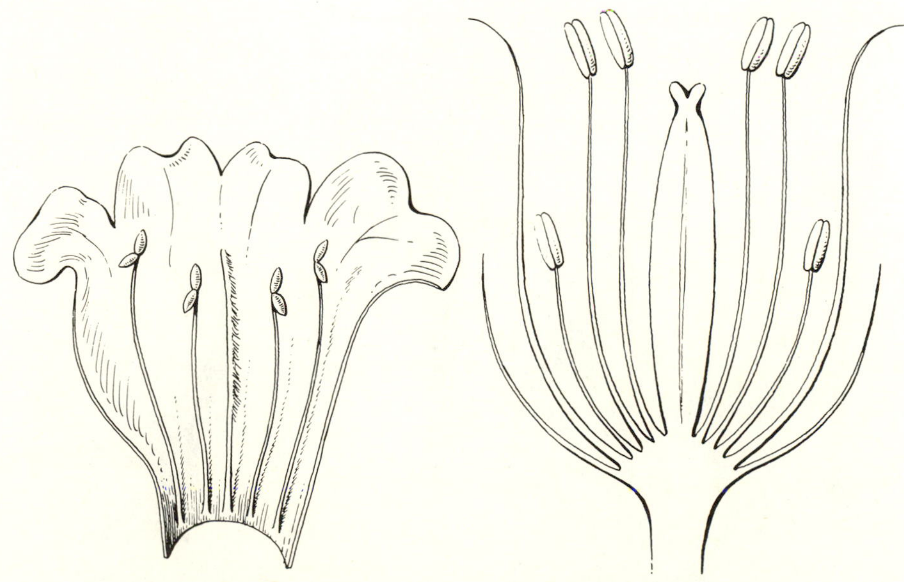

48 Didynamous fertile stamens of *Penstemon.*

49 Tetradidynamous stamens of a mustard flower.

50 Sterile stamens (staminodia) of a *Catalpa* flower.

of primitive carpels and of the way they may have evolved from folded (conduplicate) blades that are unsealed and styleless, as shown in Figure 54.

PLACENTATION TYPES. When cross sections of various ovaries are examined, it is found that the position of the ovules and placentae nearly always conforms to one of the three main patterns shown in Figures 55–57. Variations of these basic patterns are known but are uncommon.

Parietal and marginal placentation. Here the ovules are attached to the inner wall of the ovary on one or more placentae. This type occurs in simple pistils, where it is often referred to as *marginal,* and in many compound pistils, where it is called *parietal.* With rare exceptions, as in the Mustard Family, the ovary is not compartmented.

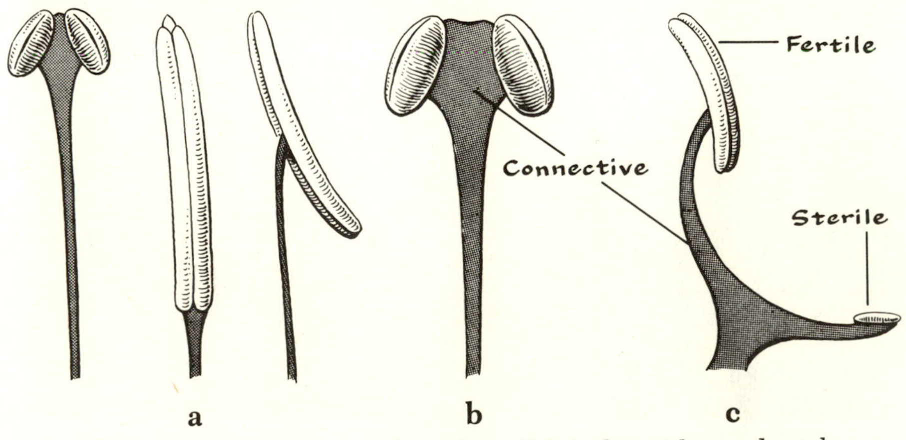

51 (**a**) Innate, adnate, and versatile anthers. (**b**) Anther with a moderately broadened connective. (**c**) Stamen of *Salvia* with an elongated connective between the anther sacs, one fertile and the other sterile.

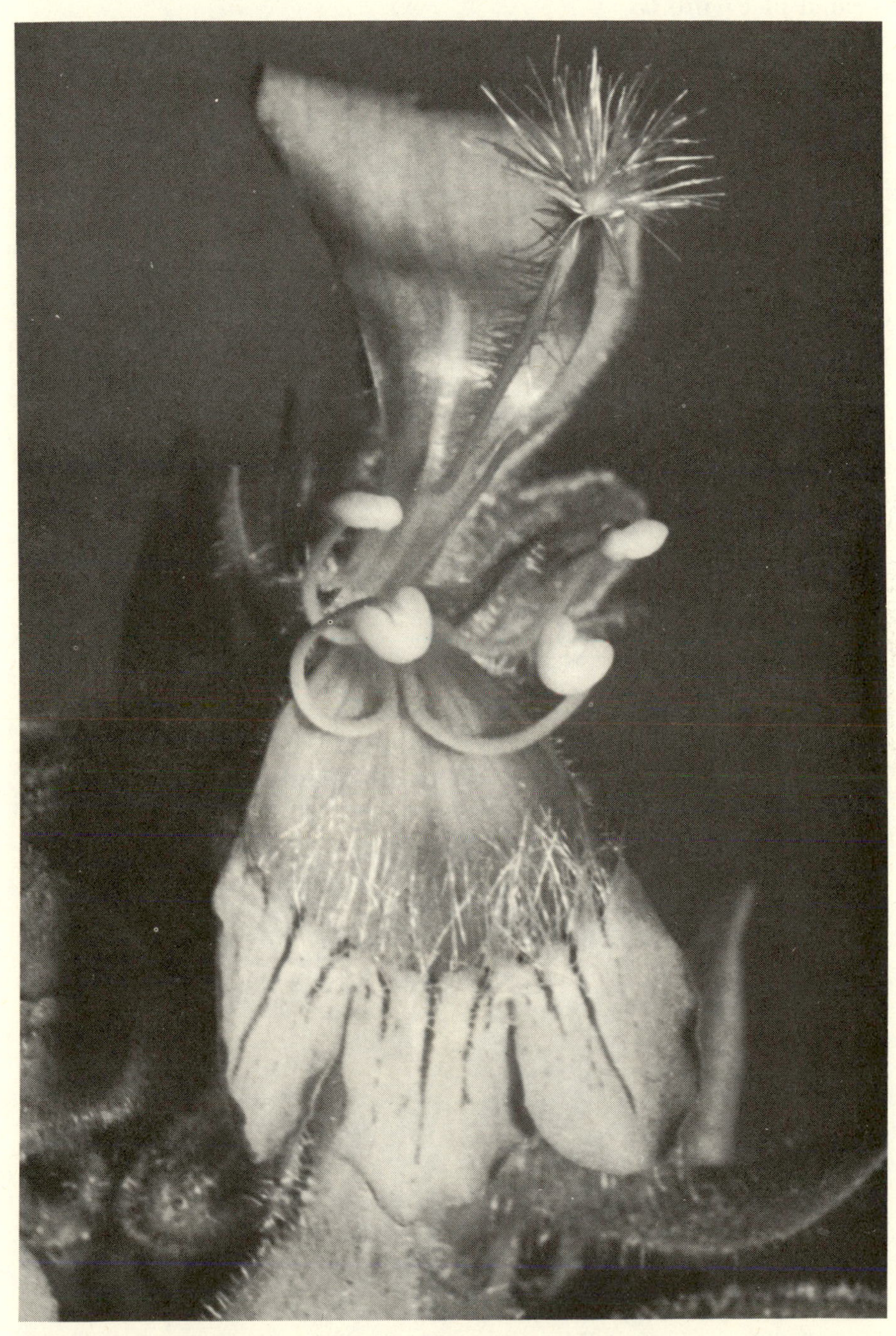

52 Flower of *Penstemon eriantherus* Pursh, with the corolla torn open to show the didynamous fertile stamens and the fifth and bearded stamen, or staminodium.

This type of placentation is generally regarded as primitive. It is illustrated in Figure 55.

Axillary (*axile*) *placentation.* Here the ovules are attached near the center of the ovary at the junction or axis of the partitions that divide the ovary into compartments. This can occur only in compound pistils and is regarded as an intermediate stage in evolution. It is illustrated by Figure 56.

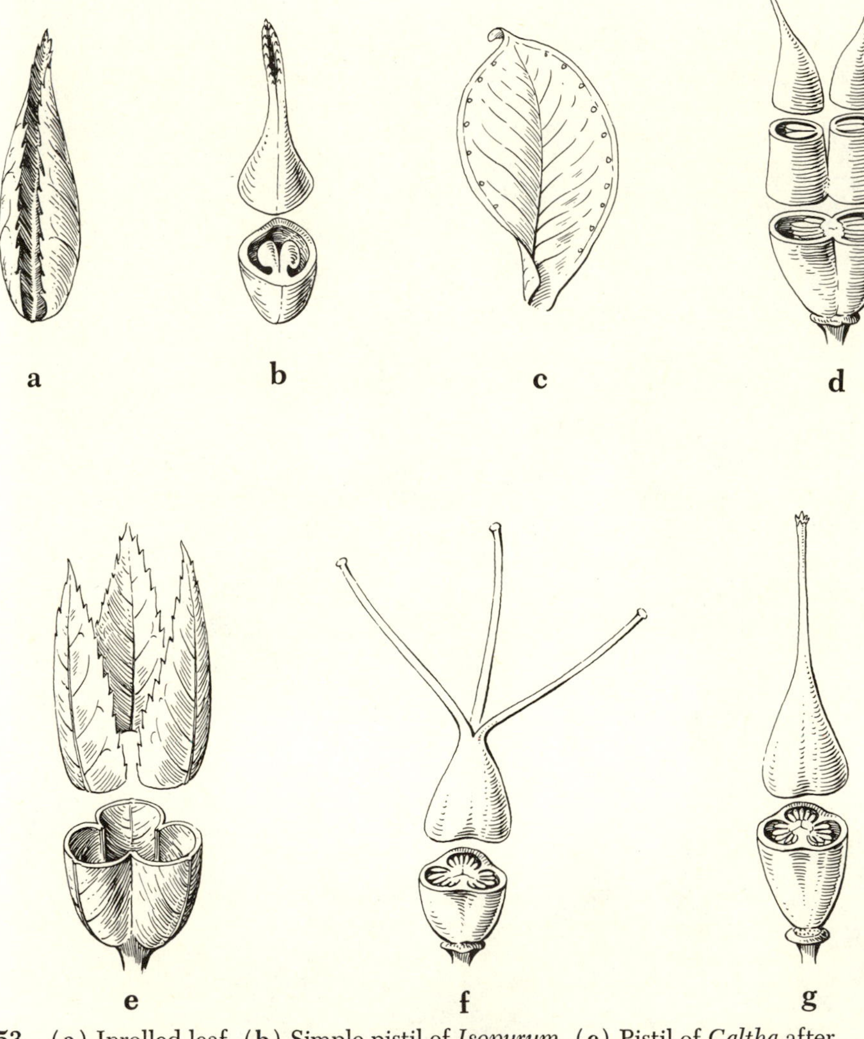

53 (**a**) Inrolled leaf. (**b**) Simple pistil of *Isopyrum.* (**c**) Pistil of *Caltha* after shedding its seeds; one carpel. (**d**) Compound pistil of two carpels of *Saxifraga.* (**e**) Three leaves forming a compound pistil. (**f**) Compound pistil of three carpels of *Hypericum,* the styles distinct. (**g**) Compound pistil of three carpels of *Hypericum,* the styles united.
(Parts *e–g* redrawn from Gray's *Lessons.*)

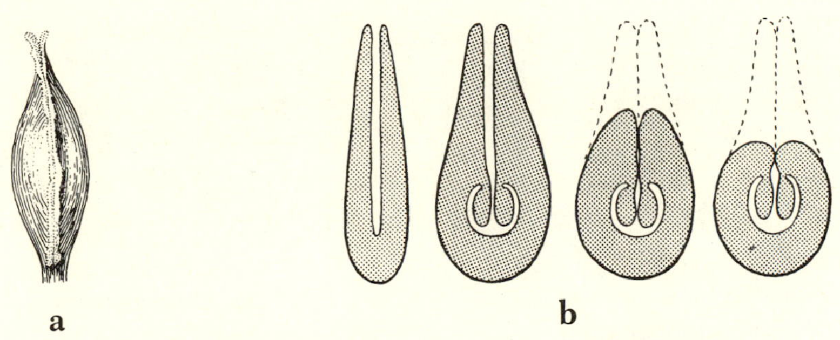

54 (**a**) Primitive conduplicate carpel in ventral view. [Redrawn from I. W. Bailey and B. G. L. Swamy in *Am. J. Bot.* **38**:374 (1951).] (**b**) Transverse sections of primitive carpels showing the development of a locule and stages in the closure of conduplicate carpels. [Redrawn from I. W. Bailey and B. G. L. Swamy in *Am. J. Bot.* **38**:377 (1951).]

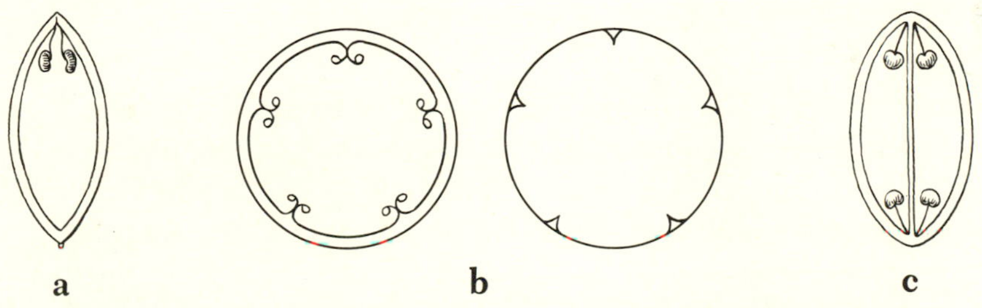

55 (**a**) Cross section of ovary of a simple pistil (Pea). (**b**) Cross section of ovary of a compound pistil of five carpels with five parietal placentae (Poppy); at right as shown diagrammatically in a floral diagram. (**c**) Cross section of ovary of a 2-carpeled pistil with two parietal placentae (Mustard).

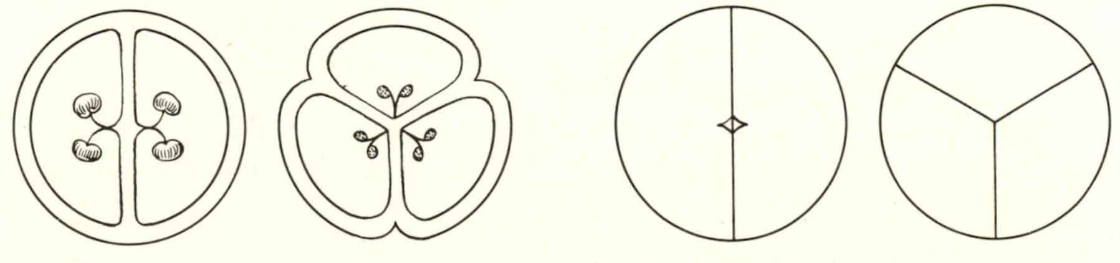

56 Two examples of axillary placentation. The left-hand figure of each pair is from a Morning-glory, that on the right from a Lily. The two right-hand figures show how these might be represented in floral diagrams.

Free-central and basal placentation. Here the ovules are produced on a projection from the base of the ovary of a compound pistil, this placenta being free from the ovary wall laterally and the ovary one-celled. *Basal placentation* is restricted to the reduced condition in which a single ovule is produced. By dissolution of the partitions in

57 Cross section of an ovary of three carpels showing free-central placentation (Chickweed).

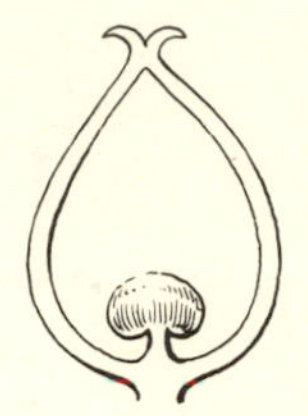

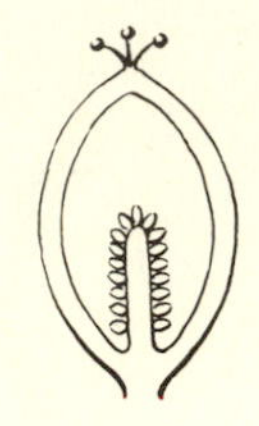

58 Longitudinal sections of ovaries showing basal placentation (left) and free-central placentation (right).

the axillary type we can derive this type of placentation, illustrated by Figures 57 and 58.

Fruits

The fruit of a flowering plant may be defined as a matured ovary and its contents, together with other flower parts that may sometimes adhere to it. The fruit is often an important diagnostic feature of a family or genus: the legume of the Pea Family, for example, or the aggregate of drupelets of the Blackberry and Raspberry (genus *Rubus*). We should keep in mind that for the taxonomist the saying "By their fruits ye shall know them" takes on real meaning!

The ovary wall, known as the *pericarp,* consists of three layers in fruits: the *exocarp,* or outer layer, which is often the skin; the *mesocarp,* or middle layer, which may become fleshy; and the *endocarp,* or inner layer, which is sometimes modified in various ways.

As yet we have only an artificial classification of fruits, based on their gross morphology rather than on their mode of origin. Until a better classification and a better nomenclature of fruits are forthcoming, the following key will distinguish the commonly recognized kinds. (It should be pointed out that the term "pod," while meaning some sort of dry and dehiscent fruit, has no precise meaning to the taxonomist.)

Key to Types of Fruits

1. Fruit derived from several flowers . . . . MUTIPLE FRUIT (Pineapple, Mulberry, Fig)
1. Fruit derived from a single flower
    2. Derived from more than one pistil . . . . AGGREGATE FRUIT (Raspberry, Strawberry)
    2. Derived from a single pistil . . . . SIMPLE FRUITS (see below)

3. Fruit fleshy, usually indehiscent (not splitting)
    4. Flesh of fruit derived from a hypanthium that surrounds the papery carpels . . . . POME
    (Apple, Pear, Quince)
    4. Flesh of fruit derived from the ovary wall
        5. Pericarp with an outer fleshy layer and an inner bony layer (the stone) . . . . DRUPE
        (Peach, Cherry)
        5. Pericarp without an inner bony layer, more or less fleshy throughout . . . . BERRY
        (Tomato, Grape)
        [Modifications of the berry type of fruit may be defined as follows]
            6. Septa evident in cross section; the outer layer leathery . . . . HESPERIDIUM
            (Orange, Lemon)
            6. Septa lacking; the outer layer leathery to hard and woody . . . . PEPO
            (Cucumber, Watermelon, Cantaloupe)
3. Fruit dry at maturity, dehiscent or indehiscent
    7. Fruit indehiscent (not splitting open)
        8. With one or more wings . . . . SAMARA
        (Maple, Ash, Elm)
        8. Without wings
            9. From a compound pistil, becoming hard and bony-shelled . . . . NUT (or NUTLET if small)
            (Oak, Walnut, Hazelnut)
            9. From a simple pistil, often with a thin shell but not bony
                10. Pericarp fused to the seed . . . . GRAIN (CARYOPSIS)
                (Wheat, Rice, Corn)
                10. Pericarp separable from the seed . . . . AKENE
                (Sedges, Sunflower, Aster)
    7. Fruit dehiscent (splitting open)
        11. From a simple pistil (1 carpel)
            12. Splitting on two sutures . . . . LEGUME
            (Pea, Bean, Alfalfa)
            12. Splitting on one suture . . . . FOLLICLE
            (Milkweed, Larkspur)
        11. From a compound pistil (carpels 2 or more, united)
            13. Carpels separating from each other but each retaining its seed . . . . SCHIZOCARP
            (Parsnip, Carrot)
            13. Carpels splitting, releasing 1 or more seeds
                14. Fruit 2-celled, the two valves splitting away from a persistent, thin partition or septum (replum) . . . . SILIQUE (or SILICLE if short)
                (Mustard)
                14. Fruit 1–several-celled, the partition not persistent if fruit 2-celled . . . . CAPSULE
                (Willow, Iris, Yucca)

[There are modifications of the capsular type of fruit (see Fig. 59) as follows]

15. Opening by a lid (circumscissile capsules)
    16. With a single seed .... UTRICLE (Pigweed)
    16. With several seeds .... PYXIS (Portulaca)
15. Opening by holes near the top ... PORICIDAL CAPSULE (Poppy)
15. Opening by splitting lengthwise
    17. Splitting on the septae ..... SEPTICIDAL CAPSULE (Yucca)
    17. Splitting between the septae and in the locules or chambers ..... LOCULICIDAL CAPSULE (Glacier Lily)

Ovules and Seeds

A seed is a matured ovule. It should not be confused with a fruit, which, though it may resemble a seed, is really a matured ovary containing one or more seeds. The typical seed consists of an *embryo* surrounded and protected by a *seed coat,* with or without a quantity of stored food, known as the *endosperm,* in which the embryo is embedded. Figures 61 and 62 illustrate the principal parts of seeds, one with and the other without endosperm.

The *embryo* is the young plant within the seed. It may be very much smaller than the seed, in which case it is likely to be surrounded by a quantity of endosperm, upon which it draws for nourishment until it is independent. It may be large enough to fill the seed cavity, in which case it usually contains reserve food within itself. The principal parts of the embryo are the *cotyledons,* which are a pair of leaf-like organs in the dicotyledons, a single one in the monocotyledons, a whorl of several in the gymnosperms; the *epicotyl,* projecting above the point of attachment of the cotyledons, which will become stem and leaf; the *hypocotyl,* a transitional stem-like portion between the attachment of the cotyledons and the root-forming part; and the *radicle,* or root-forming part. The hypocotyl may be poorly developed in some embryos and well defined in others.

Since embryos show varying degrees of specialization, they may be useful in the delimitation of genera and families, as will be seen in such families as the *Chenopodiaceae* (Goosefoot Family) and the *Brassicaceae* (Mustard Family). The primitive embryo was probably straight; more advanced types are curved or bent in the middle, and some are even spirally coiled.

The *seed coat* may be either a single or a double structure derived

59 Three types of capsules: on the left below is a septicidal capsule of *Yucca,* in the middle above is a pair of loculicidal capsules of *Iris,* and on the right below is a poricidal capsule of *Papaver,* a Poppy.

from the integument or integuments of the ovule. When there are two seed coats, the outer one (*testa*) is usually hard and tough, and the inner one (*tegumen*) is thin and delicate. Variations in the structure and markings of the seed coat may furnish useful taxonomic characters for distinguishing groups of plants. Commonly used features of this sort are the wings on the seeds of the *Bignoniaceae* (Trumpet-vine

60 Silique fruits of *Lunaria* (Money Plant) that have split and left only the partition, or replum, behind on the plant. This partition is very delicate and silvery. The inflorescences at this stage are sometimes used for winter bouquets.

Family) and the tufts of hair on the seeds of the *Salicaceae* (Willow Family) and *Asclepiadaceae* (Milkweed Family). The seed hairs of the genus *Gossypium* (Cotton) are of great economic importance.

The *endosperm* is present in the immature stages of the formation of seeds and may persist, as a cellular food mass in which the embryo is embedded, when the seed is ripe. In many plants, however, the endosperm is used up, and its space is occupied by the embryo when the seed is mature, the food stored in the endosperm tissue having been assimilated by the embryo and re-elaborated, in part, as food stored in the cotyledons. The presence or absence of endosperm in

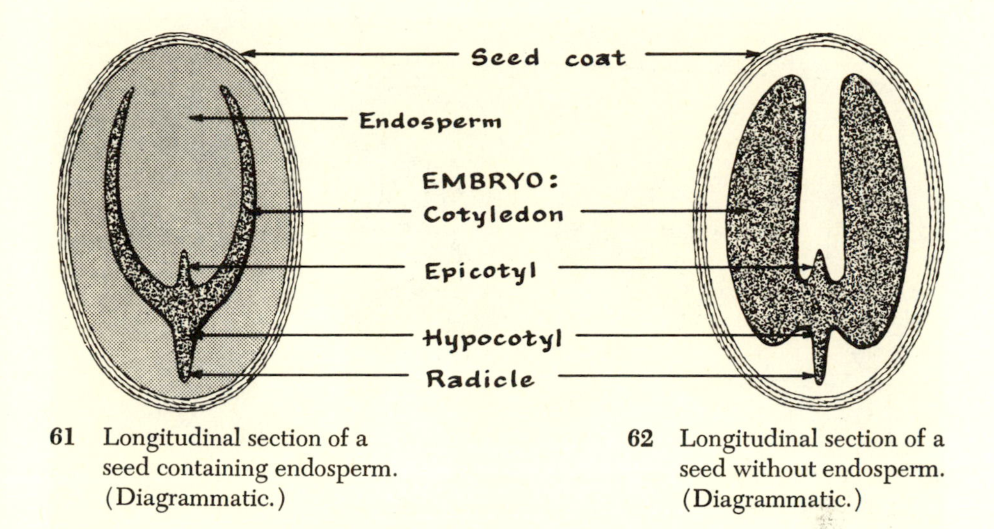

61 Longitudinal section of a seed containing endosperm. (Diagrammatic.)

62 Longitudinal section of a seed without endosperm. (Diagrammatic.)

mature seeds is a useful diagnostic character, usually at the family level.

Some seeds have an additional feature known as an *aril,* which is an outgrowth from the seed stalk (*funiculus*) or sometimes from the placenta. It is often of a mucilaginous or gelatinous texture and sometimes brightly colored. Examples of arillate seeds are the Litchi (*Sapindaceae*), the aril of which is eaten, the Yew (*Taxaceae*), in which it is red, and the spice known as mace, which is the aril of the Nutmeg (*Myristicaceae*).

External features of the seed, in addition to those mentioned above, are the *hilum,* the scar left on the seed at the point where the funiculus was detached; the *raphe,* the ridge formed by the fusion of the funiculus with the seed when the funiculus is bent; the *chalaza,* the upper portion of the raphe where the funiculus merges with the base of the ovule; and the *micropyle,* the minute pore through which the pollen tube once entered the ovule. These features are illustrated in Figure 63.

The *direction,* the way in which the ovule points in relation to its

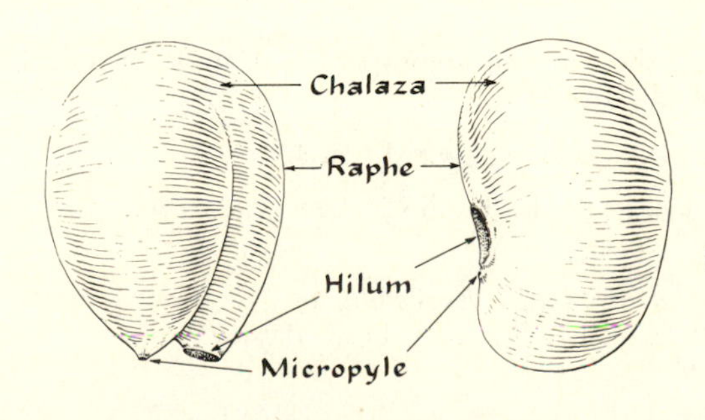

63 Side views of seeds of Violet (left) and Bean (right) showing external features.

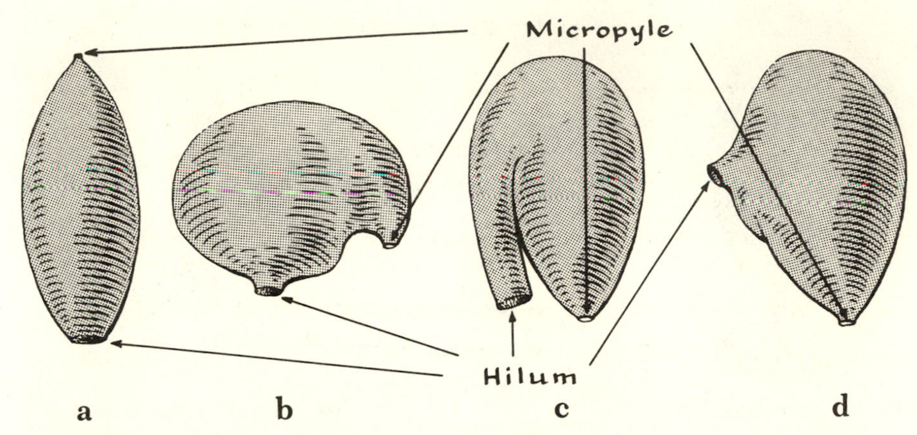

64 (**a**) Orthotropous ovule. (**b**) Campylotropous ovule. (**c**) Anatropous ovule. (**d**) Amphitropous ovule.

attachment to the ovary, is sometimes used as an aid in defining taxa. This direction or condition may take four forms. If the ovule is *orthotropous,* it is straight, and the micropylar end is directly opposite the hilum, as in Figure 64, a. If the ovule is *campylotropous,* it is curved, so that the micropylar end is close to the hilum, but the ovule as a whole is more or less horizontal, as in Figure 64, b. If the ovule is *anatropous,* it is inverted, the funiculus being curved at the apex and the body of the ovule lying against it. This is one of the commonest types and is illustrated by Figure 64, c. If the ovule is *amphitropous,* the funiculus and the raphe are short, and the ovule is attached laterally. This condition, intermediate between the campylotropous and the anatropous, is illustrated by Figure 64, d.

REFERENCES

Vegetative organs

ARBER, AGNES. 1930. "Root and shoot in the angiosperms: A study of morphological categories," *New Phytol.* **29**:297–315.

CANDOLLE, A. P. DE. 1827. *Organographie végétale*, I. Paris.

CANDOLLE, C. DE. 1868. "Théorie de la feuille," *Arch. Sci. Bibl. Universelle* **32**:32–64.

GOEBEL, K. 1905. *Organography of Plants, Especially of the* Archegoniatae *and* Spermatophyta (authorized English edition by Isaac Bayley Balfour), Part II. Oxford.

Inflorescences

CROIZAT, LEON. 1943. "The concept of the inflorescence," *Bull. Torr. Club* **70**:496–509.

PARKIN, J. 1914. "The evolution of the inflorescence," *Proc. Linn. Soc.* **42**:511–562.

PHILIPSON, W. R. 1946–1948. "Studies in the development of the inflorescence," *Ann. Not.* n.s. **10**:257–270, **11**:285–297, 409–416, **12**:65–75, 147–156.

RICKETT, H. W. 1944. "The classification of inflorescences," *Bot. Rev.* **10**:187–231.

RICKETT, H. W. 1955. "Materials for a dictionary of botanical terms: III, Inflorescences," *Bull. Torr. Club* **82**:419–445.

WOODSON, R. E., JR. 1935. "Observations on the inflorescence of *Apocynaceae*," *Ann. Mo. Bot. Gard.* **22**:1–48.

Floral diagrams

EICHLER, A. W. 1875–1878. *Blüthendiagramme*, 2 vols. Leipzig.

SCHAFFNER, J. H. 1916. "A general system of floral diagrams," *Ohio. J. Sci.* **16**:360–364.

Floral morphology

ARBER, AGNES. 1937. "The interpretation of the flower: A study of some aspects of morphological thought," *Biol. Rev.* **12**:57–84.

BAILEY, I. W., and B. G. L. SWAMY. 1951. "The conduplicate carpel of dicotyledons and its initial trends of specialization," *Am. J. Bot.* **38**:373–379.

BAILEY, I. W., and C. G. NAST. 1943. "The comparative morphology of the *Winteraceae:* II, Carpels," *J. Arnold Arb.* **24**:472–481.

BANCROFT, H. 1935. "A review of researches concerning floral morphology," *Bot. Rev.* **1**:77–99.

DOUGLAS, G. E. 1944. "The inferior ovary," *Bot. Rev.* **10**:125–186.

EAMES, A. J. 1929. "The role of flower anatomy in the determination of angiosperm phylogeny," *Proc. Int. Congr. Plant Sci. 1926,* **1**:423–427.

EAMES, A. J. 1931. "The vascular anatomy of the flower with a refutation of the theory of carpel polymorphism," *Am. J. Bot.* **18**:147–188.

GAUTHIER, ROGER. 1950. "The nature of the inferior ovary in the genus *Begonia*," *Contr. Inst. Bot. Univ. Montreal* **66**:1–91.

GUNDERSON, A. 1928. "Flower structures of dicotyledons," *Torreya* **28**:70–73.

HENSLOW, G. 1888. *The Origin of Floral Structures Through Insect and Other Agencies.* New York.

HENSLOW, G. 1891. "On the vascular systems of floral organs and their importance in the interpretation of the morphology of flowers," *J. Linn. Soc.* **28**:151–197.

HUNT, K. W. 1937. "A study of the style and stigma, with reference to the nature of the carpel," *Am. J. Bot.* **24**:288–295.

JACKSON, G. 1934. "The morphology of flowers of *Rosa* and certain closely related genera," *Am. J. Bot.* **21**:453–466.

SAUNDERS, E. R. 1937, 1939. *Floral Morphology,* I, II. Cambridge, England.

SCHAFFNER, J. H. 1937. "The fundamental nature of the flower," *Bull. Torr. Club* **64**:576–582.

THOMAS, H. H. 1934. "The nature and origin of the stigma," *New Phytol.* **33**:173–198.

VAN TIEGHEM, P. 1868. "Recherches sur la structure du pistil," *Ann. Sci. Nat. Ser. V,* **12**:127–226.

VAN TIEGHEM, P. 1868. *Recherches sur la structure du pistil et sur l'anatomie comparée de la fleur.* Paris.

WERNHAM, H. F. 1911. "Floral evolution: with particular reference to the sympetalous dicotyledons," *New Phytol.* **10**:293–305.

WILSON, C. L. 1937. "The phylogeny of the stamen," *Am. J. Bot.* **24**:686–699.

WILSON, C. L. 1942. "The telome theory and the origin of the stamen," *Am. J. Bot.* **29**:759–765.

WILSON, C. L., and THEODOR JUST. 1939. "The morphology of the flower," *Bot. Rev.* **5**:97–131.

WODEHOUSE, R. P. 1928–1936. "Pollen grains in the identification and classification of plants," *Bull. Torr. Club* **55**:181–198, 449–462, **56**:123–138, **57**:21–46, **63**:495–514; *Am. J. Bot.* **16**:297–312, **18**:749–764.

Fruits

CHUTE, HETTIE M. 1930. "The morphology and anatomy of the achene," *Am. J. Bot.* **17**:703–723.

EGLER, F. E. 1943. "The fructus and the fruit," *Chron. Bot.* **7**:391–395.

MACDANIELS, L. H. 1940. *The Morphology of the Apple and Other Pome Fruits,* Mem. Cornell Univ. Agr. Exp. Sta. 230.

MASTERS, M. T. 1871. "Classification of fruits," *Nature* **5**:6.

Ovules and seeds

JOHANSEN, D. A. 1950. *Plant Embryology.* Ronald Press, Waltham, Mass.

General

(Anonymous). 1960. "Preliminary list of works relevant to descriptive biological terminology," *Taxon* **9**:245–257.

BROWN, R. W. 1954. *Composition of Scientific Words.* Baltimore, Md.

CANDOLLE, A. P. DE. 1880. *La phytographie, ou l'art de décrire les végétaux.* Paris.

DENNIS, C. J. 1960. "The language of biology. III. Definitions," *Turtox News* 38:106–108.

FEATHERLY, H. I. 1954. *Taxonomic Terminology of the Higher Plants.* Ames, Iowa.

GRAY, ASA. 1887. *Gray's Lessons in Botany* (rev. ed.). New York.

HOUGH, J. N. 1953. *Scientific Terminology.* Holt, Rinehart, and Winston, New York.

JACKSON, B. D. 1949. *A Glossary of Botanic Terms, with Their Derivation and Accent* (4th ed.). London.

LAWRENCE, G. H. M. 1951. *Taxonomy of Vascular Plants.* Macmillan, New York.

LEE, A. E., and CHARLES HEIMSCH. 1962. *Development and Structure of Plants: A Photographic Study.* Holt, Rinehart, and Winston, New York.

CHAPTER 9

Angiosperms

The angiosperms, or *Angiospermae,* which are generally known as the flowering plants, constitute that subdivision of the seed plants (*Spermatophyta, Phanaerogamae,* or *Embryophyta Siphonogama*) whose members characteristically have stems with xylem tissue composed in part of vessels and bear their seeds within one or more closed carpels.

The angiosperms are usually called "flowering plants" because they produce what are commonly thought of as true flowers (some of which, however, are much reduced). But the gymnosperms (*Gymnospermae*), also, often produce flower-like structures, such as the male and female cones, or "flowers," of pines. Without a more precise definition of the term "flower," therefore, this feature is not entirely distinctive and reliable.

Phylogeny

Fossil evidence indicates that the angiosperms originated in the Mesozoic era, probably in the Jurassic period, of geologic time, roughly 181 million years ago and perhaps even earlier than that. But the fossil record is so incomplete that to date it has not been possible to ascertain much concerning the origin and true relationships, or phylogeny, of this vast assemblage of plants that now dominates the vegetation of the earth, including perhaps a quarter of a million species and about three hundred families.

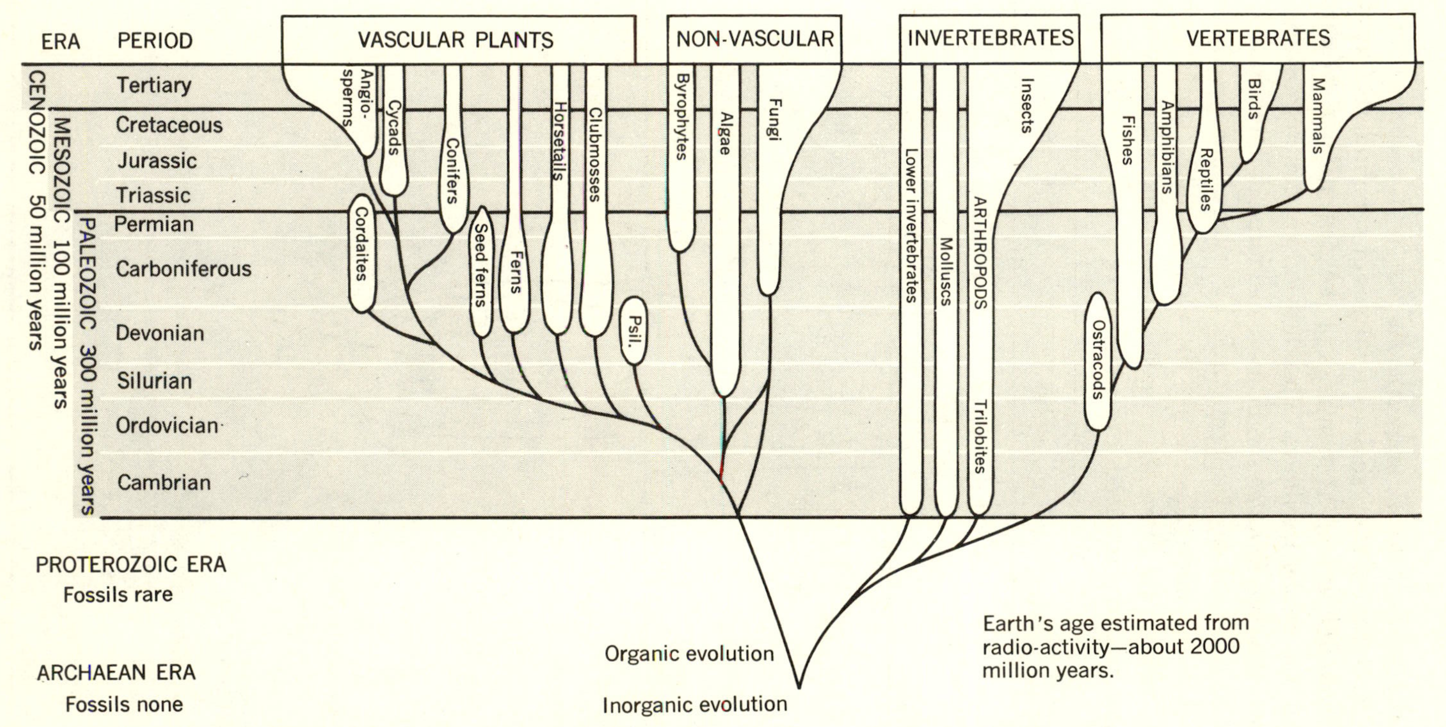

65 Diagram illustrating the dominance of plant and animal groups in the geologic periods and their relationships. [Modified from Alfred Gunderson in *Brooklyn Botanic Garden Leaflets,* Ser. 18, No. 4 (1930); used by permission.]

Among the earliest fossil angiosperms we find members of the *Ranales,* or Buttercup Order, such as relatives of the modern *Liriodendron* and *Magnolia,* as well as members of the *Amentiferae,* or catkin-bearers, like our modern willows. This has led to the supposition by some workers that the *Ranales* should be considered as the basal group in the evolution of the angiosperms, and by others that the *Amentiferae* should be considered as among the most primitive angiosperms. Nor are we able to do more than make an educated guess as to the ancestral group that produced the angiosperms: whether it was the

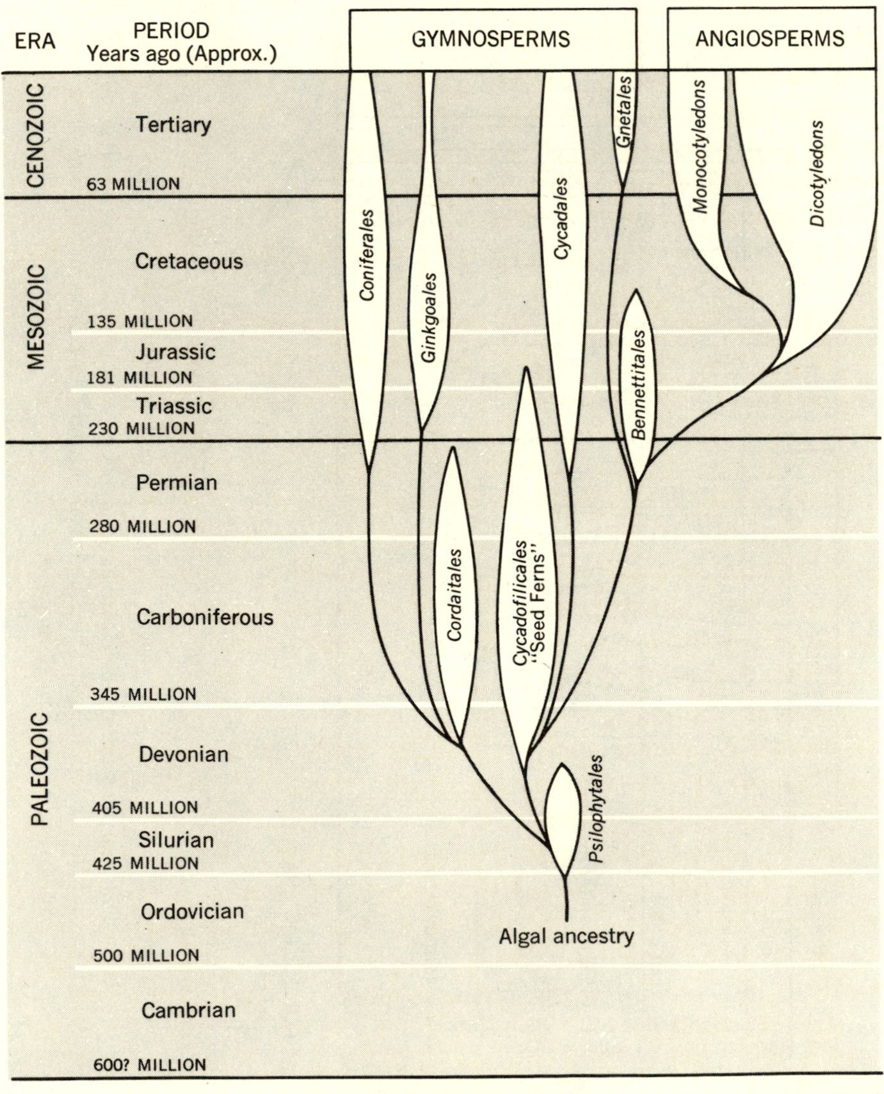

66 More detailed chart of possible relationships and evolutionary history of the seed plants.

extinct group of ancient seed ferns, some group of the cycads, or even some group of as yet undiscovered fossil gymnosperms. All we know is that suddenly, in Upper Cretaceous time, there appeared a great outburst of angiosperms and that these were already so far advanced in evolution as to be generally assignable to living genera. Figures 65 and 66 illustrate some possible relationships and details of the fossil records of various groups: the first a generalized picture of plant and animal groups, the second a more detailed chart of inferred relationships among modern and fossil seed plants.

The ultimate goal of taxonomists is to achieve a phylogenetic classification of living plants. To reach this goal, we must understand the past history of each group and must know which characters of the group are primitive and which are advanced. Because so little is known of the past history of angiosperms, we must rely on an assessment of primitive and advanced characters if we are to produce a reasonably objective and phylogenetic classification. Though this part of the problem is not easily solved, there are certain means of drawing conclusions. As pointed out by Sporne (1956) and others, certain characters, including anatomical ones, are undoubtedly primitive, as indicated by the fossil record and as determined by other means. These primitive characters, furthermore, seem to be statistically correlated with certain other characters, which therefore seem also to be primitive. It follows, then, that the absence of such primitive characters indicates an advanced condition, and that by analyzing the degree of advancement we may place various groups in a phylogenetic arrangement, each group being located by a calculated advancement index. The resulting phylogenetic scheme, as suggested by Sporne, would resemble a target with concentric rings representing degrees of advancement from center to periphery. Groups, such as families or orders, could then be pictured as variously shaped areas on the target, their radial extent indicating the range of advancement within the group, and the shape of the area indicating the distribution of the included plants (genera or species) according to their advancement. Such a scheme is shown in Figure 67, which is a hypothetical representation of this sort.

It has long been customary to represent phylogenetic classifications by a "family tree" diagram, modern groups being assigned places on the trunk and branches of the tree, as was done by Bessey (Fig. 4). But in reality we are dealing with living groups that are represented by only the twig ends of such a tree, and it is pure speculation to fill in the trunk and branches without a full knowledge of the past history. Speculation of this sort, nevertheless, is not entirely fruitless, and it does help to suggest possible ancestral relationships. Furthermore,

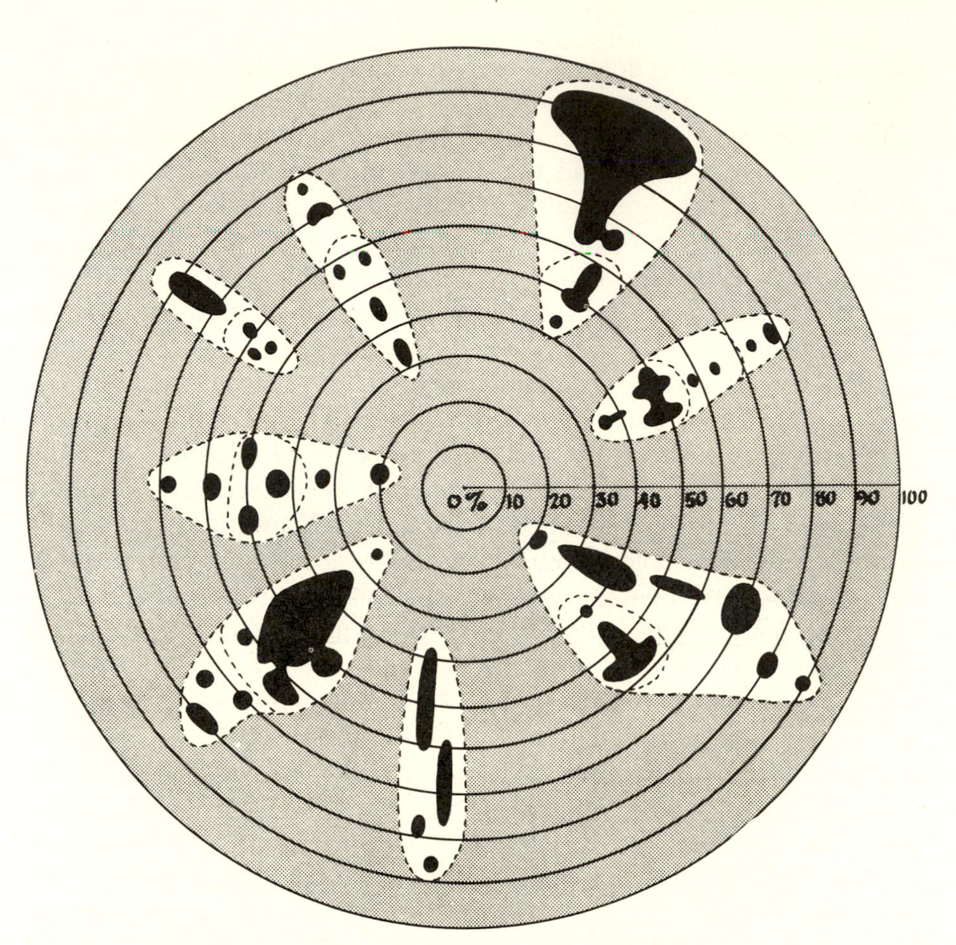

67 Scheme for showing phylogenetic relationships within a group such as the flowering plants. The black areas represent hypothetical families, and the broken lines enclosing groups of families represent areas of affinity. The concentric circles represent degrees of advancement based on a percentage. This is the sort of scheme proposed by Sporne [*Biol. Rev.* **31**:1–29 (1956)].

while lacking positive proof in all cases, and while admitting that single characters may be criteria of advancement or primitiveness in one group and not in another, taxonomists generally agree that certain criteria can be used in a phylogenetic classification. Some of these are listed below.

Criteria Used in Classification

1. *The presence or absence of petals, and, when present, whether separate or united.* This gives the primary subdivisions of dicotyledons: *apetalae,* without petals; *polypetalae,* with separate petals; and

sympetalae, with united petals. These groups, though somewhat artificial, are extremely useful. Separate petals seem to have come first, historically, and from this condition the derived conditions of apetaly and sympetaly have been evolved.

2. *The hypogynous, perigynous, or epigynous nature of the flower.* The sequence here seems to have been from hypogynous through perigynous to epigynous types. But we occasionally find epigyny in otherwise hypogynous groups, as in the Huckleberries in the otherwise hypogynous family *Ericaceae.* And perigyny seems to be totally lacking in the sympetalous groups.

3. *The number of parts.* It is believed that primitive flowers, in general, had indefinite numbers of parts, the stamens and carpels often being numerous. The general tendency of evolution has been toward fewer parts.

4. *The union of parts.* Separate parts are generally regarded as ancestral to united parts: separate petals preceded united petals, and separate carpels preceded united carpels. The condition in which the carpels are separate is *apocarpous;* the condition of united carpels is *syncarpous.* The apocarpous condition, with more than one carpel, can occur only in hypogynous or perigynous flowers, all epigynous flowers being syncarpous or else having only a single carpel.

5. *The nature of the perianth.* In the monocotyledons, particularly, the nature of the perianth distinguishes groups of orders. The calyx may be green and the perianth biseriate; these two parts may be similar and all petaloid; or the perianth may be variously reduced to bristles, scales, or hairs.

6. *The nature of the fruit.* To some extent, the nature of the fruit is determined by the nature of the gynoecium. In some families, such as the *Poaceae* (Grass Family), *Brassicaceae* (Mustard Family), and *Fabaceae* (Pea Family), the fruit type is unique and immediately distinguishes the family from all others. On the other hand, partly because of our present unsatisfactory nomenclature for fruits, the nature of the fruit may be a confusing feature to the beginner. For example, we have not distinguished between the akene of the Buckwheat Family (*Polygonaceae*), which comes from a hypogynous flower, and the fruit of the Sunflower Family (*Asteraceae*), which comes from an epigynous flower.

7. *The type of placentation.* This may be determined by inspection of either the ovary or the fruit, and it often indicates affinities. The free-central and basal placentation types, in particular, have been used to associate whole groups of families.

8. *The morphology of the seed.* Here are included such characteristics as the presence or absence of endosperm (referred to as albumen in older treatments) and the nature of the embryo. The number of cotyledons, of course, is a major means of distinguishing large groups.

9. *Anatomical characters.* One of the most dependable characters used to distinguish monocotyledons from dicotyledons is the gross anatomy of the stem. But occasionally we find peculiarities of stem anatomy that are distinctive of lesser groups, such as the occurrence of bicollateral bundles in the Pumpkin Family (*Cucurbitaceae*) and in the Gentian Order (*Gentianales*). Also included here are such peculiarities as the occurrence of specialized secretory tissue, such as that found in families having milky juice, and the presence of exceptionally long and strong phloem fibers. A fairly long list of correlated characters of vascular tissue is given by Sporne (1956), these being associated with evolutionary stages in the dicotyledons. The vascularization of flower parts, of stem nodes, and of integuments of seeds has also yielded clues to affinities.

10. *Vegetative characters.* These have to do with roots, stems, and leaves, and are sometimes highly useful as indicative of relationships. The woody or herbaceous habit, the presence or absence of stipules, the general morphology of leaves, and the peculiarities of pubescence are included here. Certain groups are well adapted to dry habitats by the succulent nature of the plants. In a few families there are plants with milky juice. In some families or lesser groups the plants are aromatic, with a characteristic odor or flavor. All such vegetative characters aid in recognition as well as in classification.

11. *Serum diagnosis.* A technique of determining relationships by the reaction to serum of members of various families has been developed by Karl Mez (see Chester in the references). Adapting the methods of bacteriologists, he based his system on the similarity of the proteins produced by the plants. This was determined by the degree of precipitation produced when the protein extracts were mixed with a serum from an animal. By 1926 enough data had been assembled to enable Mez to draw up a family tree that showed many similarities to similar schemes proposed by workers who used morphological characters. It seems, however, that a phylogenetic scheme based on serum diagnosis alone is not likely to be any closer to the truth than any other scheme based on a single morphological character. The same proteins are apparently found in very distantly related families. Plant biochemistry in general, however, is proving to be a useful tool, as pointed out earlier (pp. 75–76).

Classification of Angiosperms

Since John Ray, near the end of the seventeenth century, first established a system of classification that delimited monocotyledons and dicotyledons, there has been little disagreement with this division into two classes, the *Monocotyledoneae* (one cotyledon) and the *Dicotyledoneae* (two cotyledons). These two groups are distinguished from each other not only by the number of cotyledons, however, but also by combinations of other characters as well, as noted in the following tabular comparison. Exceptions may occur for nearly all the characters enumerated, but the exceptions are few, and they almost never involve more than a single character in any one plant. The most useful differences are those associated with stem structure, leaf venation, and number of parts in the flower.

COMPARISON OF MONOCOTYLEDONS AND DICOTYLEDONS

Monocotyledons	*Dicotyledons*
EMBRYO: with 1 cotyledon, which usually develops underground; endosperm often present in the seed.	EMBRYO: with 2 cotyledons, which often develop above the ground; endosperm present or lacking in the seed.
ROOTS: the primary root of short duration, soon replaced by adventitious roots, which form a fibrous root system or sometimes a fascicle of fleshy roots, but usually without a taproot.	ROOTS: the primary root often persistent and becoming a strong taproot, with smaller secondary roots.
GROWTH FORM: mostly herbaceous; a few are arborescent, such as Bamboos, Palms, Yuccas, and Dracaenas.	GROWTH FORM: either herbaceous or woody.
VASCULAR SYSTEM: consisting of numerous scattered bundles, without definite arrangement and in a ground parenchyma; no cambium except in a few such as *Dracaena;* no differentiation into cortical and stelar regions in stems.	VASCULAR SYSTEM: usually consisting of a definite number of primary bundles in a ring, with a cambium, and secondary growth in diameter of the stem; differentiation into cortical and stelar regions in stems.
LEAVES: usually parallel-veined, often sheathing at the base, commonly oblong or linear in shape; petiole seldom developed.	LEAVES: usually net-veined (pinnate or palmate), seldom sheathing at the base, usually broader in shape; petiole commonly developed.
FLOWERS: with their parts usually in threes or multiples of three (3-merous).	FLOWERS: with their parts usually in fours or fives (4- or 5-merous).

CLASSIFICATION OF MONOCOTYLEDONS. One of the most recent and detailed treatments of this large assemblage of plants is that of Hutchinson, a British botanist: *The Families of Flowering Plants:* II, *Mono-*

cotyledons (London, 1934). In this volume that author has divided all monocotyledons into three groups (which he calls divisions) based on the nature of the perianth. Though not yet recognized by all American taxonomists, these groups seem to form a natural basis for a primary breakdown and apparently consist of families and orders that are rather closely related. Hutchinson's grouping is given below and is followed, with minor modifications, in this text. We shall call his primary groups subclasses of the class *Monocotyledoneae.*

Subclass 1: Calyciferae. This group includes the most ancient monocotyledons, those believed to have been derived from an early pair of groups in the *Ranales,* one in which the flowers produced small, one-seeded fruits (akenes), and another in which the flowers produced fewer, larger, several-seeded fruits (follicles) with a tendency to split open at maturity. Since these two conditions still exist in the *Ranales,* the earliest monocotyledons evidently came from two sources in that dicotyledonous order, as shown on the chart of the relationships found among the chief orders of monocotyledons (Fig. 68).

The *Calyciferae* have a biseriate perianth, the sepals usually being green, and the corolla, when present, is usually colored or white. There

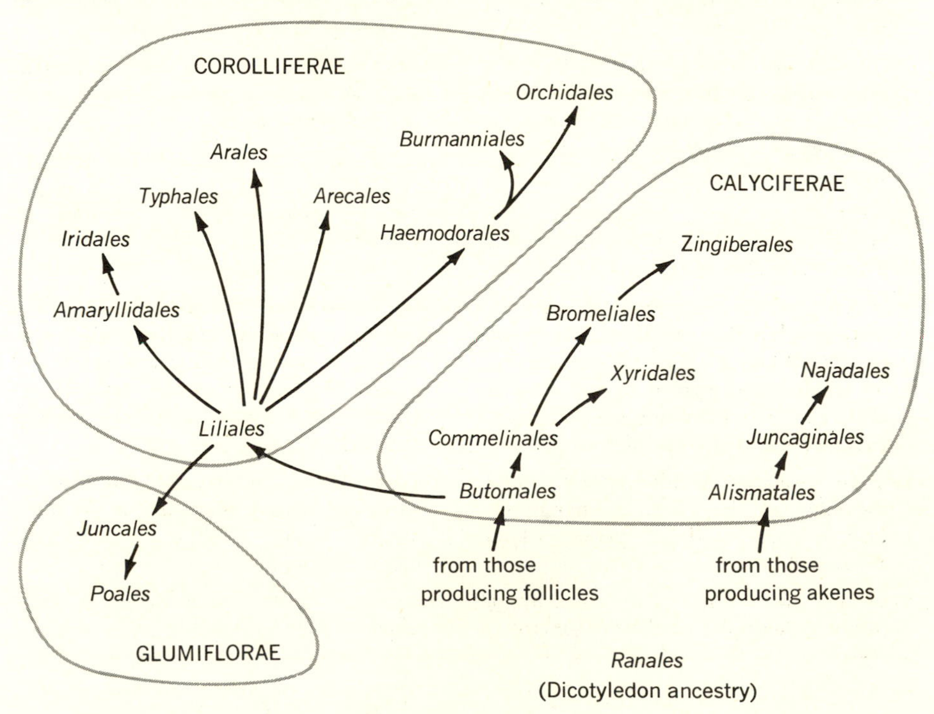

68 Chart to show possible relationships among the chief orders of monocotyledons.

is no adnation of the sepals and petals into a common perianth tube, as there sometimes is in the second group below. The plants are usually rhizomatous or annual, and bulbs or corms are never produced. Some of the families are aquatic.

Subclass 2: Corolliferae. This group is believed to have been derived from the *Calyciferae,* and it includes as its basal order the *Liliales.* Here the perianth is uniseriate and often petaloid throughout, but in derived families it may be reduced to hairs or bristles. There is a tendency for the calyx and the corolla to unite into a perianth tube. The plants are predominantly terrestrial, and bulbs and corms are fairly common. One notable exception in regard to the perianth character occurs in the genus *Trillium* (family *Liliaceae*), which has green sepals.

Subclass 3: Glumiflorae. Here there is a tendency for the perianth to be chaffy or to be reduced to scales or lodicules, and there is a further tendency among the more reduced families (*Cyperaceae* and *Poaceae*) for the flowers to be aggregated into spikelets. The plants of this group, having a general grass-like aspect, are often referred to as "the grasses and grass-like plants." Fibrous roots and rhizomes are common; only a very few plants produce poorly developed bulbs.

Figure 68 illustrates possible relationships among these three major groups and among the orders within each.

CLASSIFICATION OF DICOTYLEDONS. In order to arrive at a satisfactory classification of dicotyledons, we must first come to some agreement about the basal or most primitive group and then try to derive other groups from it. As we have noted, there has been much speculation about the way in which the dicotyledons have evolved and about the group or groups within them that should be regarded as primitive and consequently ancestral to the others.

One of the theories, followed by Endlicher, Eichler, Wettstein, and Engler, suggests that the dicotyledons were derived from gnetalian, or conifer-like, ancestors among the gymnosperms. The primitive dicotyledons, then, would be woody, without petals, with unisexual, wind-pollinated flowers having few parts, and those grouped in a cone-like inflorescence such as a catkin. The catkin-bearers, or *Amentiferae,* accordingly, were regarded as primitive and were supposed to have given rise to groups that had a perianth and often had insect-pollinated flowers. This view has the corroboration of some of the fossil evidence, for we find catkin-bearing plants such as willows among the earliest angiosperm fossils.

Another theory, supported and elaborated in various ways by Bentham and Hooker, Bessey, Hallier, Arber and Parkin, and Hutchinson, might be called the "ranalian theory" because it suggested that the dicotyledons were derived from gymnospermous ancestors such as the cycads of the order *Bennettitales,* which gave rise to the *Ranales,* or Magnolia-Buttercup Order. The primitive dicotyledon, then, would have perfect flowers with numerous separate parts spirally arranged on an elongated axis, as in *Magnolia,* and simpler flowers, without a perianth, would be derived by reduction. The fossil evidence also lends support to this view, for relatives of the modern *Magnolia* are also among the earliest dicotyledons known. This is the theory most generally accepted at the present time.

Authorities also differ about the subsequent evolution of the dicotyledons. One useful scheme divides them into three major groups: the *Polypetalae* (those with separate petals), including the *Ranales;* the *Apetalae* (those without petals), which have been derived from various sources in the *Polypetalae* by reduction and consequently are not a natural group; and the *Sympetalae* (those with united petals, also called the *Metachlamydeae*), which have also been derived from the *Polypetalae.* The *Apetalae* are subdivided into the *Amentiferae* (those with catkins) and the *Floriferae* (those without catkins). The *Polypetalae* are subdivided into hypogynous, perigynous, and epigynous groups; and the *Sympetalae* are subdivided into hypogynous and epigynous groups, perigyny being lacking in the *Sympetalae.*

This arrangement, with representative orders of each group, is shown in Figure 69. By utilizing the simple characters on which this chart is based, the student will be able to place any unknown plant known to be a dicotyledon among its near relatives. Most manuals and floras follow the sequence given, from *Apetalae* through *Polypetalae* to *Sympetalae,* and also for the secondary groups within those groups. Monocotyledons precede dicotyledons in most manuals.

Another attempt to show the possible relationships among the dicotyledonous orders is presented in Figure 70. This is based on Bessey's system and retains his ideas about the evolution of a basically hypogynous line, shown on the right, and of a perigynous-epigynous line, shown on the left. The apetalous orders are doubtfully assigned places within this scheme. The orders in the shaded section are sympetalous, and according to this scheme parallel evolution produced separate hypogynous sympetalous groups and epigynous sympetalous groups. We cannot be sure, however, that this actually took place. The broken lines connecting various orders indicate considerable doubt about the relationships.

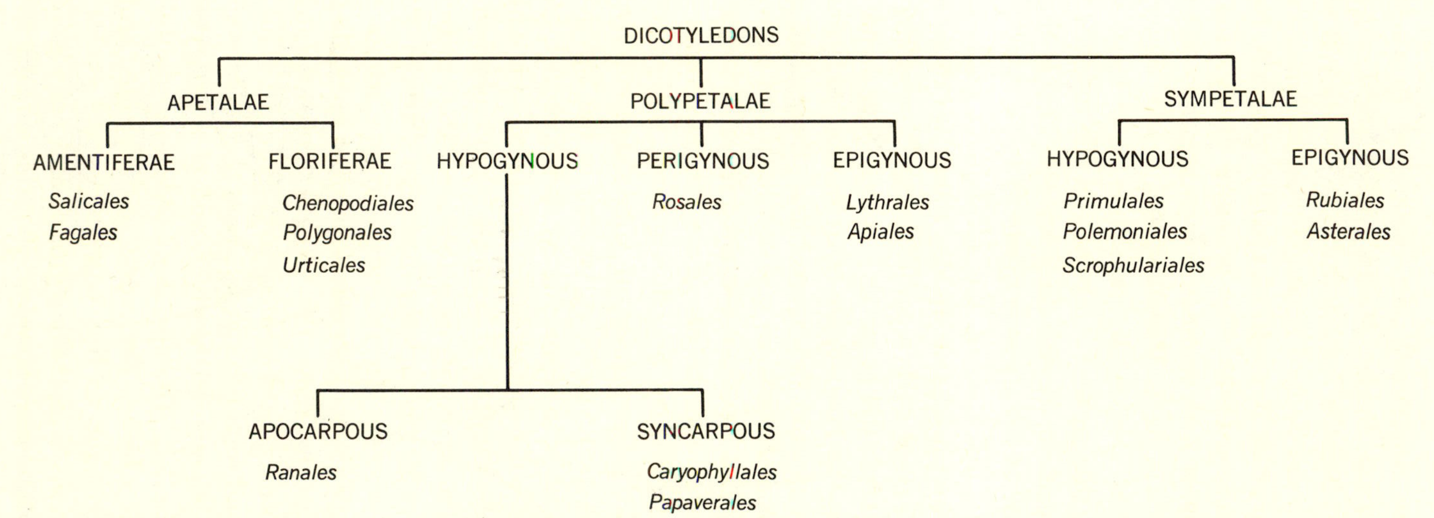

69 Generalized scheme of classification of the dicotyledons, showing some of the characteristic orders represented. The *Ranales*, the most primitive order, or their precursors gave rise to other orders either by reduction or by specialization in various other ways, the *Apetalae* being considered as reduced members, and the *Sympetalae* as members that became specialized by the development of united petals, which is believed to have aided in pollination by insects.

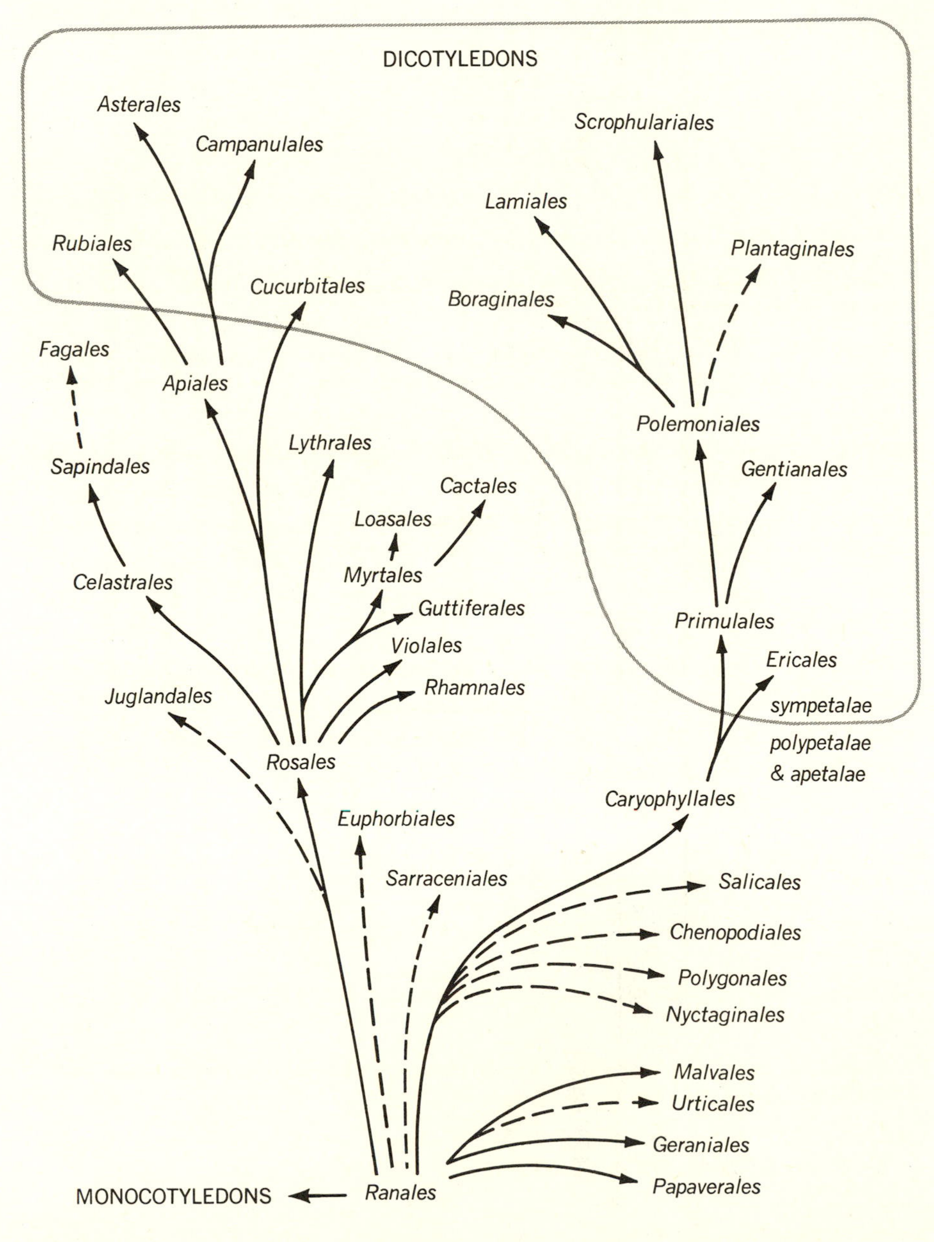

70 Chart showing possible relationships among dicotyledonous orders.

REFERENCES

ALSTON, R. E., T. J. MABRY, and B. L. TURNER. 1963. "Perspectives in chemotaxonomy," *Science* **142**:545–552.

ANDREWS, H. N., JR. 1947. *Ancient Plants and the World They Lived In.* Cornell University Press, Ithaca, N.Y.

ANDREWS, H. N., JR. 1963. "Early seed plants," *Science* **142**:925–931.

ARNOLD, C. A. 1947. *An Introduction to Paleobotany.* McGraw-Hill, New York.

AXELROD, DANIEL I. 1952. "A theory of angiosperm evolution," *Evolution* **6**:29–60.

BAILEY, I. W. 1949. "Origin of the angiosperms: Need for a broadened outlook," *J. Arnold Arb.* **30**:64–70.

BAILEY, I. W. 1957. "The potentialities and limitations of wood anatomy in the study of phylogeny and classification of angiosperms," *J. Arnold Arb.* **38**:243–254.

BERTRAND, P., and P. CORSING. 1938. "Phylogénie des végétaux vasculaires," *Bull. Soc. Bot. Fr.* **85**:331–348.

BERTRAND, P., and P. CORSING. 1938. "Phylogénie des végétaux vasculaires: Observations complémentaires," *Bull. Soc. Bot. Fr.* **85**:503–504.

BESSEY, C. E. 1895. "The point of divergence of monocotyledons and dichotyledons," *Bot. Gaz.* **22**:229–232.

BESSEY, C. E. 1897. "Phylogeny and taxonomy of the angiosperms," *Bot. Gaz.* **24**:145–178.

BESSEY, C. E. 1915. "The phylogenetic taxonomy of flowering plants," *Ann. Mo. Bot. Gard.* **2**:109–164.

BOWER, F. O. 1935. *Primitive Land Plants.* Cambridge, England.

BROWNE, I. M. P. 1935. "Some views on the morphology and phylogeny of the leafy vascular sporophyte," *Bot. Rev.* **1**:383–404.

CAMPBELL, D. H. 1929. "The phylogeny of the angiosperms," *Bull. Torr. Club* **55**:479–497.

CAMPBELL, D. H. 1930. "The phylogeny of monocotyledons," *Ann. Bot.* **44**:311–331.

CAMPBELL, D. H. 1940. *The Evolution of the Land Plants* (Embryophyta). Stanford University Press, Stanford, Calif.

CHALK, L. 1937. "The phylogenetic value of certain anatomic features of dictoyledonous plants," *Ann. Bot.* n.s. **1**:429–437.

CHAMBERLAIN, C. J. 1929. "An evaluation of the structural evidence for genetical relationships in plants: Some evidence for vascular plants," *Proc. Int. Congr. Pl. Sci. 1926,* **1**:473–480.

CHEADLE, V. I. 1942. "The role of anatomy in phylogenetic studies of the monocotyledoneae," *Chron. Bot.* **7**:253–254.

CHESTER, K. S. 1937. "A critique of plant serology: A review in English for the serodiagnostic work of Karl Mez," *Quart. Rev. Biol.* **12**:19–46, 165–190, 294–321.

COPELAND, H. F. 1940. "The phylogeny of the angiosperms," *Madroño* **5**:209–218.

DARRAH, W. C. 1939. *Principles of Paleobotany.* Leyden.

DAVY, J. B. 1937. "On the primary groups of dicotyledons," *Ann. Bot.* n.s. **1**:429–437.

DEBEER, C. R. 1936. *Embryology and Evolution.* Oxford.

DITTMER, H. J. 1964. *Phylogeny and Form in the Plant Kingdom.* Van Nostrand, Princeton, N.J.

EAMES, A. J. 1911. "On the origin of the herbaceous type in the angiosperms," *Ann. Bot.* **25**:215–224.

EAMES, A. J. 1929. "The role of flower anatomy in the determination of angiosperm phylogeny," *Proc. Int. Congr. Pl. Sci. 1926,* **1**:423–427.

FAEGRI, K. 1939. "Some fundamental problems of taxonomy and phylogenetics," *Bot. Rev.* **3**:400–423.

FOSTER, A. S. 1936. "Leaf differentiation in the angiosperms," *Bot. Rev.* **2**:349–372.

FOSTER, A. S. 1952. "Foliar venation in angiosperms from an ontogenetic standpoint," *Am. J. Bot.* **39**:752–766.

FRENGUELLI, J. 1946. "El origen de las angiospermas," *Bol. Soc. Argent. Bot.* **1**:169–208.

GIBBS, R. D. 1954. "Comparative chemistry and phylogeny of flowering plants," *Trans. Roy. Soc. Canada III,* **48**:1–47.

GILBERT, S. G. 1940. "Evolutionary significance of ring porosity in woody angiosperms," *Bot. Gaz.* **102**:105–120.

GILMOUR, J. S. L., and W. B. TURRILL. 1941. "The aim and scope of taxonomy," *Chron. Bot.* **6**:217–219.

GUNDERSON, A. 1939. "Flower buds and phylogeny of dicotyledons," *Bull. Torr. Club* **66**:287–295.

GUNDERSON, A. 1939. "The classification of dicotyledons," *Torreya* **39**:108–110.

GUNDERSON, A. 1941. "Flower structure and the classification of dicotyledons," *Brooklyn Bot. Gard. Rec.* **30**:93–98.

GUNDERSON, A. 1943. "Flower forms and groups of dicotyledons," *Bull. Torr. Club* **70**:510–516.

HALLIER, H. 1905. "Provisional scheme of the natural (phylogenetic) system of flowering plants," *New Phytol.* **4**:151–162.

HARRIS, T. M. 1935. "The ancestry of the angiosperms," *Proc. Sixth Int. Congr. Amsterdam* **2**:230–231.

HILL, A. W. 1906. "The morphology and seedling structure of geophilous species of *Peperomia,* together with some views on the origin of monocotyledons," *Ann. Bot.* **20**:395–427.

HOEG, O. A. 1937. "The Devonian floras and their bearing upon the origin of vascular plants," *Bot. Rev.* **3**:563–592.

HUTCHINSON, J. 1923–1924. "Contributions toward a phylogenetic classification of flowering plants," *Kew Bull.* **1923**:65–89, 241–261, **1924**:49–66, 114–134.

HUTCHINSON, J. 1929. "The phylogeny of flowering plants," *Proc. Int. Congr. Pl. Sci. 1926,* **1**:413–421.

JOHNSON, J., and I. A. HOGGAN. 1931. "The challenge of plant virus differentiation and classification," *Science* n.s. **73**:29–32.

JUST, T. 1948. "Gymnosperms and the origin of angiosperms," *Bot. Gaz.* **110**:91–103.

KULP, J. L. 1961. "Geologic time scale," *Science* **133**:1105–1114.

KUNKEL, L. C. 1935. "Possibilities in plant virus classification," *Bot. Rev.* **1**:1–17.

LEWIS, D. 1942. "The evolution of sex in flowering plants," *Biol. Rev.* **17**:46–67.

LOTSY, P. 1910. "Phylogeny of Plants," *Bot. Gaz.* **49**:460–461.

MAHESHWARI, P. 1950. *An Introduction to the Embryology of Angiosperms.* McGraw-Hill, New York.

MARTIN, A. C. 1946. "The comparative internal morphology of seeds," *Am. Midl. Nat.* **36**:513–660.

MATTHEWS, J. R. 1941. "Floral morphology and its bearing on the classification of angiosperms," *Trans. Bot. Soc. Edinb.* **23**:60–82.

MCNAIR, J. B. 1934. "The evolutionary status of plant families in relation to some chemical properties," *Am. J. Bot.* **21**:427–452.

MCNAIR, J. B. 1935. "Angiosperm phylogeny on a chemical basis," *Bull. Torr. Club* **62**:515–532.

METCALFE, C. R. 1946. "The sytematic anatomy of the vegetative organs of the angiosperms," *Biol. Rev.* **21**:159–172.

PENNELL, F. W. 1948. "The taxonomic significance of an understanding of floral evolution," *Brittonia* **6**:301–308.

POPE, M. A. 1925. "Pollen morphology as an index to plant relationship," *Bot. Gaz.* **80**:63–73.

PULLE, A. 1938. "The classification of the spermatophytes," *Chron. Bot.* **4**:109–113.

REICHERT, E. T. 1919. *A Biochemic Basis for the Study of Problems of Taxonomy, Heredity, Evolution, etc. with Special Reference to the Starches.* Washington, D.C.

RUSBY, H. H. 1929. "The value and limitations of histology in vegetable taxonomy," *Proc. Int. Congr. Pl. Sci. 1926,* **2**:1356–1360.

SARGANT, E. 1908. "The reconstruction of a race of primitive angiosperms," *Ann. Bot.* **22**:121–186.

SCHAFFNER, J. H. 1938. "The importance of phylogenetic taxonomy in systematic botany," *Ecology* **19**:296–300.

SCHUBERT, C. 1921. "The evolution of primitive plants from the geologist's point of view," *New Phytol.* **19**:272–275.

SINNOTT, E. W. 1914. "Investigations on the phylogeny of the angiosperms," *Am. J. Bot.* **1**:303–322.

SINNOTT, E. W., and I. W. BAILEY. 1914–1915. "Investigations on the phylogeny of the angiosperms," *Am. J. Bot.* **1**:441–453; *Ann. Bot.,* **28**:547–600; *Am. J. Bot.* **2**:1–22.

SPORNE, K. R. 1949. "A new approach to the problem of the primitive flower," *New Phytol.* **48**:259–276.

SPORNE, K., R. 1956. "The phylogenetic classification of the angiosperms," *Biol. Rev.* **31**:1–29.

SPRAGUE, T. A. 1940. "Taxonomic botany, with special reference to the angiosperms," in J. Huxley, *The New Systematics,* pp. 435–454. London.

STEBBINS, G. L. 1938. "Cytological characteristics associated with the different growth habits in the dicotyledons," *Am. J. Bot.* **25**:189–197,

TAKHTAJAN, A. L. 1953. "Phylogenetic principles of the system of higher plants," *Bot. Rev.* **19**:1–45.

THOMAS, H. H. 1931. "The early evolution of the angiosperms," *Ann. Bot.* **45**:647–672.

THOMAS, H. H. 1936. "Paleobotany and the origin of the angiosperms," *Bot. Rev.* **2**:397–418.

THORNE, R. F. 1958. "Some guiding principles of angiosperm phylogeny," *Brittonia* **10**:72–77.

THORNE, R. F. 1963. "Some problems and guiding principles of angiosperm phylogeny," *Am. Nat.* **97**:287–305.

TIPPO, O. 1942. "A modern classification of the plant kingdom," *Chron. Bot.* **7**:203–206.

TURRILL, W. B. 1942. "Taxonomy and phylogeny," *Bot. Rev.* **8**:247–270, 473–532, 655–707.

TURRILL, W. B. 1951. "Modern trends in the classification of plants," *Taxon* **1**:17–19.

VESTAL, P. A. 1940. "Wood anatomy as an aid to classification and phylogeny," *Chron. Bot.* **6**:53–54.

WEEVERS, T. 1943. "The relation between taxonomy and chemistry of plants," *Blumea* **5**:412–422.

WERNHAM, H. F. 1911. "Floral evolution," *New Phytol.* **10**:78–83, 109–120, 145–159, 217–226, 293–307.

WIELAND, G. R. 1929. "Antiquity of the angiosperms," *Proc. Int. Congr. Pl. Sci. 1926*, **1**:429–456.

WIELAND, G. R. 1929. "Views of higher seed plant descent since 1879," *Science* n.s. **70**:223–228.

WIELAND, G. R. 1931. "Why the angiosperms are old," *Science* n.s. **74**:219–221.

WILSON, C. L. 1937. "The phylogeny of the stamen," *Am. J. Bot.* **24**:686–699.

WODEHOUSE, R. P. 1936. "Evolution of pollen grains," *Bot. Rev.* **2**:67–84.

YAMPOLSKY, C. 1925. "Origin of sex in the phanaerogamic flora," *Genetica* **7**:521–532.

Part II

SELECTED ORDERS AND FAMILIES OF MONOCOTYLEDONS

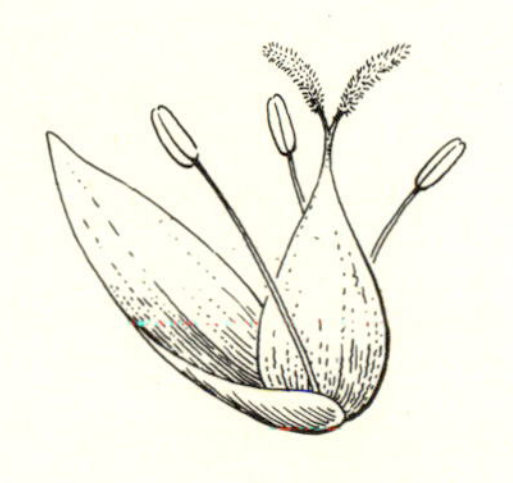

The Subclass Calyciferae

The perianth is biseriate (in two series). *The calyx is green* and is attached below the corolla, not in the same whorl. The corolla, when present, is usually petaloid and colored or white, contrasting with the calyx. There is no union of the calyx and corolla into a common perianth tube. The gynoecium may be apocarpous or syncarpous. The plants are mostly rhizomatous or annual, without bulbs or corms. Some of the families include moisture-loving or definitely aquatic plants (*Alismataceae, Juncaginaceae, Najadaceae, Butomaceae,* and *Hydrocharitaceae*); one family includes mostly epiphytes (*Bromeliaceae*); and some families are tropical or subtropical (*Musaceae, Strelitziaceae,* and *Zingiberaceae*).

Key to the Principal Families of Calyciferae

Fruit 1-seeded, or sometimes 2-seeded

Plants scapose, the leaves basal

Leaves broad; inflorescence a bracteate panicle or open raceme; inner perianth segments petaloid . . . . ALISMATACEAE

Leaves linear, grass-like; inflorescence a narrow, spike-like, ebracteate raceme; inner perianth segments sepaloid . . . . JUNCAGINACEAE

Plants with leafy stems . . . . NAJADACEAE

Fruit 3–many-seeded

Gynoecium of separate carpels . . . . BUTOMACEAE

Gynoecium of united carpels

Submerged aquatics with small epigynous flowers and often with whorled leaves . . . . HYDROCHARITACEAE

Plants not submerged aquatics; flowers hypogynous or epigynous; leaves not whorled

Plants mostly epiphytic or xerophytic, with densely clustered linear and often spiny-toothed leaves; flowers in terminal, spike-like inflorescences (or in *Tillandsia* the leaves entire and well spaced) . . . . BROMELIACEAE

Plants terrestrial, seldom xerophytic, their leaves not spiny-toothed

Flowers hypogynous

Plants with leafy stems; flowers in cymes . . . . COMMELINACEAE

Plants scapose; flowers in heads . . . . XYRIDACEAE

Flowers epigynous

Plants not aromatic; fertile stamens 5–6

Leaves and bracts spirally arranged; fruit fleshy . . . . . . MUSACEAE

Leaves and bracts 2-ranked; fruit capsular . . . . . . STRELITZIACEAE

Plants aromatic; fertile stamen 1 . . . . ZINGIBERACEAE

ORDER ALISMATALES

ALISMATACEAE: Water-plantain Family

Scapose marsh or aquatic herbs with large paniculate or racemose and bracteate inflorescences. Flowers perfect or unisexual, hypogynous, regular, 3-merous, the calyx of 3 green sepals, the corolla of 3 white or sometimes lavender petals, which fall early. Stamens 6 or more. Carpels (pistils) usually numerous, forming a head or whorl of akenes in fruit.

The family includes about 13 genera and 50 species, widely distributed in temperate and tropical regions.

EXAMPLES: *Alisma:* Water plantain.
Sagittaria: Arrowhead.
Echinodorus: Burhead.

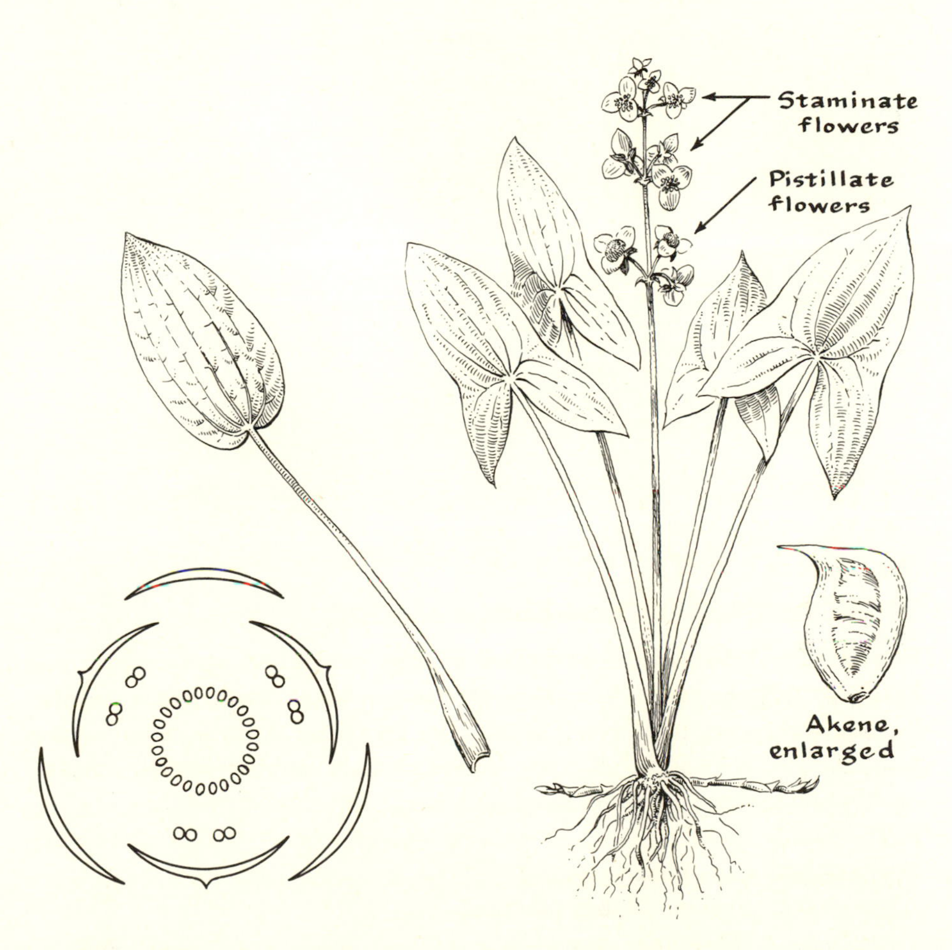

71 Floral diagram and leaf of *Alisma*.

72 Plant of *Sagittaria latifolia*.

ORDER JUNCAGINALES

JUNCAGINACEAE: Arrowgrass Family

Marsh herbs from rhizomes, with long, narrow, basal leaves and spike-like, scapose, ebracteate inflorescences. Flowers small, greenish, usually perfect, hypogynous, regular, the perianth of 6 concave segments. Stamens 6 or 3, the anthers nearly sessile. Carpels 6 or 3, weakly united, each containing 1 or 2 ovules.

The family includes 4 genera and about 10 species, mainly in temperate regions, and often in brackish or saline areas.

EXAMPLES: *Triglochin:* Arrowgrass, a common stock-poisoning plant.
Scheuchzeria: a plant commonly found in cold bogs.

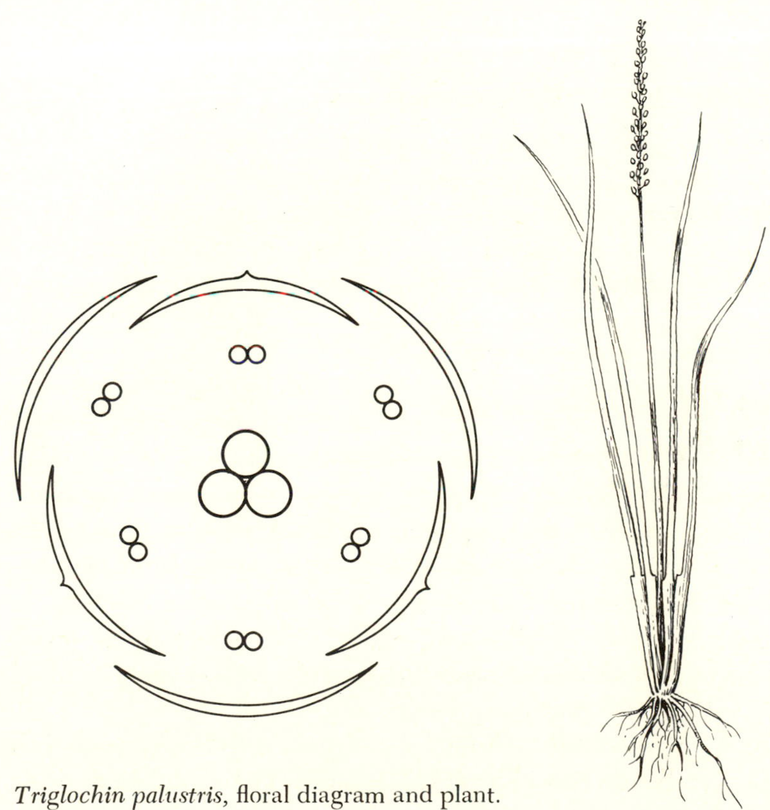

73 *Triglochin palustris,* floral diagram and plant.

ORDER NAJADALES

NAJADACEAE: Pondweed Family

Submerged or partly submerged aquatic herbs, with slender, leafy stems, which root in the mud on the bottom of fresh-water or saline streams, ponds, or shores. Flowers perfect or unisexual, in spikes or axillary clusters, sometimes solitary. True perianth none, or in *Potamogeton* consisting of 4 green parts that are probably sepals but are variously interpreted as bracts or dilated connectives of stamens. Stamens 1–4. Carpels 1 or more, usually 1-seeded in fruit and becoming akenes.

The family includes about 10 genera and 105 species, widely distributed. As treated here, it includes the families *Potamogetonaceae, Ruppiaceae, Zannichelliaceae,* and *Najadaceae* of some authors.

EXAMPLES: *Potamogeton:* Pondweed, a genus of some 50 or more species, furnishing good wild-fowl food in fresh water.

Najas: Naiad, with about 30 species in fresh and brackish water.

Ruppia: Ditch Grass, with a single polymorphic species.

Zostera: Eel Grass, 6 species of marine plants, both hemispheres.

Zannichellia: Horned Pondweed, with 2 species of wide distribution in fresh or brackish water.

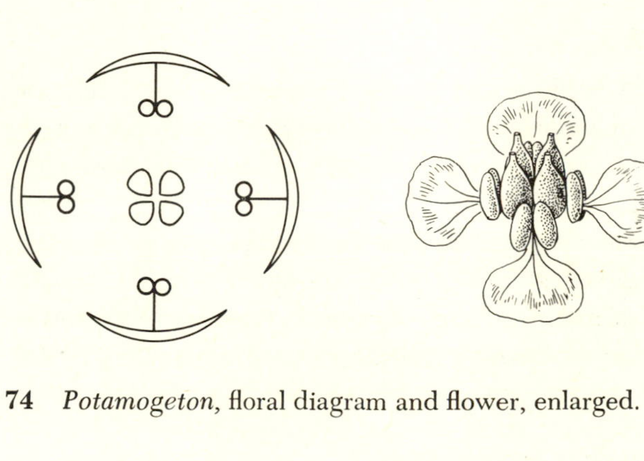

74 *Potamogeton*, floral diagram and flower, enlarged.

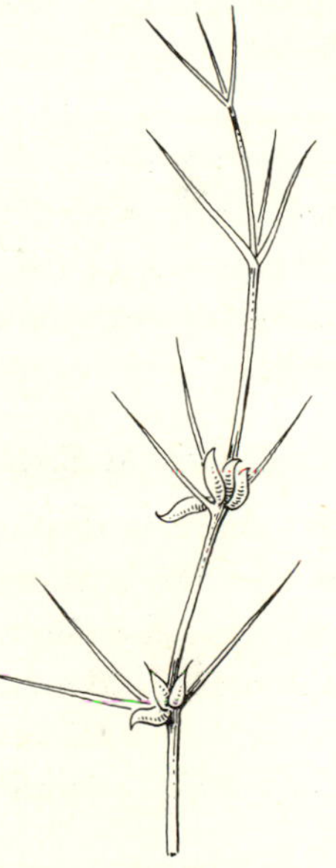

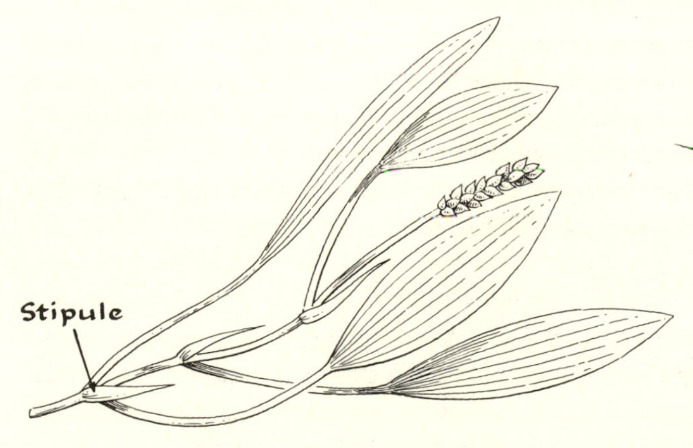

75 Portion of plant of *Potamogeton nodosus*.

76 Portion of plant of *Zannichellia*. The leaves are opposite or often apparently whorled.

ORDER BUTOMALES

Perennial aquatic herbs of fresh or salt water. Leaves in a basal tuft, or cauline and then alternate or whorled. Flowers showy or small, perfect or unisexual, hypogynous or epigynous, the perianth mostly biseriate, the outer segments green and the inner petaloid. Stamens many to 3. Carpels free and the flowers hypogynous, or united and the flowers epigynous, each carpel many-ovuled, the ovules scattered over the interior face of the carpel.

This is one of the most primitive groups of monocotyledons, showing a relationship with the dicotyledons through follicle-producing members of the *Ranales*.

BUTOMACEAE: Flowering Rush Family

The family is distinguished in the order by having showy, hypogynous flowers with separate carpels. It includes 4 genera and 6 species, in Europe, Asia, and America, mainly temperate and tropical in distribution.

EXAMPLES: *Butomus umbellatus:* Flowering Rush.
Hydrocleis nymphoides: of Brazil, resembling Waterlilies and sometimes cultivated.

HYDROCHARITACEAE: Frog's Bit Family

The family is distinguished in the order by having inconspicuous, epigynous flowers with united carpels. It includes 6 genera and about 60 species, widely distributed.

EXAMPLES: *Elodea:* a common freshwater weed with whorled leaves.
Vallisneria: Tape Grass, a widely distributed dioecious freshwater plant with ribbon-like leaves.

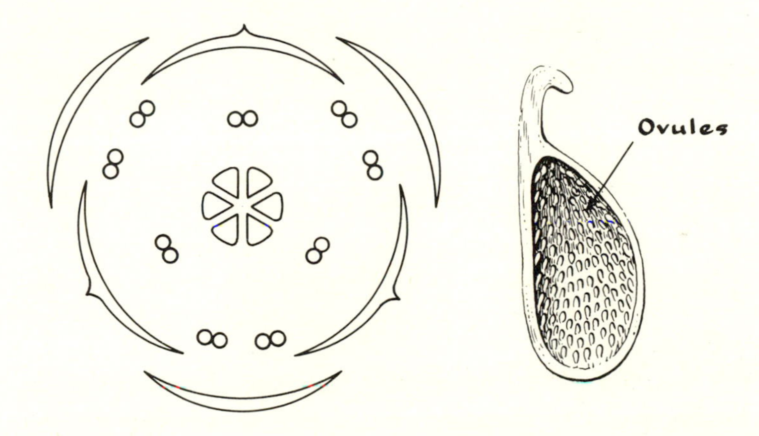

77 *Butomus*, floral diagram and longitudinal section of a carpel.

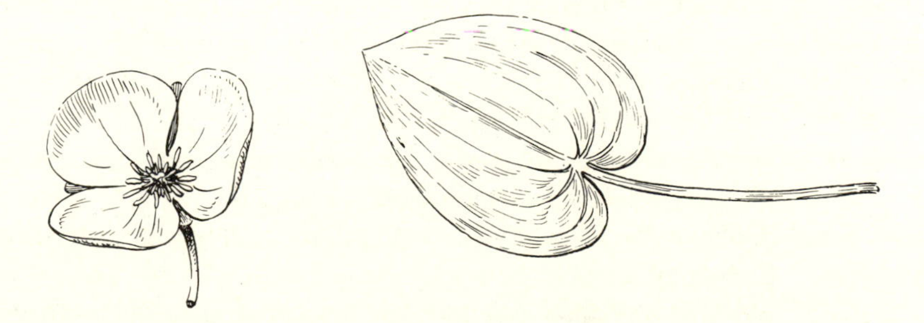

78 *Hydrocleis*, flower and leaf.

ORDER COMMELINALES

COMMELINACEAE: Spiderwort Family

Terrestrial herbs with jointed stems bearing alternate sheathing leaves. Flowers hypogynous, perfect, regular to somewhat irregular, 3-merous, in axillary cymes. Sepals green. Petals often blue, delicate in texture. Stamens 6 and all fertile, or sometimes 3 or 2 fertile and the others sterile or lacking, the filaments often hairy and brightly colored. Pistil 1, of 3 united carpels, the style single and the stigma capitate, or sometimes the ovary 2-celled by abortion of one carpel. Fruit usually a few-seeded and loculicidal capsule.

The family includes about 26 genera and 600 species, mainly in warm regions, but extending into north temperate latitudes.

EXAMPLES: *Tradescantia:* Spiderwort, an American genus of some 40 species, often in sandy areas.

Commelina: Dayflower, with about 100 species in warm regions.

Zebrina pendula: Wandering Jew, a commonly cultivated foliage plant for window boxes or hanging baskets.

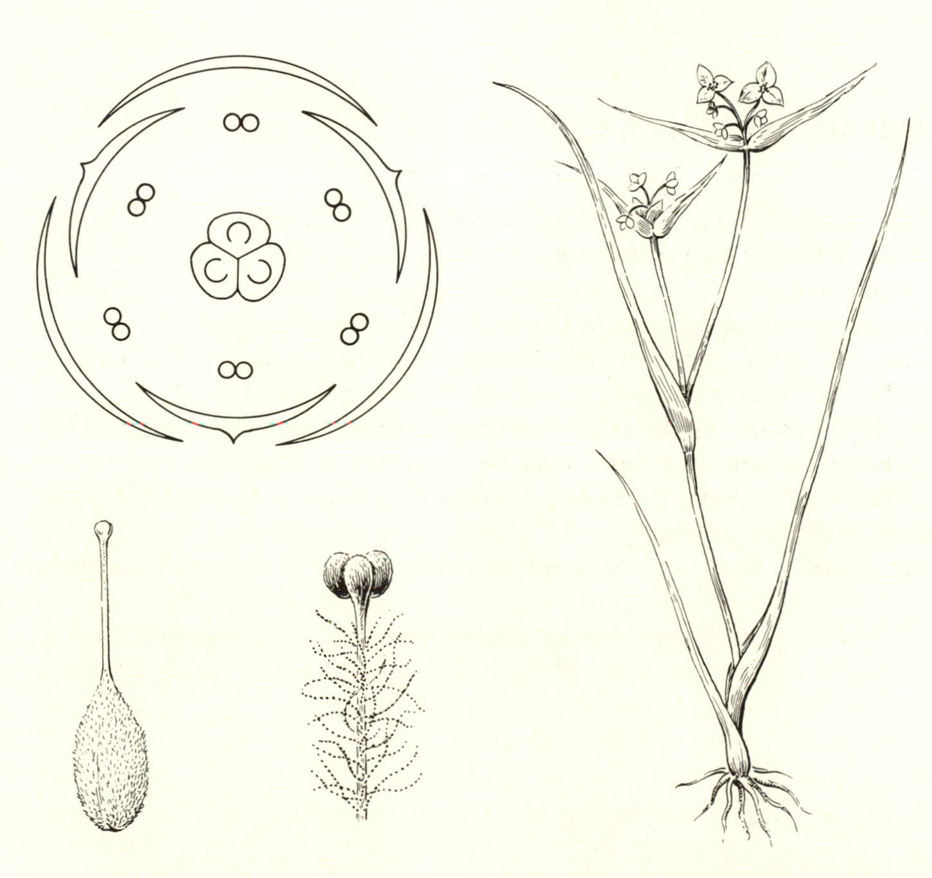

79 *Tradescantia,* floral diagram, pistil, and stamen, enlarged.

80 Plant of *Tradescantia occidentalis.*

ORDER XYRIDALES

XYRIDACEAE: Yellow-eyed Grass Family

Scapose marsh herbs with linear leaves often sheathing at the base, and small, perfect, hypogynous, somewhat irregular flowers in bracteate heads. Perianth biseriate, the outer segments hyaline or chaffy, the inner petaloid, usually yellow, and united below into a tube. Fertile stamens 3 and opposite the petals, and often with 3 outer staminodia in the normal position. Pistil 1, of 3 united carpels, the ovary 1-celled with 3 parietal placentae, or partly 3-celled with basal placentation, the ovules numerous or few. Fruit a 3-valved capsule enclosed within the persistent corolla tube.

The family includes 2 genera and about 40 species, in warm regions and often in saline marshes.

EXAMPLES: *Xyris:* Yellow-eyed Grass, with about 33 species, 15 of these in Florida.
Abolboda: of tropical America, with about 7 species.

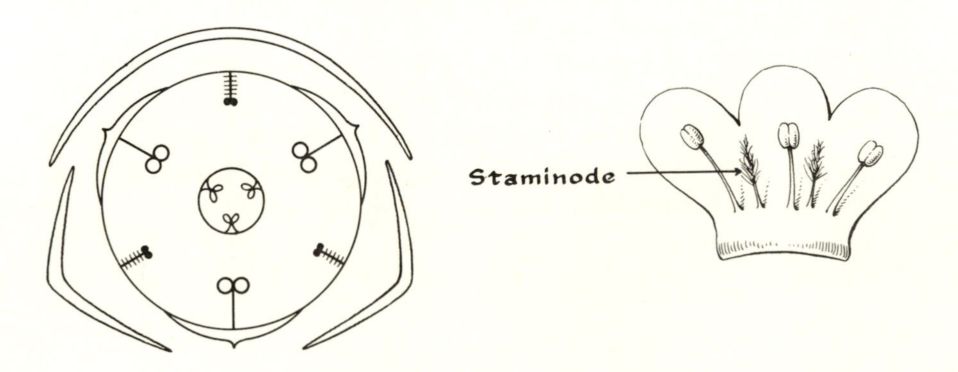

81 *Xyris,* floral diagram and corolla opened to show the stamens, enlarged.

ORDER BROMELIALES

BROMELIACEAE: Pineapple Family

Mostly epiphytic, tropical or subtropical, xerophytic plants with densely clustered linear and usually spiny-toothed leaves. Inflorescence terminal, spike-like, often with colored bracts. Perianth of free or united segments, consisting of green sepals and corolline petals, and usually with appendages within forming a corona. Stamens 6, usually inserted on the perianth. Pistil 1, the ovary superior or inferior, 3-celled. Fruit often fleshy.

The family includes 4 tribes, 45 genera, and about 1,000 species, confined to tropical and subtropical America.

EXAMPLES: *Tillandsia usneoides:* Spanish Moss, common on trees in the Gulf states, and used in the upholstering of furniture.

Ananas sativus: Pineapple, a valuable fruit crop, especially in the Hawaiian Islands. The fiber from the leaves is used to make piña cloth.

82 *Tillandsia usneoides* L., Spanish Moss, hanging in festoons from a Live Oak in Florida. Like most members of the *Bromeliaceae,* the plant is epiphytic.

83 *Tillandsia fasciculata* Sw., a member of the Bromeliaceae, called Wild Pineapple or Air-plant. It grows attached to tree trunks in swamps and hammocks of peninsular Florida, Mexico, and the West Indies.

ORDER ZINGIBERALES

Tropical or subtropical herbs of moist places, with rhizomes and fibrous or tuberous roots. Pseudostems often formed by overlapping sheaths of the leaves. Flowers epigynous, the perianth biseriate, the corolla regular or irregular, often inconspicuous. Stamens 5–6, or reduced to 1 and then with or without staminodia, the staminodia sometimes very conspicuous and petaloid. Fruit a capsule, or indehiscent and fleshy.

The order represents a high degree of specialization within the *Calyciferae* and somewhat parallels that of the Orchids in the *Corolliferae*. The exact interpretation of the morphology of the flowers has been the subject of much discussion.

Six families are included in the order, the following being representative.

MUSACEAE: Banana Family

A small family of 1 genus and some 45 species, in the tropics of the Old World, distinguished in the order by having spirally arranged leaves and bracts, irregular flowers with 5–6 stamens, often unisexual by abortion, and indehiscent fleshy fruits.

EXAMPLES: *Musa paradisiaca* var. *sapientum:* the common Banana.
Musa textilis: Manila Hemp, used in making rope.

STRELITZIACEAE: Strelitzia Family

A small family of 4 genera and about 38 species, in tropical America, South Africa, and Madagascar. It is distinguished in the order by having 2-ranked leaves and bracts, flowers with 3 distinct sepals, 3 irregular petals, 5–6 stamens, and capsular fruits.

EXAMPLES: *Strelitzia reginae:* Bird of Paradise Flower, often cultivated.
Ravenala madagascariensis: Travelers' Tree.

ZINGIBERACEAE: Ginger Family

A family of about 24 genera and 300 species, in tropical and subtropical regions, especially in Asia. It is distinguished in the order by the plants' being aromatic, with flowers having the inner perianth segments regular, but both the inner and outer segments united, and the androecium of 1 fertile stamen, the others being reduced to petaloid staminodia or lacking. The fruit is either a 3-valved capsule or fleshy and indehiscent.

EXAMPLES: *Zingiber officinale:* Ginger, a native of tropical Asia.
Curcuma angustifolia: Arrowroot, of the East Indies, yielding a valuable starch.
Amomum subulatum: Cardamom, of Bengal, a common spice.

The Subclass Corolliferae

The perianth is usually uniseriate (in one series), petaloid throughout (except in the genus *Trillium* of the family *Liliaceae,* which has green sepals), or it is reduced to hairs or scales, *the calyx and corolla thus similar in appearance.* There is a tendency toward a union of the calyx and corolla into a common, petaloid perianth tube. The gynoecium is syncarpous or reduced to a single carpel. Bulbs and corms are fairly common, though not universal, and some plants are rhizomatous. There are some families of aquatic plants (*Lemnaceae, Typhaceae,* and *Sparganiaceae*); and some groups are mainly tropical or subtropical (*Araceae, Arecaceae, Haemodorales, Burmanniales,* and *Orchidaceae*).

Key to the Principal Families of Corolliferae

Plants thalloid and aquatic, without true stems and leaves . . . . . . LEMNACEAE

Plants with stems and leaves, sometimes aquatic

Herbs with minute flowers on a dense, fleshy spike (spadix), which is usually subtended by a conspicuous bract (spathe) . . . . ARACEAE

Herbs, shrubs, or trees; if herbaceous the flowers not as above

Trees or shrubs with large pinnate or palmate leaves in a terminal tuft; fruit a large nut, drupe, or berry; flowers mostly unisexual . . . . ARCACEAE

Mostly herbs, but if woody the leaves simple, the flowers perfect, and the fruit a capsule

Flowers perfect, the perianth petaloid and often showy; plants seldom aquatic

Ovary superior, the flowers hypogynous . . . . LILIACEAE

Ovary inferior, the flowers epigynous

Flowers regular; stamens 6 or 3; placentation axillary

Stamens 6; plants scapose . . . . AMARYLLIDACEAE

Stamens 3; plants with equitant-leafy stems . . . . IRIDACEAE

Flowers irregular; stamens 2 or 1; placentation parietal . . ORCHIDACEAE

Flowers unisexual, minute, in dense heads or spikes; perianth not petaloid; plants aquatic

Inflorescence a continuous or interrupted and cylindrical spike; fruit disseminated by silky hairs . . . . TYPHACEAE

Inflorescence of few to many globose heads; fruit without hairs . . . . SPARGANIACEAE

ORDER LILIALES

LILIACEAE: Lily Family

Perennial herbs, or sometimes woody, from rhizomes, bulbs, or fleshy roots. Flowers often in racemes, regular, hypogynous, usually perfect, 3-merous (except *Maianthemum,* which is 4-merous), and often showy, the perianth usually petaloid (except *Trillium,* which has green sepals). Stamens commonly 6. Pistil 1, of 3 united carpels, with 1 or 3 styles and usually 3 stigmas, the ovary 3-celled, with axillary placentation and an indefinite number of ovules. Fruit a capsule or berry.

The family includes about 200 genera and 2,500 species, world-wide in distribution, including ornamentals and a few poisonous plants. As treated here, it includes the families *Trilliaceae, Smilacaceae, Melanthaceae, Convallariaceae, Dracaenaceae, Calochortaceae,* etc. of various American authors. It thus includes 11 subfamilies for the world.

EXAMPLES: *Lilium:* Lily, with about 45 species.
Calochortus: Mariposa or Sego Lily, with about 57 species.
Allium: Onion, with about 250 species.
Zigadenus: Death Camas, some species being very poisonous.
Yucca: Yucca or Soapweed, a genus of xerophytic shrubs.
Dracaena: Dragon Tree, a genus of small trees of dry warm areas.
Camassia: Camas, with an edible bulb.

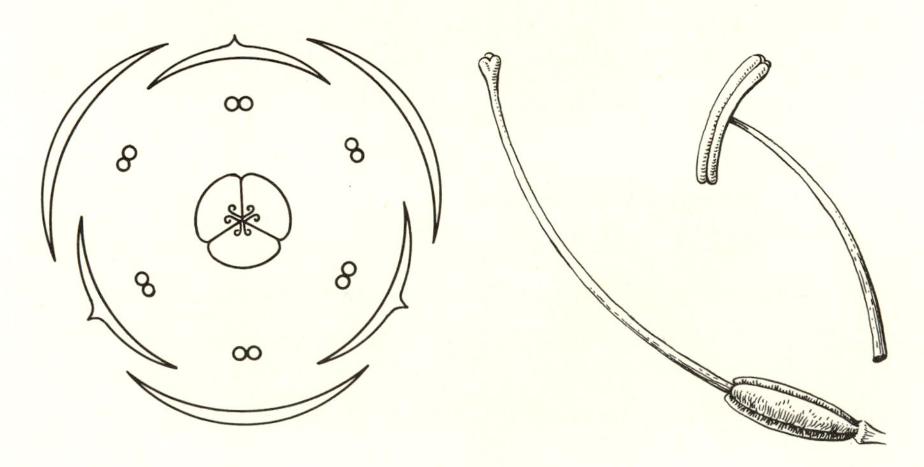

84 *Lilium,* floral diagram, stamen, and pistil.

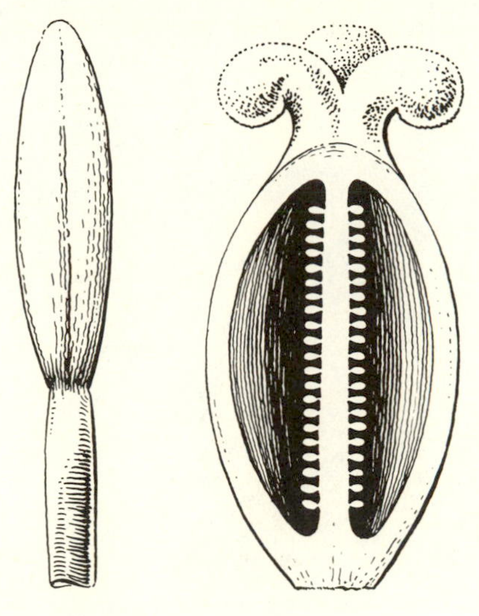

85 *Tulipa,* stamen and section of pistil.

86 (Left) Perianth segment of a species of *Calochortus.* (Right) Pistil of *Zigadenus.*

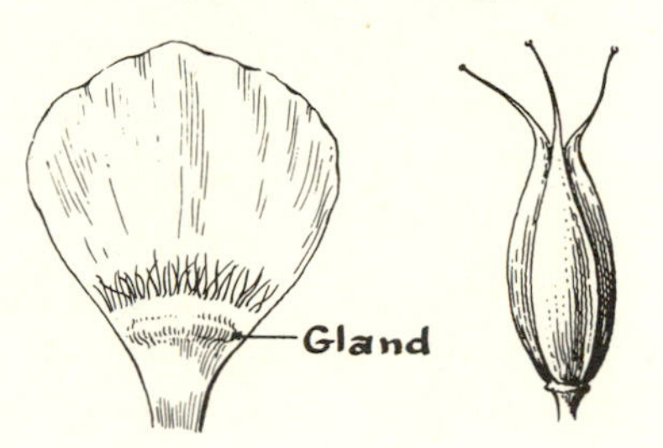

87 Flower of *Lilium philadelphicum* L. The sepals and petals are all alike and all petaloid, typical of many members of the *Corolliferae.*

88 *Yucca glauca* Nutt., a common plant, of western plains and hills, having a woody base.

89 *Yucca glauca* Nutt., detailed view of a fruit, which is a septicidal capsule.

ORDER AMARYLLIDALES

AMARYLLIDACEAE: Amaryllis Family

Perennial scapose herbs or woody plants from bulbs or rhizomes. Flowers solitary, umbellate, or in elongated racemes or panicles, epigynous, with a tubular epigynous hypanthium, regular or nearly so, the perianth petaloid and often with appendages within forming a corona. Stamens 6. Pistil 1, of 3 united carpels, the style single and stigmas 3. Fruit a capsule or berry.

As treated here, the family includes about 65 genera and nearly 900 species, mainly in warm regions and often xerophytic.

As defined by Hutchinson,[1] this family would include those liliaceous groups, such as Onions (*Allium*), that have a scapose habit and umbellate inflorescence subtended by one or more spathaceous bracts. He considers this type of inflorescence to be of greater taxonomic importance than the character of superior or inferior ovary, which is usually used to distinguish the *Liliaceae* from the *Amaryllidaceae*. The Century Plants, moreover, are often treated as a separate family.

EXAMPLES: *Amaryllis:* a South African genus, often cultivated.
Narcissus: Narcissus, Jonquil, Daffodil.
Agave: Century Plant, of desert areas.

[1] J. Hutchinson: *The Families of Flowering Plants:* II, *Monocotyledons* (1934).

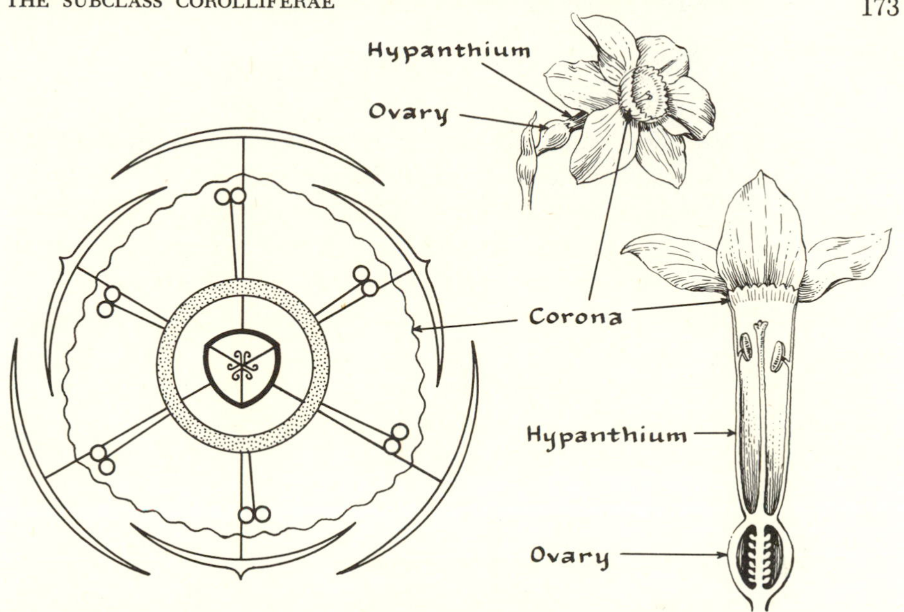

90 *Narcissus,* floral diagram, flower, and longitudinal section of flower.

91 *Narcissus pseudo-narcissus* L., the cultivated Daffodil, showing the conspicuous corona.

ORDER IRIDALES

IRIDACEAE: Iris Family

Mostly perennial herbs from rhizomes or corms, the leaves equitant and conduplicate (2-ranked and folded lengthwise at the base). Flowers epigynous, 3-merous, regular, with a short or long solid epigynous hypanthium, the perianth petaloid. Stamens 3. Pistil 1, of 3 united carpels, the styles often petaloid, the placentation axillary. Fruit a loculicidal capsule.

The family includes about 57 genera and 800 species, in temperate and tropical regions.

EXAMPLES: *Iris:* Iris, Blue Flag.
Gladiolus: Gladiolus, often cultivated.
Sisyrinchium: Blue-eyed Grass.
Crocus: about 75 species of Old World plants often cultivated for their early blooms.

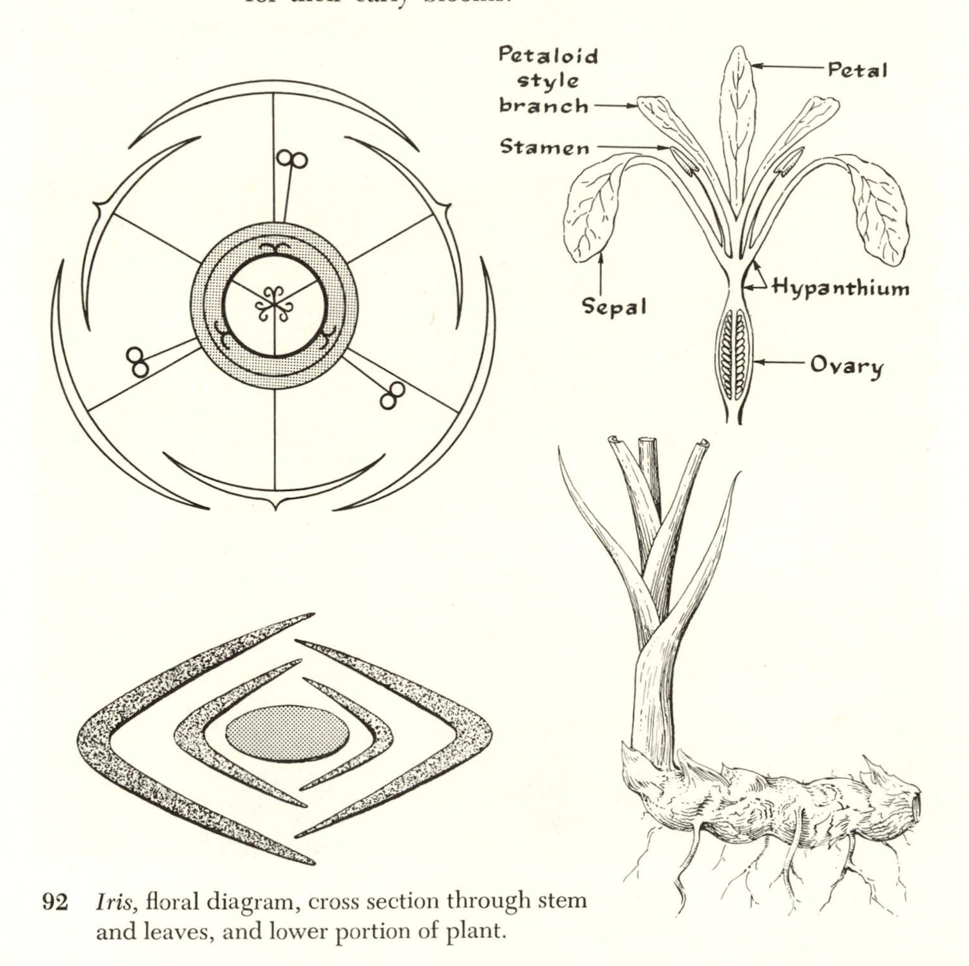

92 *Iris,* floral diagram, cross section through stem and leaves, and lower portion of plant.

93 *Iris missouriensis* Nutt., flowers and fruit.

ORDER TYPHALES

Aquatic herbs from rhizomes, the leaves sheathing at the base. Flowers anemophilous, unisexual, small, produced in dense spikes or heads, the perianth reduced to hairs or scales. The order is probably a reduced group, the culmination of a line of evolution derived from the *Liliales,* and includes the following two families.

TYPHACEAE: Cattail Family

Tall aquatic or marsh herbs from rhizomes, with 2-ranked linear sheathing leaves. Flowers monoecious, in dense terminal spikes in which the upper part is staminate and the lower pistillate. Staminate flowers of 2–5 (usually 3) stamens, each flower surrounded by a hair-like or membranaceous perianth of indefinite parts. Pistillate flowers with a single pistil having a long-stipitate ovary, the stipe bearing numerous perianth hairs. Fruit 1-seeded and dehiscent at maturity.

The family includes the single genus *Typha,* with about 9 species, in temperate and tropical regions but not occurring south of the equator in the Americas and Africa.

EXAMPLES: *Typha latifolia:* Broad-leaved Cattail.
Typha angustifolia: Narrow-leaved Cattail.

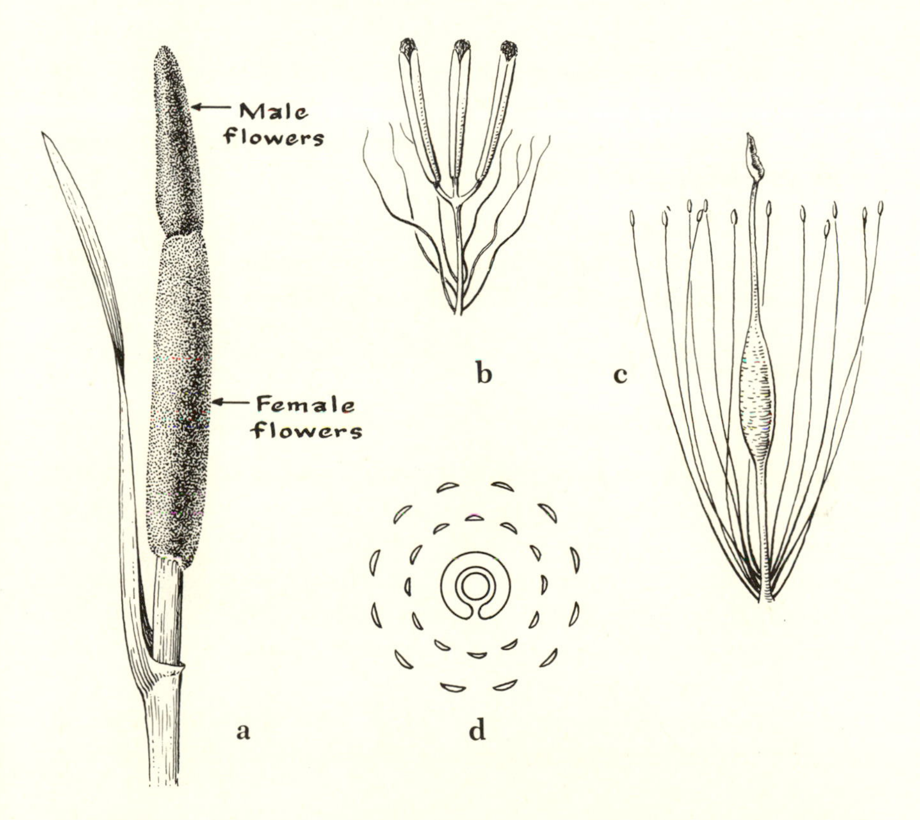

94 *Typha,* (**a**) upper part of flowering plant; (**b**) male flower, enlarged; (**c**) female flower, enlarged; (**d**) floral diagram of female flower.

SPARGANIACEAE: Bur-reed Family

Rather low (mostly 1 meter or less high), grass-like, aquatic or marsh herbs from rhizomes, with 2-ranked linear sheathing leaves. Flowers monoecious, in dense globose heads, the upper heads staminate and the lower pistillate. Perianth composed of a few narrow scales. Stamens 3 or more, their filaments free or partly united near the base. Pistil 1, the ovary narrowed at the base, 1-celled or sometimes 2-celled. Fruit an akene.

The family includes the single genus *Sparganium*, with about 15 species, in temperate and cool regions of the Northern Hemisphere, Australia, and New Zealand.

EXAMPLES: *Sparganium eurycarpum:* Broad-fruited Bur-reed, transcontinental in North America.

Sparganium angustifolium: Narrow-leaved Bur-reed, of North America and Australia.

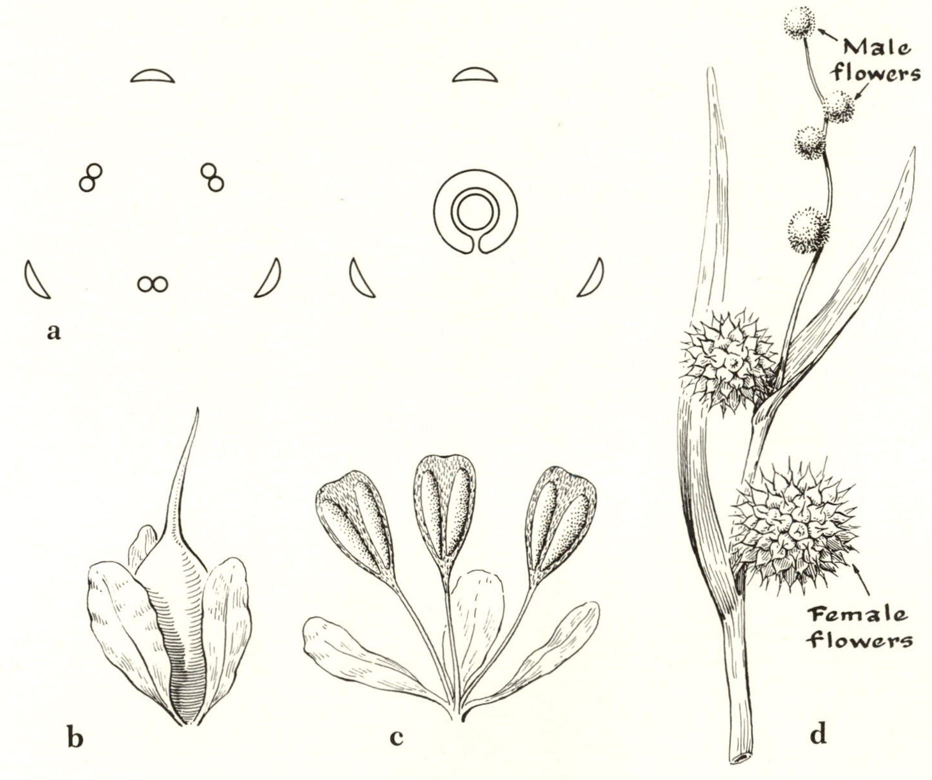

95 *Sparganium*, (**a**) floral diagrams of staminate (left) and pistillate (right) flowers; (**b**) pistillate flower, enlarged; (**c**) staminate flower, enlarged; (**d**) upper portion of flowering plant.

96 *Sparganium multipedunculatum* (Morong) Rydb., showing fruiting heads. The staminate flowers have dropped off.

ORDER ARALES

Terrestrial or sometimes aquatic and mostly herbaceous plants with minute flowers produced on a dense spike (spadix), which is usually subtended by a large and sometimes ornamental bract (spathe). Fruit a berry.

The group probably was derived from the *Liliales,* representing a line of evolution parallel to that of the *Arecales* and *Typhales.*

ARACEAE: Arum Family

Herbs, climbers, or rarely shrubs, of various habit and size, the leaves usually large, simple or compound, net-veined, and with sheathing petioles. Inflorescence characteristically a spathe (a more or less showy or petaloid bract) and spadix (a spike of numerous small flowers, borne on a fleshy axis, subtended or enclosed by the spathe). Flowers hypogynous, unisexual or perfect, 2–3-merous or naked, crowded, ebracteate, the stamens 1 or more, the anthers opening by terminal pores, the filaments free or united; the pistil 1, of 1 or more carpels, nearly always forming a berry in fruit.

The family is mostly tropical, with only a few representatives in temperate North America. It includes about 100 genera and 1,500 species, some ornamental and others producing food for human consumption.

EXAMPLES: *Zantedeschia aethiopica:* Calla-lily.
Arisaema triphyllum: Jack-in-the-pulpit.
Symplocarpus foetidus: Skunk-cabbage.
Colocasia esculenta: Taro, or Poi.
Pistia: Water-lettuce.
Philodendron: about 220 species of tropical climbers or sometimes shrubs.
Orontium: Golden-club.

LEMNACEAE: Duckweed Family

Small to minute, floating or submerged herbs, which are more or less thalloid (without stems or leaves, and the roots, if any, thread-like). Flowers monoecious and naked; stamens 1–2; pistil 1, 1-celled; ovules 1–7.

The family includes 2–3 genera and about 19 species, in fresh water of temperate and tropical regions.

EXAMPLE: *Lemna:* Duckweed, often found in ponds and swamps.
Spirodela: Great Duckweed.
Wolffia: plants about 1 mm long, the smallest of flowering plants.

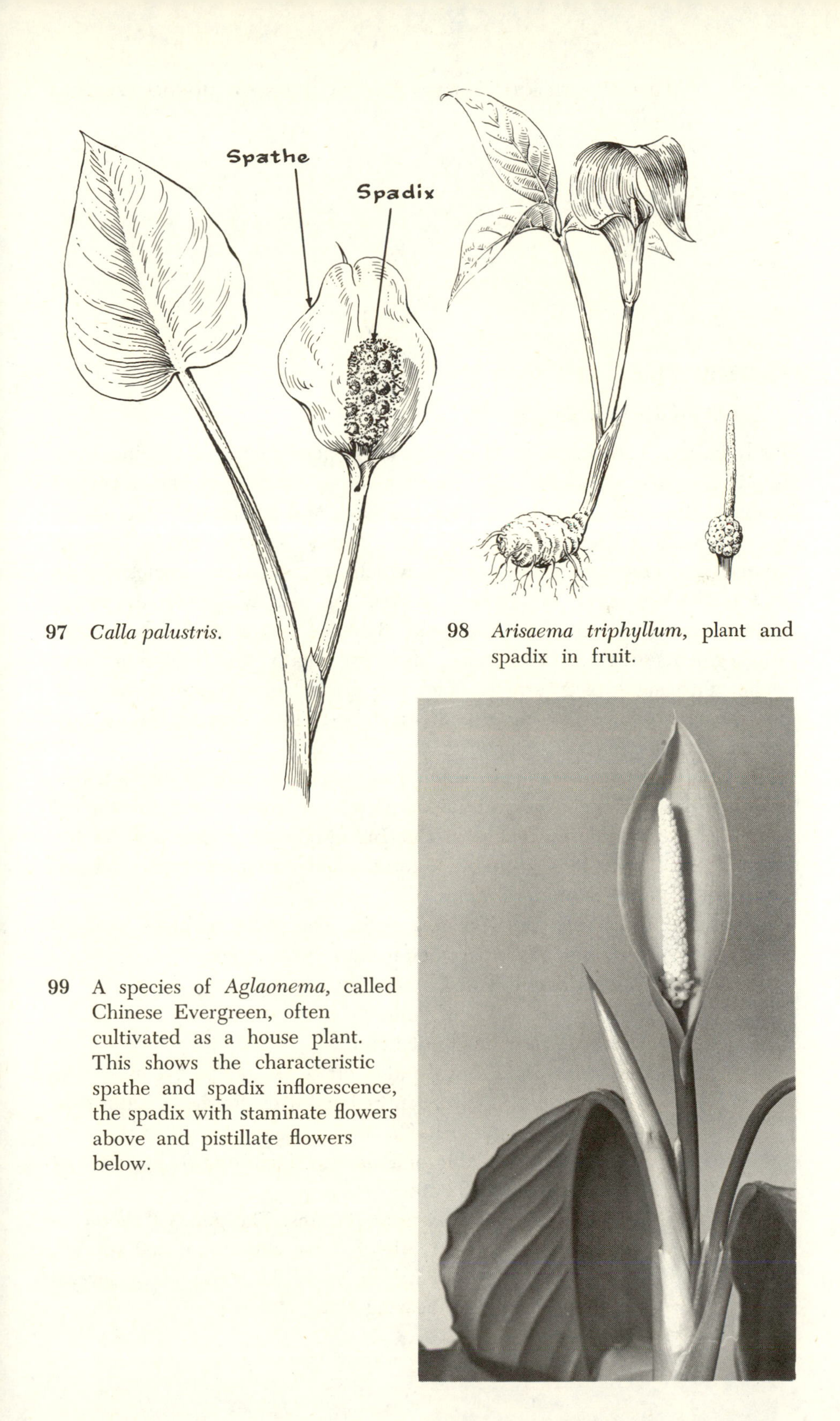

97 *Calla palustris.*

98 *Arisaema triphyllum,* plant and spadix in fruit.

99 A species of *Aglaonema,* called Chinese Evergreen, often cultivated as a house plant. This shows the characteristic spathe and spadix inflorescence, the spadix with staminate flowers above and pistillate flowers below.

ORDER ARECALES

ARECACEAE: Palm Family

Mostly unbranched shrubs or trees with large, persistent, mostly pinnate or digitate leaves in a terminal tuft, the stem often covered by the persistent leaf-bases, the stem sometimes much reduced, and sometimes vine-like or rope-like, with scattered leaves and stout spines. Flowers small, in large and usually paniculate inflorescences, which are sometimes enclosed in a large spathe-like bract, hypogynous, perfect, or, usually, unisexual (polygamous, monoecious, or dioecious), with 3 sepals and 3 petals, these perianth segments distinct or connate. Stamens usually 6, rarely numerous. Gynoecium usually of 3 carpels, which are distinct or connate at the base, each carpel producing a single seed. Fruit berry-like, drupaceous, or nut-like.

The family, of considerable economic importance, is divided into 8 tribes, including about 170 genera and 1,500 species, in tropical and subtropical regions of the world, especially in the Indo-Malayan region and in the Guianas and Brazil. It is probably a terminal evolutionary group derived from woody members of the *Liliales*.

EXAMPLES: *Cocos nucifera:* Coconut Palm. The genus includes some 30 species, the others being entirely American.

Phoenix dactylifera: Date Palm.

Elaeis guineensis: Oil Palm.

Areca catechu: Betel-nut Palm.

Roystonea regia: Royal Palm, often planted in avenues.

Washingtonia filifera: Sentinel Palm of southern California.

Raphia pedunculata: Raffia.

Sabal: Cabbage Palm, or Palmetto, including 8 species, 4 of them in the Gulf states.

Calamus and *Daemonorops:* Rattans. The genus *Calamus* is one of the largest in the family, with some 200 species, mostly in Asia, and having rope-like stems often several hundred feet long, bearing stout spines.

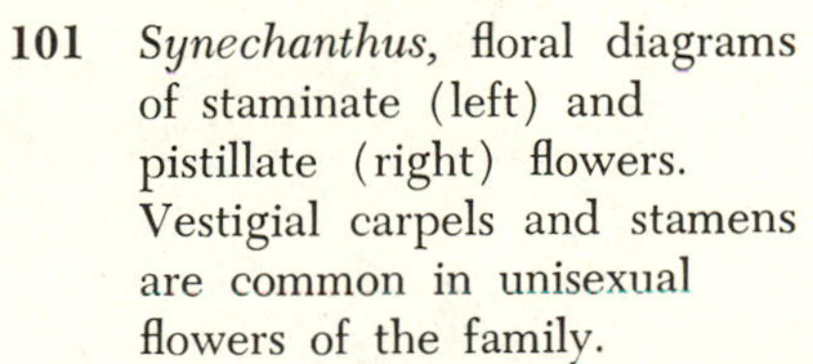

100 Tree of the Coconut Palm, *Cocos nucifera.*

101 *Synechanthus,* floral diagrams of staminate (left) and pistillate (right) flowers. Vestigial carpels and stamens are common in unisexual flowers of the family.

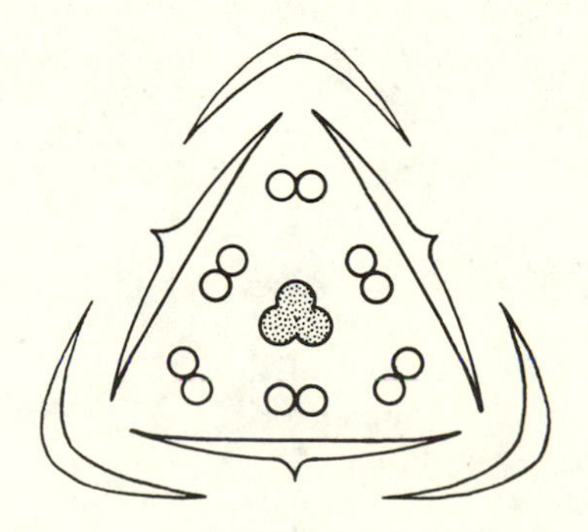

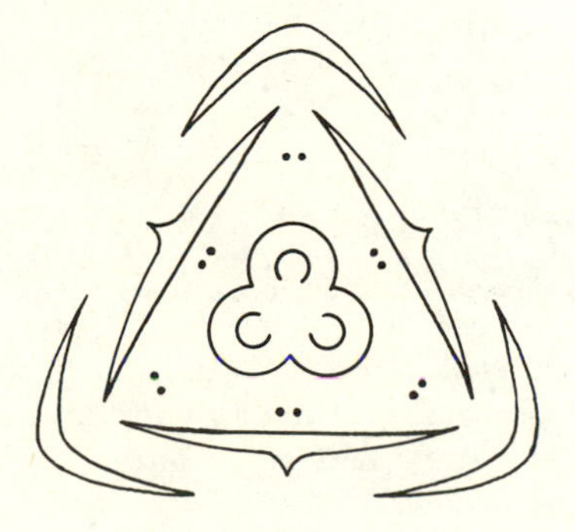

102 *Serenoa repens* (Bartr.) Small, the Saw Palmetto of the southeastern United States, here forming an understory in a stand of *Pinus elliottii* Engelm., Slash Pine.

103 *Roystonea regia* (H. B. K.) O. F. Cook, the Royal Palm. Shown here are the crown of the tree, an inflorescence, and part of the trunk.

ORDER HAEMODORALES

Perennial herbs, or sometimes woody plants, from rhizomes or corms. Leaves mostly entire, rarely lobed, in a basal tuft in herbaceous members and in tufts at the ends of branches in woody ones (*Velloziaceae*). Flowers perfect, solitary, in panicles, or in umbels, varying from regular to irregular, hypogynous or, usually, epigynous. Stamens numerous to 6, or sometimes reduced to 3, 2, or 1. Pistil 1, of 3 united carpels, the ovary either 3-celled with axillary placentation or 1-celled with parietal placentation.

The order, as defined by Hutchinson,[1] includes 6 families, which are mainly tropical and subtropical in the Southern Hemisphere. It includes plants that are somewhat intermediate in character between the *Liliales* and the *Orchidales*.

ORDER BURMANNIALES

Small annual or perennial herbs, often saprophytic. Leaves usually reduced to scales. Flowers solitary and terminal, or in racemes or cymes, perfect or unisexual, regular to irregular, epigynous. Stamens 6 or 3. Pistil 1, of 3 united carpels, the ovary either 3-celled with axillary placentation or 1-celled with parietal placentation. Ovules and seeds very numerous and small.

The order[1] includes 3 families, which are tropical and subtropical in distribution. It probably represents a development parallel to that of the *Orchidales*, as evidenced by the tendency of the plants to become saprophytic and to produce numerous minute seeds, but the degree of specialization in the flower parts has not equaled that of the Orchids.

ORDER ORCHIDALES

ORCHIDACEAE: Orchid Family

Perennial herbs, often epiphytic and sometimes saprophytic, from rhizomes or fleshy roots. Flowers strikingly irregular, often spurred, epigynous, 3-merous, the pedicel and ovary twisted through a semicircle in anthesis. Stamens 1 or 2, attached to the style (gynandrous), and together with the style and stigmas forming the *column*. Pollen often cohering in waxy masses or pollinia. Pistil 1, of 3 united carpels, the ovary usually 1-celled with 3 parietal placentae, the style single and the stigmas 3 and all functional or 2 functional and 1 sterile, the latter often produced upward into a beak, or *rostellum*. Fruit a capsule. Seeds very minute and very numerous.

This is one of the largest families of flowering plants, including about 500 genera and perhaps 15,000 species, widely distributed, but mainly in tropical forests.

[1] J. Hutchinson: *The Families of Flowering Plants:* II, *Monocotyledons* (1934).

EXAMPLES: *Cypripedium:* Lady's Slipper.
Habenaria: Rein Orchis, with about 500 species.
Corallorhiza: Coral-root, the plants saprophytic.
Vanilla planifolia: Vanilla Orchid, which yields the extract from immature pods.
Cattleya: one of the chief horticultural or corsage orchids.
Bulbophyllum and *Dendrobium:* each with about 1,000 species in the Malaysian region alone.

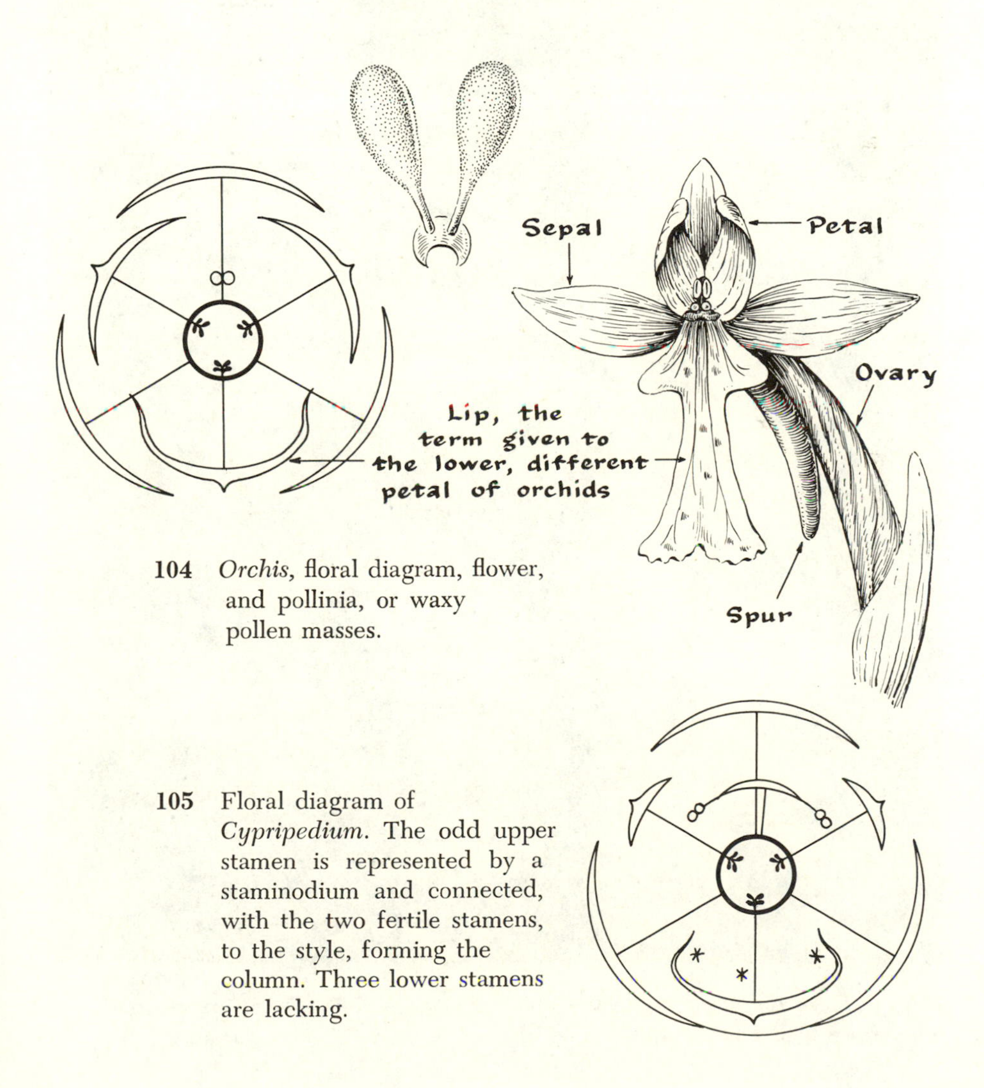

104 *Orchis,* floral diagram, flower, and pollinia, or waxy pollen masses.

105 Floral diagram of *Cypripedium.* The odd upper stamen is represented by a staminodium and connected, with the two fertile stamens, to the style, forming the column. Three lower stamens are lacking.

106 *Calypso bulbosa* (L.) Oakes, the Fairy Slipper. The lip is inflated and resembles a little slipper. The plants are 6–22 centimeters tall.

107 Detail of flowers of *Calypso bulbosa* (L.) Oakes, the Fairy Slipper. The petaloid part in the center, above the lip, is the column.

The Subclass Glumiflorae

These are the grasses and grass-like plants. The flowers are small and scarcely showy, and the perianth is reduced to scale-like or hair-like parts or sometimes lacking. In many members (*Cyperaceae* and *Poaceae*) the flowers are subtended by chaffy scales (bracts), and these are aggregated into spikelets. Fibrous roots occur in most members. Some are annuals, some are rhizomatous, and very few (some species of *Melica* of the family *Poaceae*) have poorly developed bulbs or corms. Some members are aquatic, and some (such as the Bamboos) are tropical or subtropical.

Key to the Families of Glumiflorae

Fruit a 3-valved capsule containing 3–many seeds; perianth of 6 chaffy and similar segments; flowers not associated with scale-like bracts (scales, lemmas, or glumes) . . . . JUNCACEAE

Fruit an akene or a grain (caryopsis), indehiscent and 1-seeded; perianth inconspicuous, of scales, bristles, or hairs, or sometimes lacking; flowers associated with scale-like bracts (scales, lemmas, or glumes)

Stems not jointed; leaves when present 3-ranked; fruit a lenticular or trigonous akene . . . . CYPERACEAE

Stems jointed; leaves always present and 2-ranked; fruit a grain (caryopsis), seldom angled . . . . POACEAE

ORDER JUNCALES

Mostly grass-like plants with small and chaffy liliaceous flowers, which are anemophilous.

The order includes 4 families, typified by the following, which is the only one represented in North America.

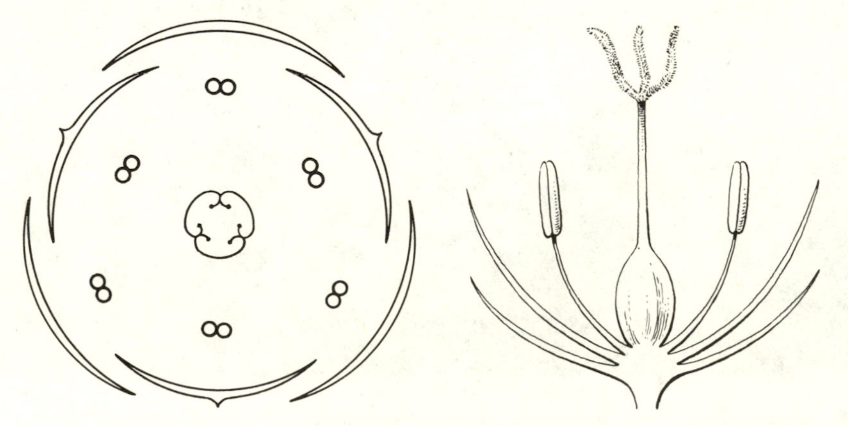

108 *Juncus,* floral diagram and longitudinal section of flower.

JUNCACEAE: Rush Family

Perennial or annual herbs, rarely shrubby (*Prionium*), usually with fibrous roots and erect or horizontal rhizomes. Stems commonly leafy only at the base, the leaves in a basal tuft or sometimes reduced to sheaths, which may be open or closed, the blades, when present, cylindric to flat and grass-like, mostly linear or filiform. Flowers mostly anemophilous, solitary or usually in panicles, corymbs, or heads, often very small, regular, 3-merous, perfect or unisexual, the perianth usually of 6 segments in two series, rarely of only 3 segments, chaffy, greenish, brownish, or black, rarely white or yellowish. Stamens 6 or 3, the anthers basifixed. Pollen in tetrads. Pistil 1, of 3 united carpels, the ovary superior and 1-celled with parietal placentation or, rarely, 3-celled with axillary placentation (*Prionium*), the style long and single to almost none, the stigmas 3. Fruit a loculicidal capsule with few to many seeds.

The family includes 8 genera and about 300 species, widely distributed in temperate and cold regions but especially well represented in the Andes of South America.

EXAMPLES: *Juncus:* Rush
Luzula: Wood Rush

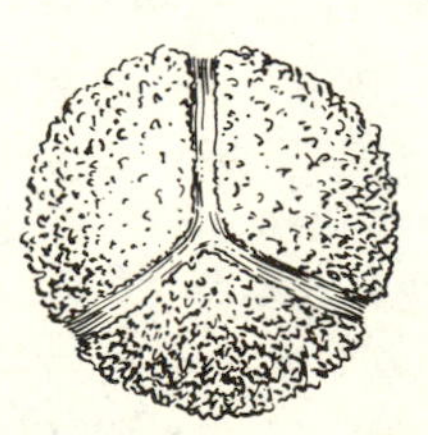

109 Pollen tetrad of *Luzula*, greatly magnified.

110 Cross section of ovary of *Prionium*, enlarged.

ORDER POALES

Annual or perennial, mostly grass-like herbs, sometimes woody, with alternate and mostly linear leaves composed of a sheath and a blade. Flowers small, naked or with a scale-like or hair-like perianth, enclosed between scale-like bracts. Stamens 6, 3, or fewer. Pistil 1, the ovary superior, 1-celled and 1-seeded. Fruit a grain (caryopsis) or an akene.

The order probably represents an early modification from the *Liliales* through the *Juncales*. It includes two important families, the Sedges and the Grasses. These two families may be distinguished as follows:

	Cyperaceae: Sedges	*Poaceae:* Grasses
Stems	Herbaceous Not jointed Usually solid Often triangular in cross-section	Herbaceous, or sometimes woody Jointed Usually hollow in internodes Circular in cross-section
Leaves	3-ranked when present Sheaths usually closed Ligule usually not present	2-ranked Sheaths usually open Ligule usually present
Fruit	An akene	A grain (caryopsis)
Habitat	Usually in wet places	Usually in drier places, but some aquatic

CYPERACEAE: Sedge Family

Grass-like herbs, often in wet places, from rhizomes or fibrous roots, the stems not jointed, often with solid pith, and frequently triangular in cross-section. Leaves usually 3-ranked, with a closed sheath; ligule rarely present. Flowers small, perfect, monoecious or dioecious, arranged in spikelets, and each flower usually solitary within a bract (*glume* or *scale*). Bracts 2-ranked or spiral in arrangement in the spikelet, the inflorescence composed of one or more spikelets and commonly subtended by one or more leaf-like bracts. Perianth reduced to hypogynous scales, bristles, or hairs, rarely somewhat petaloid, and often absent. Stamens 3 or fewer, rarely more, the anthers basifixed. Ovary superior, 1-celled, the style with 2 or 3 branches or teeth. Ovule solitary. Fruit a lenticular or trigonous akene.

The family includes 75 genera and about 3,500 species, mostly in temperate and cold regions, many extending into the Arctic. It is important as a source of forage and of food for wild fowl.

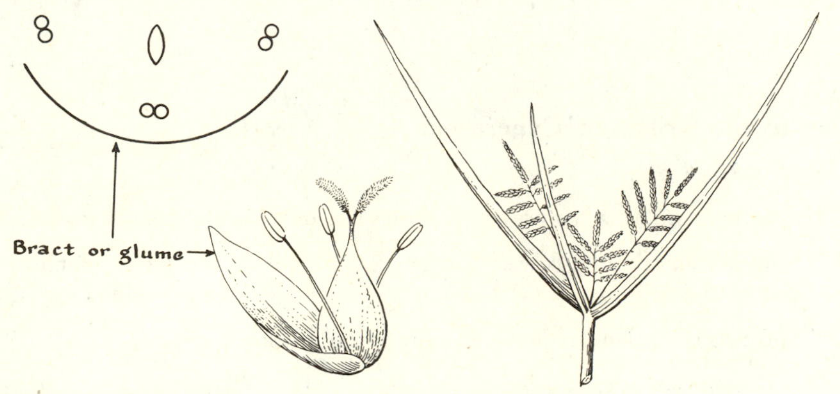

111 *Cyperus*, floral diagram, flower, and inflorescence.

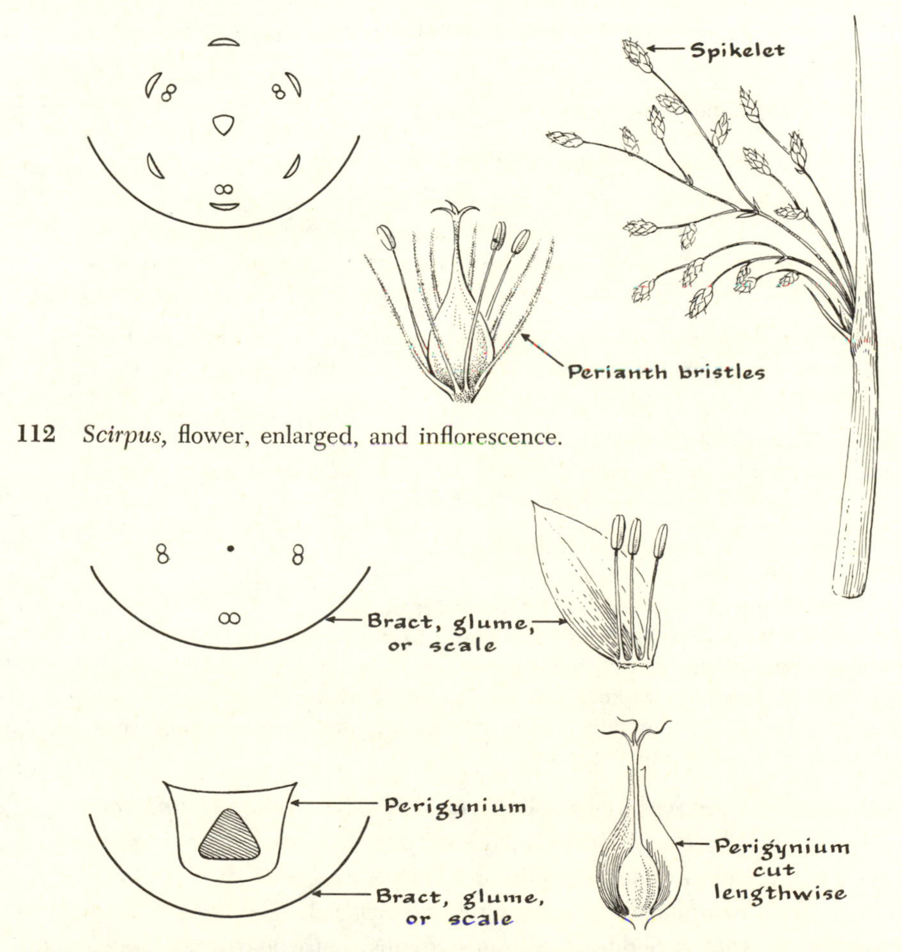

112 *Scirpus*, flower, enlarged, and inflorescence.

113 (Left) Floral diagrams of male (above) and female (below) flowers of *Carex*. (Right) Male (above) and female (below) flowers of *Carex*.

Key to the Tribes of Cyperaceae

Flowers perfect

Hypogynous scales, when present, filiform, flat, or perianth-like, not folded

Spikelets mostly 1–2-flowered, often 2 or more of the lower glumes empty . . . . RHYNCHOSPOREAE

Spikelets several–many-flowered, only 1–2 of the lower glumes empty

Glumes not 2-ranked . . . . SCIRPEAE

Glumes 2-ranked . . . . CYPEREAE

Hypogynous scales 2, folded or keeled, spikelets several–many-flowered . . . . HYPOLYTREAE

Flowers unisexual, no hypogynous setae present

Female flower not enclosed by a modified glume

Pistillate flower solitary at the base of an androgynous spikelet or the spikelets unisexual, the female spikelets 1-flowered and in the lower part of the panicle, the male spikelets in the upper part and 2- or more-flowered . . . . SCLERIEAE

Pistillate flower terminal in a unisexual spikelet or in the upper part of the panicle, the male spikelets produced lower and 2- or more-flowered . . . . CRYPTANGIEAE

Female flower enclosed by a modified glume (*perigynium*), female spikelets 1-flowered, spicate, male spikelets 2- or more-flowered and terminal or, rarely, continuous at the base of the female spike . . . . CARICEAE

Rhynchosporeae: cosmopolitan in distribution.
Scirpeae: cosmopolitan in distribution.
Cypereae: chiefly tropical, some in temperate regions.
Hypolytreae: tropical and subtropical; the most primitive groups.
Sclerieae: cosmopolitan in distribution.
Cryptangieae: chiefly in tropical America, some in Africa.
Cariceae: chiefly in temperate and cold regions, rare in the tropics, and common in mountainous regions.

EXAMPLES: *Cyperus papyrus:* Papyrus, of the Nile valley (*Cypereae*).
Eleocharis: Spike Rush (*Scirpeae*).
Scirpus validus: Bulrush (*Scirpeae*).
Eriophorum: Cotton Grass (*Scirpeae*).
Carex: Sedge, the largest genus, with nearly 800 species, in temperate, alpine, and arctic regions (*Cariceae*).

114 *Carex rostrata* Stokes, a common sedge, showing unisexual spikes, staminate at the top, and below them thicker, pistillate spikes.

POACEAE (GRAMINEAE): Grass Family

Annual or perennial herbs, or sometimes woody in warm regions. Stems often branched at the base, in perennials forming flowering stems (*culms*) and sterile shoots (*innovations*), the culms cylindrical, jointed, usually hollow in the internodes and closed in the nodes. Leaves sometimes crowded at the base of the plants, alternate and 2-ranked, each consisting of *sheath, ligule,* and *blade:* sheaths encircling the stem with the margins free and overlapping or sometimes united; ligule placed at the junction of sheath and blade, membranaceous or reduced to a fringe of hairs, rarely absent; blades usually long and narrow, often involute, parallel-veined. Flowers usually perfect, small and inconspicuous, consisting of stamens and pistil and 2–3 minute, hyaline or fleshy scales (*lodicules*) representing the perianth, each such flower subsessile between two bracts (*lemma* and *palea,* the palea sometimes wanting), the whole forming a *floret,* or false flower. Florets 1–many, 2-ranked, sessile on a short or minute and slender axis (*rachilla*) and bearing at the base 2 empty bracts (the first and second *glumes*), the florets and glumes together forming a *spikelet.* Spikelets pediceled in open or contracted panicles or racemes, or sessile in spikes. Stamens usually 3, sometimes 6, hypogynous, opening by longitudinal slits. Ovary 1-celled, with 1 ovule, which is usually adnate to one side of the carpel; styles 2, stigmas plumose. Fruit a grain, or caryopsis, with a thin pericarp adnate to the seed, with starchy endosperm and a small embryo at the base.

The family, one of the most important in the whole plant kingdom, includes about 525 genera and 5,000 species, world-wide in distribution. It has usually been subdivided into two subfamilies and as many as 27 tribes. Fourteen tribes, including about 1,500 species, occur in the United States. Of these species, about 140 are important native forage plants, and about 60 are cultivated in the United States.

A recent trend is to divide the family into six subfamilies and 26 tribes, at least regarding North American plants. For details of this system the reader is referred to G. L. Stebbins and Beecher Crampton: "A Suggested Revision of the Grass Genera of Temperate North America," in *Recent Advances in Botany* 1:133–145 (Toronto, Canada, 1961).

Comparison of subfamilies

I. *Festucoideae*	II. *Panicoideae*
Spikelets 1–several-flowered, the sterile florets, if any, usually above the fertile florets in the spikelet.	Spikelets with 1 perfect terminal floret and 1 sterile floret below it, the sterile floret usually represented by an empty lemma. Some genera are monoecious, and some produce staminate or empty florets.
Articulation usually above the glumes and between the florets, the glumes remaining on the plant when the seed is shed.	Articulation below the glumes, the whole spikelet (including the glumes) falling when the seed is shed.
Spikelets laterally compressed, the glumes and lemmas usually folded lengthwise, or sometimes the spikelets terete.	Spikelets dorsally compressed, the glumes and lemmas flat, or sometimes the spikelets terete.

Note: Any two of the above three characteristics will place the grass in the proper subfamily.

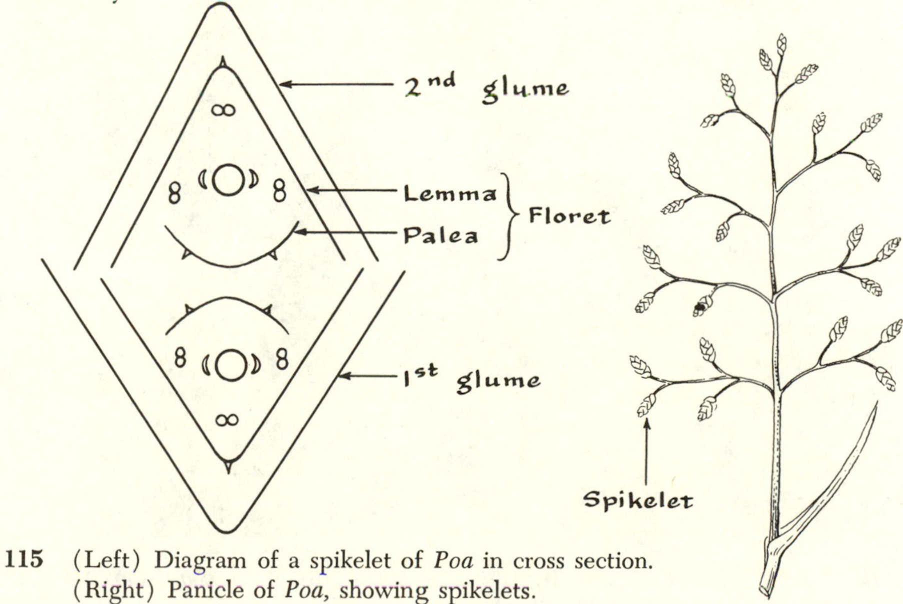

115 (Left) Diagram of a spikelet of *Poa* in cross section. (Right) Panicle of *Poa*, showing spikelets.

116 *Uniola latifolia* Michx., Broad-leaf Uniola, tribe *Festuceae*. On the ends of the very slender panicle branches are the spikelets, each with several florets. Some of the spikelets have started to disarticulate.

117 *Triticum aestivum* L., Wheat. Shown here is a bearded variety with long awns. It is a member of the *Hordeae,* or Barley Tribe, having flowers in single, symmetrical spikes.

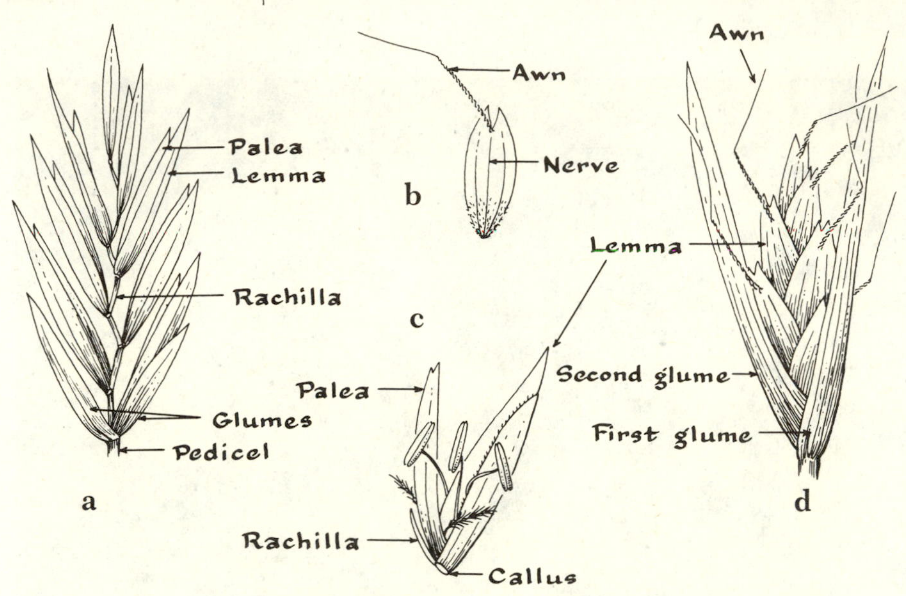

118 (a) Spikelet of bromegrass. (b) Single floret of oat grass. (c) Single floret of a grass at blooming time. (d) Spikelet of oat grass.

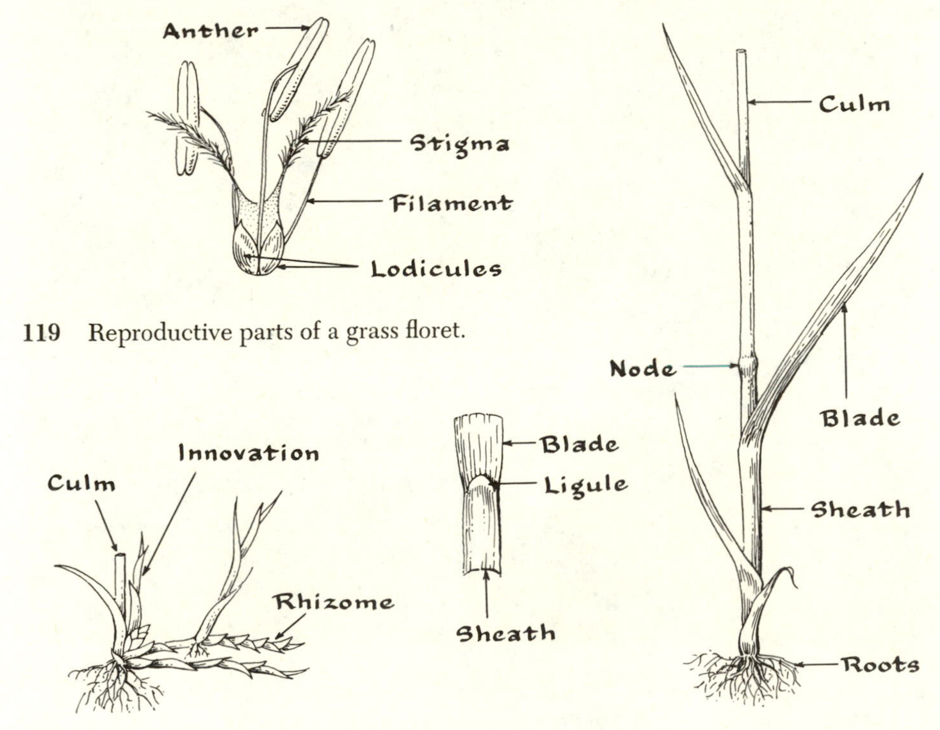

119 Reproductive parts of a grass floret.

120 Vegetative parts of grasses.

Key to the Tribes of Festucoideae[1]

Plants woody, mostly large trees or shrubs . . . . BAMBUSEAE (I)

Plants herbaceous, sometimes tall

Inflorescence of 1 or more spikes

Spikes 1-sided . . . . CHLORIDEAE (VI)

Spikes symmetrical

Articulation above the glumes, or the rachis sometimes disarticulating . . . . HORDEAE (IV)

Articulation below the glumes, the rachis not disarticulating . . . . ZOYSIEAE (VII)

Inflorescence a raceme or panicle, sometimes narrow and spike-like but never a true spike

Spikelets with 1 perfect terminal floret and 2 sterile or empty lemmas below it . . . . PHALARIDEAE (VIII)

Spikelets not as above, the sterile florets, if any, above the fertile

Spikelets unisexual, falling entire, terete; panicles with pistillate spikelets above and staminate spikelets below, these spikelets unlike . . . . ZIZANIEAE (X)

Spikelets usually perfect, if unisexual then not as above

Spikelets strictly 1-flowered

Articulation usually above the glumes, the glumes not reduced in size . . . . AGROSTIDEAE (V)

Articulation below the glumes, the glumes reduced or lacking . . . . ORYZEAE (IX)

Spikelets with 2 or more florets

Glumes longer than the first lemma . . . . AVENEAE (III)

Glumes shorter than the first lemma . . . . FESTUCEAE (II)

[1] Keys to tribes are for the United States only, and are somewhat simplified. Rare exceptions may occur although not indicated.

Key to the Tribes of Panicoideae[1]

Plants with at least some perfect florets, the inflorescence essentially uniform

Spikelets in pairs, one sessile and fertile, the other pedicellate and sterile (sometimes obsolete), rarely both pedicellate; glumes indurate . . . . ANDROPOGONEAE (XII)

Spikelets not paired, in racemes or in open or spike-like panicles; glumes herbaceous or membranaceous, not indurate . . . . PANICEAE (XI)

Plants monoecious, the male and female spikelets in separate inflorescences or in different parts of the same inflorescence . . . . TRIPSACEAE (XIII)

Tribe I. *Bambuseae* (Bamboo Tribe): Woody plants with the spikelet characters of the tribe *Festuceae.* Chiefly tropical plants, many of which are of great economic importance. The genus *Arundinaria* (Cane) is a native of the southeastern United States, and several species of various genera of Bamboos are cultivated in the warmer states, the commonest being *Sasa japonica* (Sieb. & Zucc.) Makino.

Tribe II. *Festuceae* (Fescue Tribe): Spikelets 2–several-flowered, in open, narrow, or spike-like panicles, rarely in racemes. Lemmas awnless, awned from the tip, or awned from between two minute teeth near the apex, the awns straight. Glumes shorter than the first lemma. Articulation above the glumes.

A large tribe, chiefly in the cooler parts of the world, including many important forage grasses. There are 37 genera in the United States.

EXAMPLES: *Poa:* Bluegrasses.
Festuca: Fescue grasses.
Bromus: Bromegrasses.

[1] The tribe *Melinideae* is represented in the United States by two grasses, both introduced: *Melinis minutiflora* Beauv., in southern Florida, and *Thysanolaena maxima* (Roxb.) Kuntze, in southern Florida and southern California. For characteristics of this tribe and these genera see A. S. Hitchcock, *Manual of the Grasses of the United States.* The tribe *Melinideae* has been omitted above for purposes of simplicity and brevity.

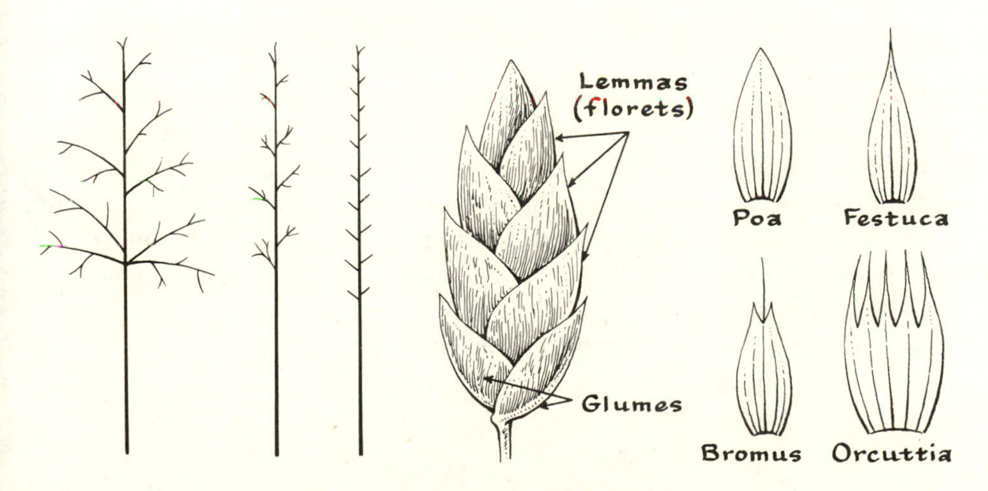

121 Tribe *Festuceae,* (left) inflorescence types, (middle) spikelet type, (right) lemma types.

Tribe III. *Aveneae* (Oat Tribe): Spikelets 2–several-flowered, with glumes as long as the first lemma or, usually, equaling or exceeding all the florets in the spikelet. Lemmas awned from the back (dorsally) or from between the teeth of a bidentate apex; the awn straight or, more commonly, bent and twisted; rarely awnless. Inflorescence a panicle, but sometimes narrow and spike-like. Articulation below the glumes in a few genera. Callus and rachilla usually hairy.

There are 10 genera in the United States, some of which are good forage grasses.

EXAMPLES: *Avena:* Oat.
Koeleria: June grass.
Deschampsia: Hairgrass.

Tribe IV. *Hordeae* (Barley Tribe): Spikelets 1–several-flowered, sessile on alternate sides of a single symmetrical spike, the articulation above the glumes. The rachis continuous and persistent or readily disarticulating at maturity. Glumes often variously modified into bristles or awns.

A large tribe, including many valuable forage grasses and several common small-grain cereals, such as Wheat, Rye, and Barley. There are 12 genera in the United States.

EXAMPLES: *Agropyron:* Wheatgrass, one of the large genera.
Elymus: Wild-rye.
Secale: Rye.
Hordeum: Barley.
Lolium: Ryegrass.
Triticum: Wheat.

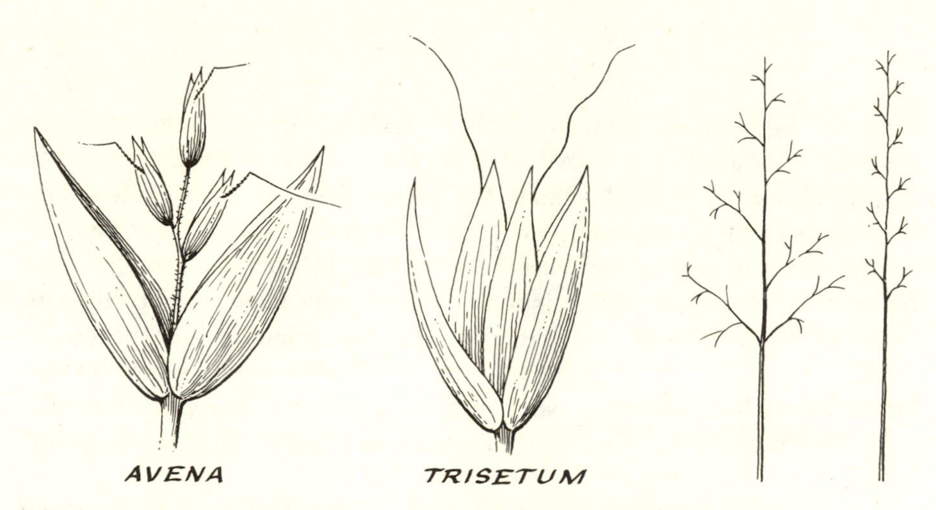

122 Tribe *Aveneae*, spikelet types and inflorescence types.

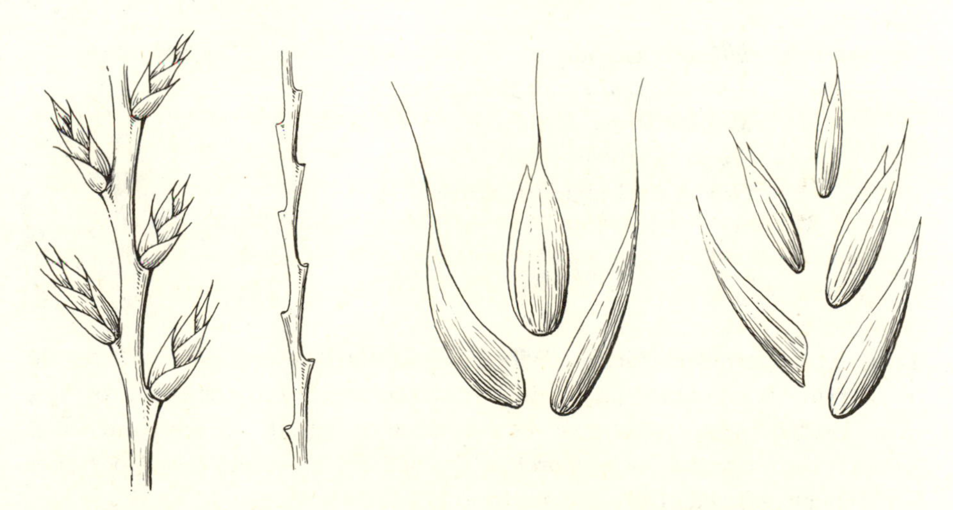

123 Tribe *Hordeae*, (left) portions of spikes showing rachis with and without spikelets, (right) generalized types of spikelets.

Tribe V. *Agrostideae* (Timothy Tribe): Spikelets strictly 1-flowered, the lemmas either awnless or awned from the tip or from the back. Articulation usually above the normal glumes but below them in a few genera. Inflorescence an open or narrow panicle, or sometimes a spikelike raceme.

The tribe, as here defined, has been divided by some authors into two tribes, those members having indurate (hardened) lemmas that fall with the fruit being split off into the tribe *Stipeae,* including such genera as *Stipa* and *Aristida.* The *Agrostideae* are probably a tribe of polyphyletic origin, having been derived in part from the *Festuceae* (those with terminally awned lemmas) and in part from the *Aveneae* (those with dorsally awned lemmas).

A large tribe, with about 25 genera in the United States, including many valuable forage grasses. Some species of *Stipa* cause mechanical injury to grazing animals.

EXAMPLES: *Phleum:* Timothy.
Agrostis: Redtop, or Bentgrass.
Muhlenbergia: Muhly.
Stipa: Needlegrass.
Oryzopsis: Indian Ricegrass.

Tribe VI. *Chlorideae* (Grama Tribe): Spikelets 1–several-flowered, sessile on one side of a continuous rachis, the inflorescence thus composed of 1 or more 1-sided spikes, which may be racemose or digitate on the culm when several. Typically the lowest floret in the spikelet is fertile, the upper ones being much reduced in size and sterile.

A large tribe, found chiefly in warm dry areas, especially in the Southwest. It includes many very valuable forage grasses. There are 18 genera in the United States.

EXAMPLES: *Bouteloua:* Grama, valuable forage grasses.
Buchloë: Buffalo Grass.
Beckmannia: Sloughgrass, a source of food for ducks.[1]

[1] See A. E. Porsild in *Sargentia* 4:9 (1943).

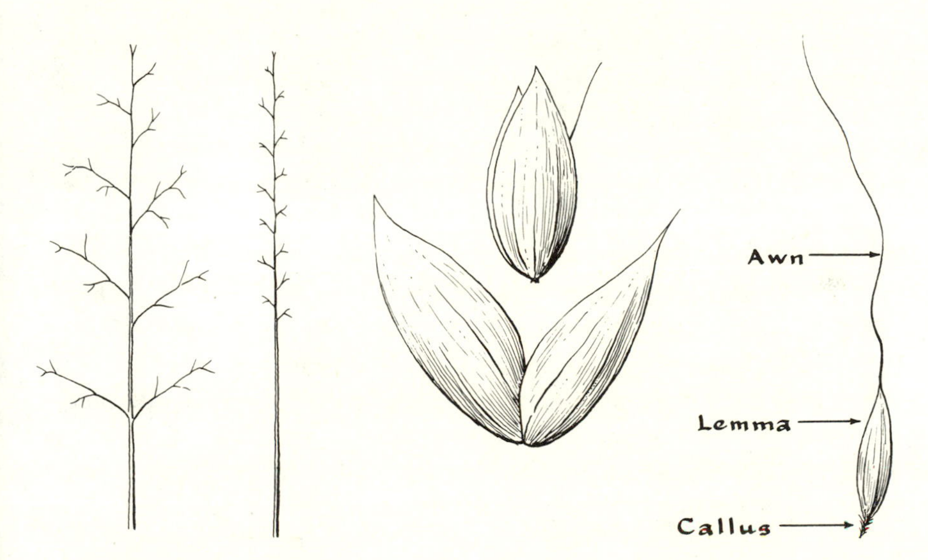

124 Tribe *Agrostideae,* (left) inflorescence types, (middle) spikelet type, (right) indurate lemma and fruit of *Stipa.*

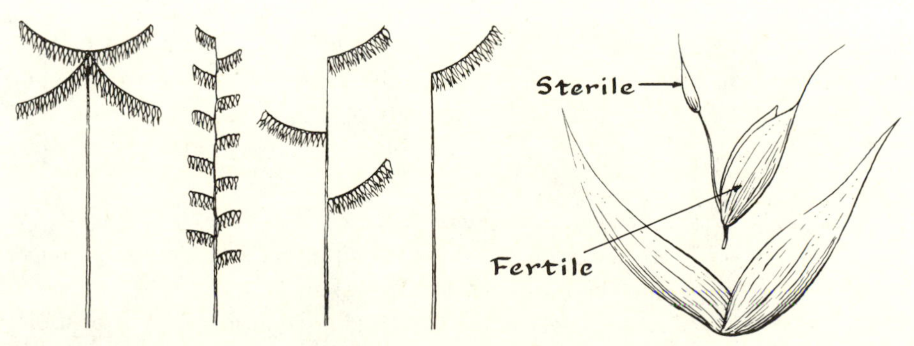

125 Tribe *Chlorideae,* inflorescence types and spikelet type.

Tribe VII. *Zoysieae* (Curly Mesquite Tribe): Spikelets sessile or subsessile, in groups, the whole group falling entire from the rachis at maturity, the articulation thus below the glumes and the inflorescence a symmetrical spike. The typical arrangement is to have the central spikelet of each group fertile while the two lateral spikelets are sterile.

This is a small tribe, probably not a natural one, and is found chiefly in desert regions of the Southwest. It includes some valuable forage grasses. There are 5 genera in the United States.

EXAMPLES: *Hilaria belangeri:* Curly Mesquite.
Hilaria jamesii: Galleta.

Tribe VIII. *Phalarideae* (Canary Grass Tribe): Spikelets with 1 perfect terminal floret and 2 sterile florets below it (these may be either staminate or empty lemmas, sometimes normal in size and sometimes much reduced). The inflorescence is an open or narrow panicle, sometimes spike-like, but never a true spike. Articulation above the glumes.

A small tribe of 6 genera, 3 of them in the United States. None of the species is abundant enough to furnish much forage.

EXAMPLES: *Phalaris canariensis:* Canary Grass, sometimes grown for ornament and for bird seed.
Hierochloë odorata: Sweet Grass, Vanilla Grass, or Seneca Grass, with an odor like vanilla and sometimes used by Indians in the making of baskets.

Tribe IX. *Oryzeae* (Rice Tribe): Spikelets 1-flowered, falling entire from the pedicels, the articulation thus below the glumes. Glumes very small or lacking. Inflorescence a panicle. Stamens 6. Plants chiefly aquatic.

A small tribe, of doubtful affinities, including 2 genera in the United States.

EXAMPLES: *Oryza sativa:* cultivated Rice, one of the most important food plants of the world, being the chief food of more people than any other single plant.
Leersia oryzoides: Rice Cutgrass, a common species of moist places in North America.

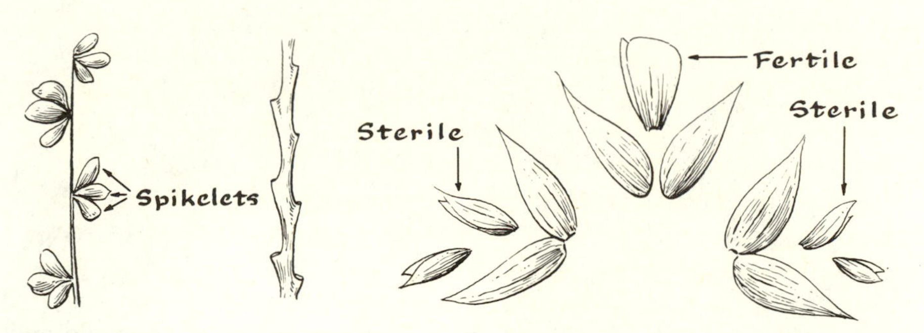

126 Tribe *Zoysieae,* (left) inflorescence, (middle) rachis, (right) spikelet group.

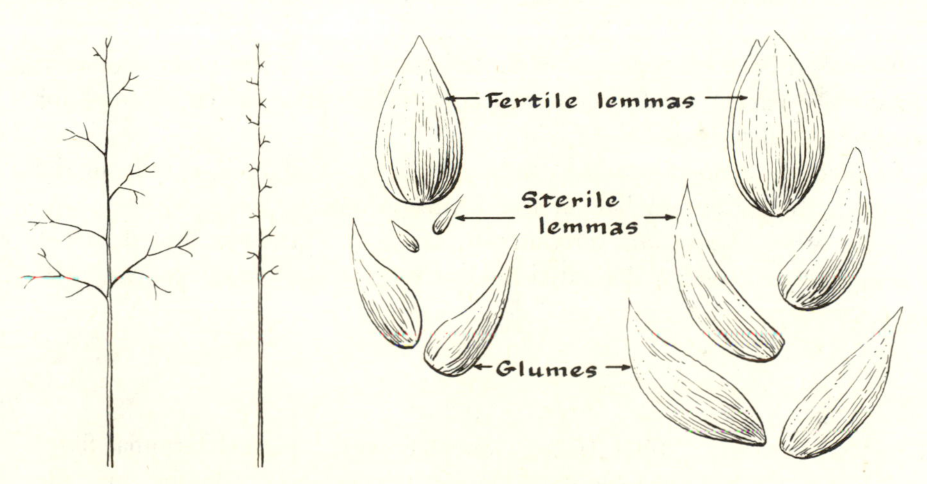

127 Tribe *Phalarideae,* inflorescence types and spikelet types.

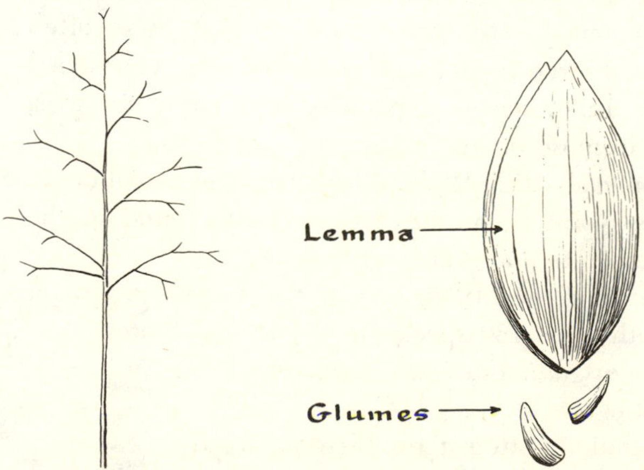

128 Tribe *Oryzeae,* inflorescence type and spikelet type.

Tribe X. *Zizanieae* (Indian Rice Tribe): Plants monoecious, the male and female spikelets on the same plant but in different parts of the panicle. Spikelets 1-flowered, falling entire, the articulation thus below the glumes. Glumes small or obsolete. Plants aquatic, mostly rather tall.

A small tribe, with 5 genera in the United States, of no great economic significance except for the genus *Zizania,* which is a source of food for wild fowl and Indians.

EXAMPLES: *Zizania aquatica:* Indian Rice, or Wild Rice, mainly in the eastern half of the United States.

Hydrochloa caroliniensis: of the southeastern United States, said to be eaten by livestock.

Tribe XI. *Paniceae* (Millet Tribe): Spikelets with 1 perfect terminal floret and 1 sterile floret below it. Rachilla joints very short. Glumes thin, the first very small (suppressed in some genera), the second glume normal and similar to the sterile lemma in size and texture so that these often appear like a pair of glumes. Fertile lemma and palea indurate (hardened and attached to the grain). Inflorescence a panicle, or 1–many racemes, which may be digitate or racemose on the main axis.

A large and often taxonomically difficult tribe, with a wide distribution. There are 80 genera in the world, some 20 of these being found in the United States, mainly in temperate and warm regions.

EXAMPLES: *Panicum:* Panicum, or Millet, one of the largest genera of grasses, with some 160 species in the United States.

Echinochloa crusgalli: Barnyard grass.

Setaria: Bristlegrass.

Cenchrus: Sandbur, often a troublesome weed.

Digitaria: Crabgrass, weedy plants, sometimes cut for hay.

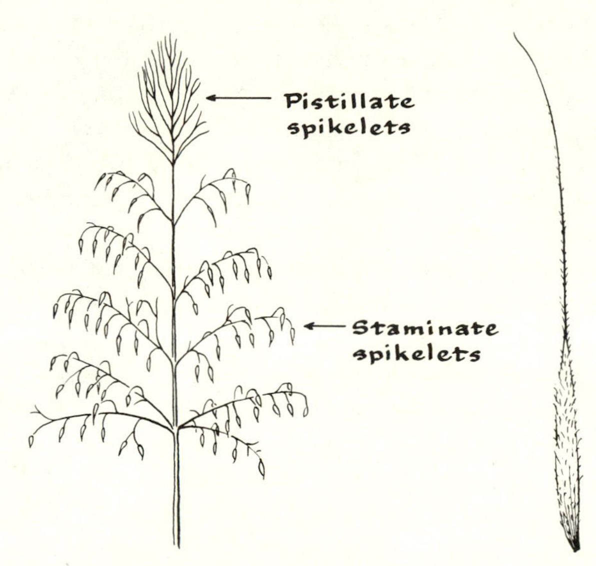

129 Tribe *Zizanieae*, inflorescence and pistillate floret.

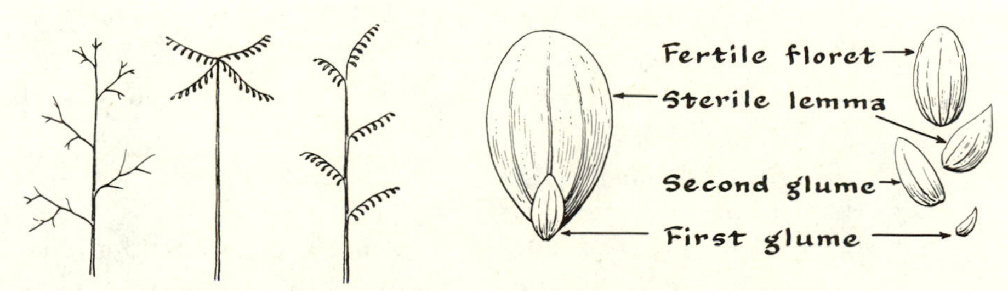

130 Tribe *Paniceae*, (left) inflorescence types, (middle) spikelet closed, (right) spikelet opened to show structure.

Tribe XII. *Andropogoneae* (Sorghum Tribe): Spikelets in pairs, the usual arrangement being as follows: one perfect and sessile, the other sterile or reduced and pedicellate, both borne on a jointed rachis, each pair falling together with a joint of the rachis. Fertile spikelets with 1 perfect terminal floret and a sterile floret below it. Glumes indurate, enclosing the florets. Lemmas very thin. Palea suppressed. (The genus *Imperata*, with narrow silky panicles, is exceptional in having all spikelets alike and fertile.)

A tribe of about 80 genera in the world, with 16 genera in the United States. It includes some important forage grasses, several cultivated crops, and a few ornamental grasses.

EXAMPLES: *Andropogon:* Beardgrass, Bluestem, or Turkeyfoot, some species being the characteristic ingredients of tall-grass prairies.

Saccharum officinarum: Sugar cane.

Sorghum vulgare: Sorghum.

Miscanthus sinensis: Eulalia, a cultivated ornamental.

Tribe XIII. *Tripsaceae* (Corn Tribe): Plants monoecious. Spikelets unisexual, the male and female spikelets in separate inflorescences or in different parts of the same inflorescence. Staminate spikelets in 2's or 3's, 2-flowered. Pistillate spikelets in 2's or, usually, single, 2-flowered, the lower floret sterile, embedded in a thickened axis (cob) or enclosed within a thickened sheath. Glumes and lemmas awnless.

A small tribe, related to the *Andropogoneae*, including 7 genera, 4 of them in the United States. Some are good forage grasses, though seldom common enough to be of significance, but Maize, or Indian Corn, is an important field crop in the United States.

EXAMPLES: *Zea mays:* Maize, or Indian Corn, with several varieties that are termed "agricultural species" by Sturtevant.[1]

Euchlaena: Teosinte, of interest as a possible ancestor of Indian Corn.

Tripsacum: Gama grass, another possible ancestor of Indian Corn.

Coix lacryma-jobi: Job's-tears, sometimes cultivated for ornament.

[1] E. L. Sturtevant: *Varieties of Corn,* U.S.D.A. Exp. Sta. Bull. 57 (1899).

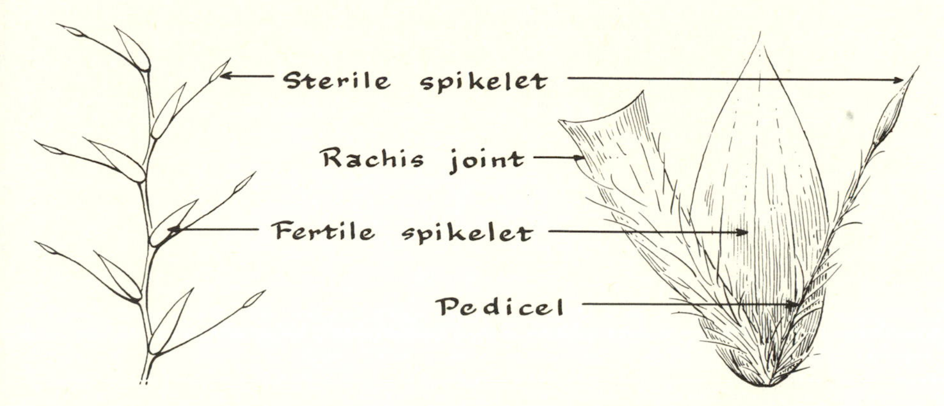

131 Tribe *Andropogoneae,* diagram of arrangement of spikelets of *Andropogon* and spikelet pair of *Andropogon.*

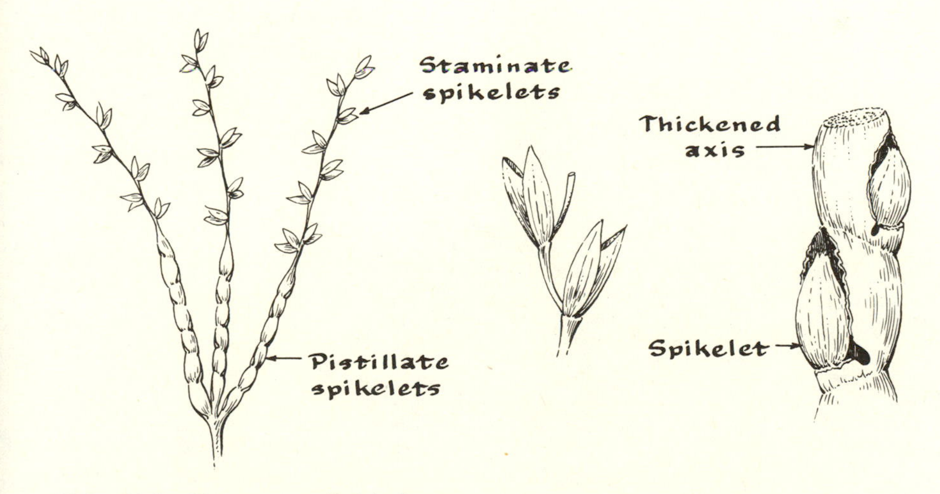

132 Tribe *Tripsaceae,* (left) diagrammatic representation of inflorescence of *Tripsacum,* (middle) staminate spikelets of *Tripsacum,* (right) pistillate spikelets of *Tripsacum.*

Part III

SELECTED ORDERS AND FAMILIES OF DICOTYLEDONS

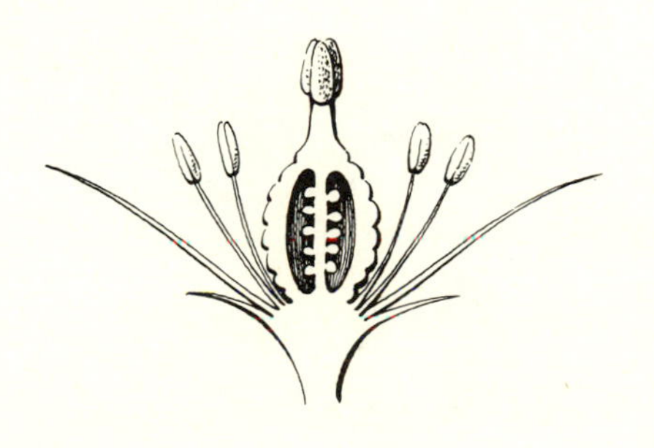

THE APETALAE

This is generally regarded as an artificial group of orders and families, probably derived from various members of the *Polypetalae*. The flowers have no petals, with the possible exception of certain members of the Polygonales such as *Rumex*, in which there are two sets of perianth segments, an outer and an inner set, both sepaloid, but the inner set probably better called petals. The calyx may or may not be present, and it is sometimes petaloid.

The Amentiferae

This group includes all dicotyledons having some or all of their flowers produced in catkins, or aments. The plants are woody, varying from large trees to ordinary or diminutive shrubs.

Key to the Families of Amentiferae

Leaves pinnately compound, without stipules; plants aromatic; fruit a bony nut completely enclosed in a leathery husk . . . . JUGLANDACEAE

Leaves simple and stipulate; plants not aromatic; fruit naked or partly or completely enclosed in an involucre

Fruit a capsule; seeds hairy; plants dioecious . . . . SALICACEAE

Fruit a nut or nutlet; seeds not hairy; plants monoecious

Male and female flowers in catkins (in *Corylus* the female catkins very small and bud-like); leaves often thin and often doubly serrate; fruit a nut or nutlet enclosed within or subtended by foliaceous or herbaceous bracts, the nutlets often winged . . . . BETULACEAE

Male flowers in catkins, female flowers in clusters or solitary; leaves leathery, not doubly serrate but often lobed; fruit a nut enclosed or partly enclosed within a cup or bur of hard, prickly or scaly bractlets, the nut never winged . . . . FAGACEAE

ORDER SALICALES

SALICACEAE: Willow Family

Trees or shrubs (rarely herbaceous) with simple, alternate, stipulate leaves. Flowers dioecious, in erect or pendulous catkins (aments), the calyx much reduced or lacking, petals none, and stamens 2 or more. Ovary 1-celled, with parietal placentation and numerous ovules. Seeds covered with long hairs. Fruit a 2–4-valved capsule.

There are 4 genera and some 200 species in the world, widely distributed; common in northern and mountainous areas, but absent from the Malay Archipelago and Australia. Only two genera, *Salix* and *Populus,* occur in North America.

EXAMPLES: *Salix babylonica:* Weeping Willow, an ornamental.
Populus alba: White or Silver Poplar.
Populus nigra var. *italica:* Lombardy Poplar.
Populus tremuloides: Quaking Aspen.

NOTE: Winter twigs are useful in the identification of woody plants and are often distinctive as to genus and species. A comparison between twigs of *Salix* and *Populus* will show that *Populus* has a terminal bud that produces ring-like scars at intervals on the twigs, marking each year's growth; and the buds of *Populus* are protected by several scales. Twigs of *Salix* have no terminal buds and no terminal bud scars, and the buds are protected by a single scale.

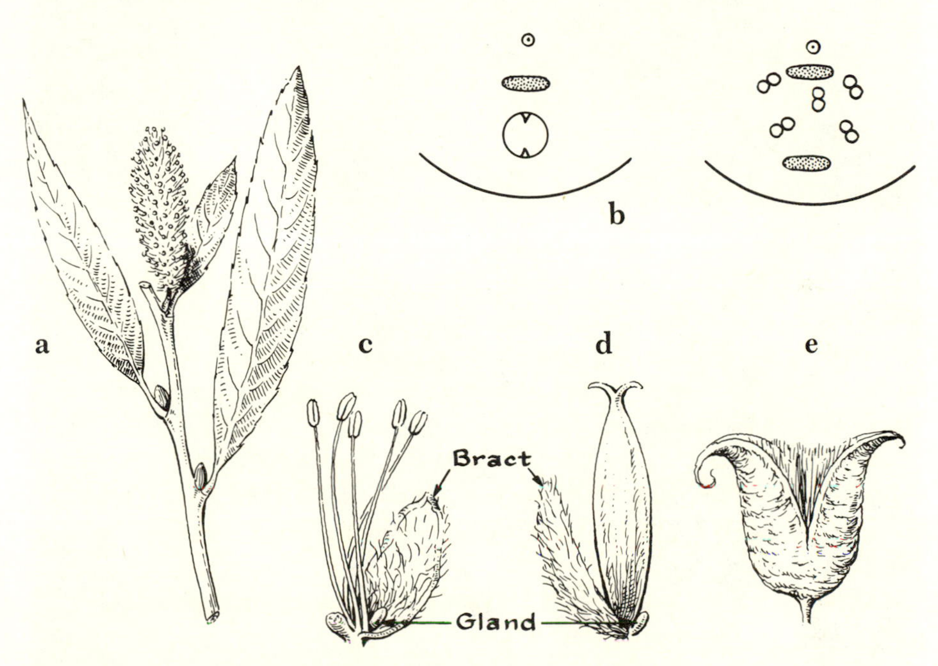

133 *Salix,* (**a**) twig with catkin, (**b**) floral diagrams of male and female flowers, (**c**) male flower, (**d**) female flower, (**e**) capsule.

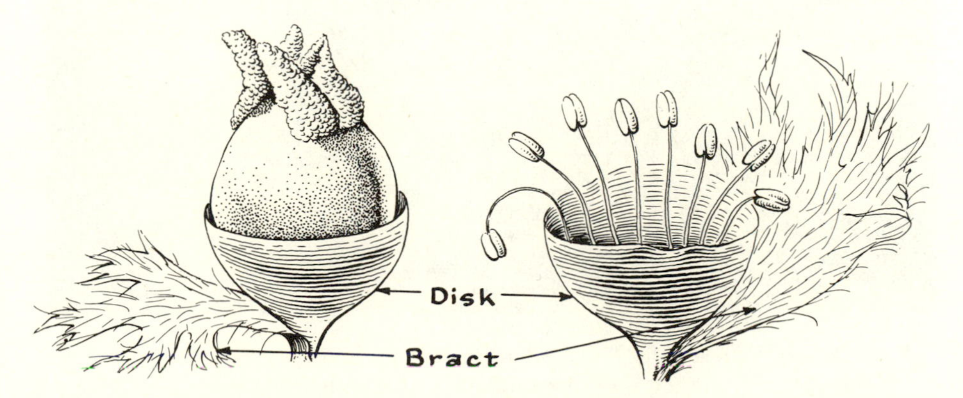

134 *Populus,* (left) female flower, (right) male flower.

135 *Salix ligulifolia* (Ball) Ball, catkins in flowering stage, with the staminate on the left and the pistillate on the right. This is a precocious willow; that is, it blooms before the leaves appear.

136 *Populus tremuloides* Michx., Trembling Aspen, showing pendulous staminate catkins.

ORDER JUGLANDALES

JUGLANDACEAE: Walnut Family

Mostly resinous and aromatic trees with alternate, pinnately compound leaves without stipules. Flowers monoecious, the male in pendulous catkins, the female in few-flowered clusters or, less commonly, in pendulous catkins. Male flower subtended by a bract and a pair of bracteoles, and often there are several scale-like parts that may represent a reduced perianth; the stamens indefinite in number and arrangement, 3–100. Female flowers perigynous, with an adnate hypogynous hypanthium, the sepals 0–4, the pistil single, with a 1–4-celled ovary, short style, and usually 2 plumose stigmas. Fruit a single-seeded nut surrounded by the fleshy or leathery hypanthium husk, which may be indehiscent or may split longitudinally into 4 valves.

The family includes about 6 genera and 60 species, mostly in north temperate regions and in the mountains of the northern tropics, but the fossil record indicates that in the Cretaceous and Tertiary periods the family had a much wider distribution, extending into Greenland and Alaska. Two genera and about 19 species are native to the United States.

EXAMPLES: *Juglans regia:* English Walnut, widely distributed in Eurasia.
Juglans nigra: Black Walnut, native to the United States and a valuable timber tree.
Juglans cinerea: Butternut, or White Walnut.
Carya ovata: Shagbark Hickory.
Carya illinoensis: Pecan, cultivated for the fruit.

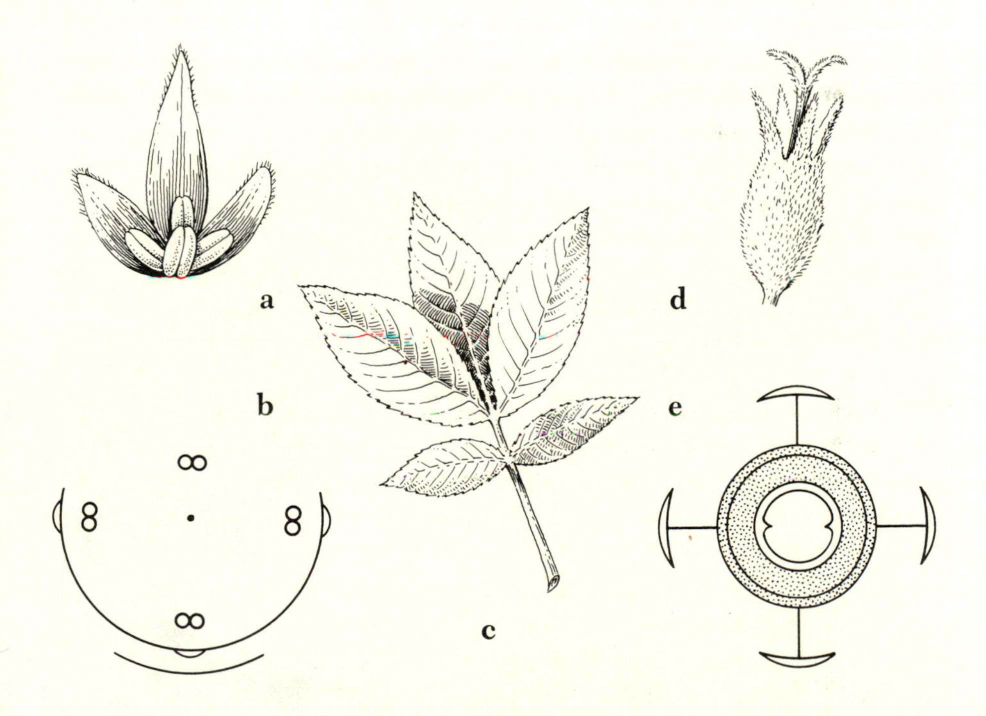

137 *Carya ovata,* (**a**) male flower, (**b**) floral diagram of male flower, (**c**) leaf, (**d**) female flower, (**e**) floral diagram of female flower.

ORDER FAGALES

Trees or shrubs with simple, alternate, stipulate leaves. Flowers monoecious, at least the staminate in catkins. Fruit a nut or nutlet.

BETULACEAE: Birch Family

Trees or shrubs with simple, alternate, stipulate leaves, which are generally thin and often doubly serrate. Flowers monoecious, both sexes usually in catkins (the inflorescence is theoretically composed of an elongated axis along which there are spirally arranged condensed dichasia including primary, secondary, and tertiary bracts), the female catkins few-flowered in *Corylus*. Perianth rudimentary or none; when present in female flowers, adnate to the 2-celled, 4-ovuled ovary with 2 styles. Fruit a 1-seeded nut or nutlet, often winged, enclosed or subtended by foliaceous bracts.

The family includes 6 genera and about 110 species, mostly in the cooler parts of the northern hemisphere but in earlier geologic time more numerous and widespread. Five genera and about 25 species occur in the United States.

EXAMPLES: *Betula lutea:* Yellow Birch, an important timber tree.
Betula papyrifera: Paper Birch, or Canoe Birch.
Alnus: Alder.
Corylus: Hazelnut.
Carpinus caroliniana: Hornbeam, or Blue Beech.
Ostrya virginiana: Hop Hornbeam, or Ironwood.

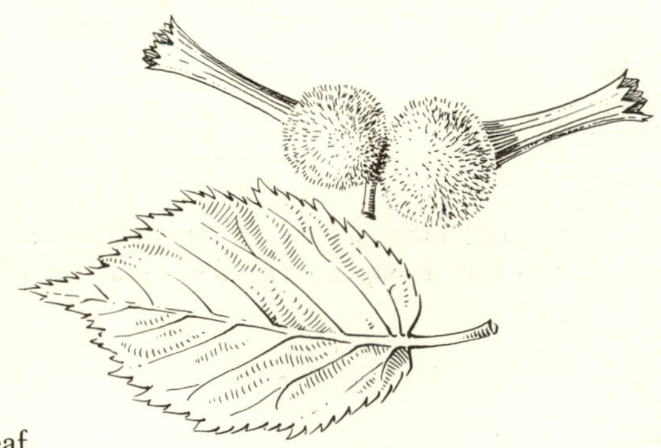

138 *Corylus cornuta,* leaf and pair of fruits inside an involucre.

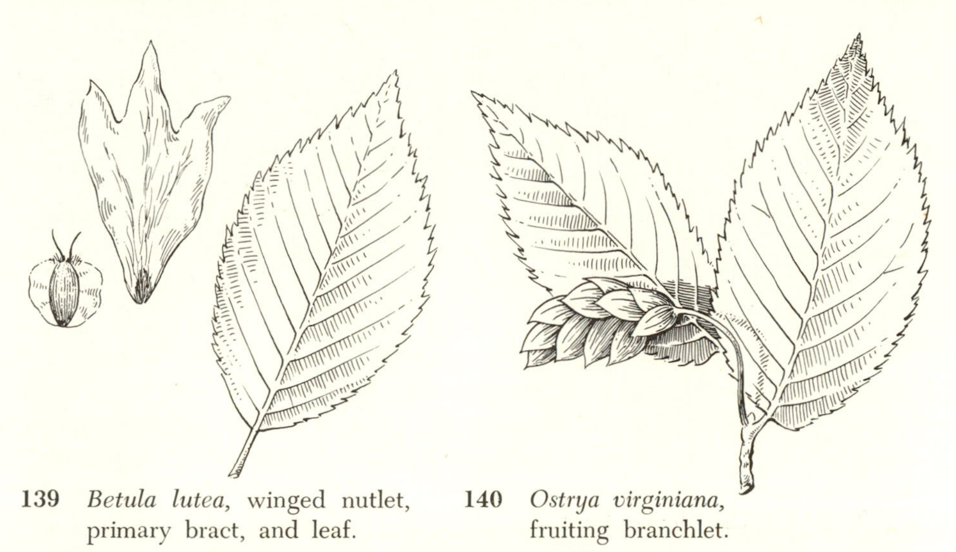

139 *Betula lutea,* winged nutlet, primary bract, and leaf.

140 *Ostrya virginiana,* fruiting branchlet.

141 *Betula occidentalis* Hook. Portion of a branch showing pistillate catkin on left, and pendulous staminate catkins lying on top of another pistillate catkin on the right.

FAGACEAE: Beech Family

Trees or shrubs with simple, alternate, stipulate leaves, which are often strongly ribbed and may be deeply lobed, the texture usually leathery. Flowers monoecious, the male usually in catkins (in globose heads in *Fagus*), the female in few-flowered clusters. Male flowers with a 4–7-lobed calyx and 4 or more stamens. Female flowers with a 4–8-lobed calyx and a single pistil, usually of 3 carpels, with 3 styles and stigmas. Fruit a nut, 1-seeded by the abortion of the other ovules, and partially or completely enclosed by an involucre of scales or a spiny bur.

The family includes 9 genera and about 600 species, widely distributed, but mostly in temperate regions. Five genera and about 90 species occur in the United States, some of them valuable timber trees.

EXAMPLES: *Fagus:* Beech.

Quercus: Oak, a large and taxonomically difficult genus of over 200 species, some of them of great economic impor tance for lumber and cork.

Castanea: Chestnut.

Castanopsis chrysophylla: Giant Chinquapin, of the West Coast of the United States.

Lithocarpus densiflora: Tanbark Oak, of the West Coast of the United States.

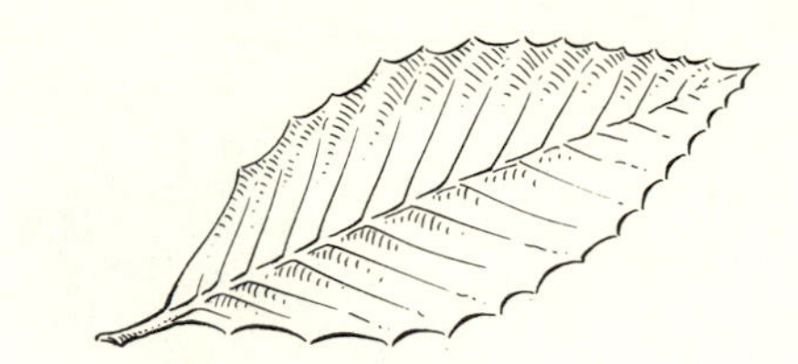

142 *Fagus grandifolia,* leaf.

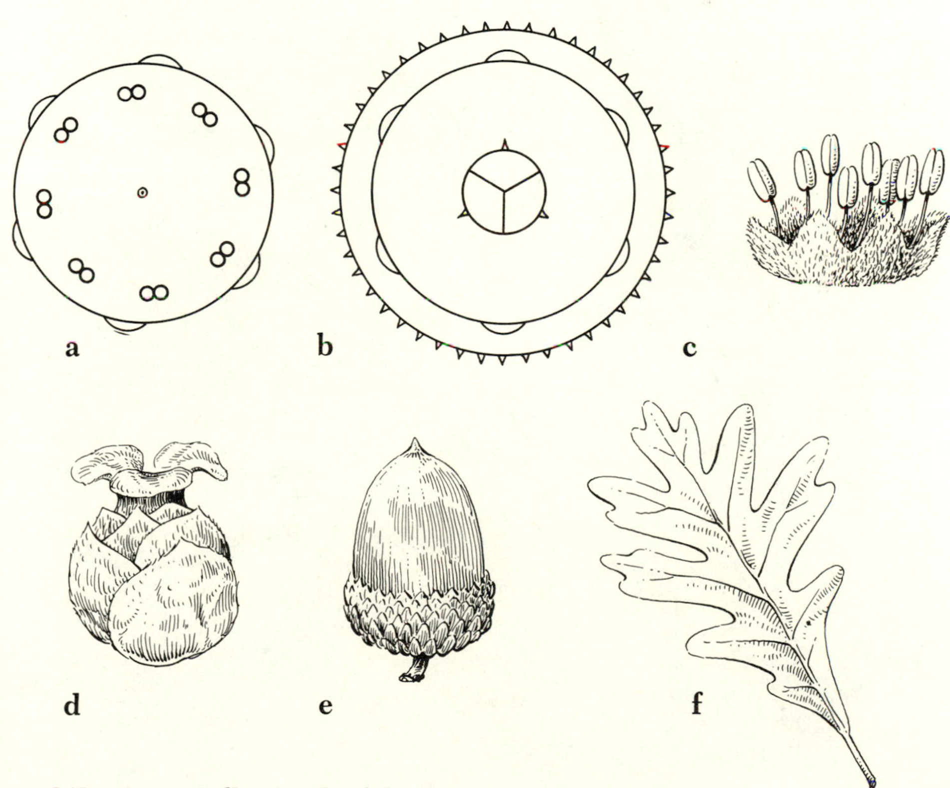

143 *Quercus alba,* (**a**) floral diagram of male flowers, (**b**) floral diagram of female flower, (**c**) male flower, enlarged, (**d**) female flower, enlarged, (**e**) fruit, (**f**) leaf.

144 *Castanea dentata* Borkh., American Chestnut, showing staminate and pistillate flowers.

The Floriferae

This group includes those members of the *Apetalae* whose flowers are produced, not in catkins, but in various other types of inflorescences. The plants may be herbaceous or woody.

Key to the Families of Floriferae

Juice milky; fruit a multiple . . . . MORACEAE

Juice not milky; fruit simple

Trees with unsymmetrical leaves; fruit a samara or a drupe . . . . ULMACEAE

Herbs, vines, or shrubs with symmetrical leaves; fruit otherwise

Plants aromatic . . . . CANNABINACEAE

Plants not aromatic

With stinging hairs or at least mostly opposite leaves; inflorescence of numerous, minute, green flowers, axillary . . . . URTICACEAE

Without stinging hairs, the leaves alternate; inflorescence sometimes of green flowers but not all axillary

Fruit not an anthocarp, not surrounded by the withered remains of the perianth tube; stamens mostly opposite the sepals or the sepals lacking

Embryo straight; plants not mealy; perianth petaloid or else in two series (*Rumex*); stipules sheathing the stem or the flowers usually involucrate (*Eriogonum*); fruit usually a triangular akene, rarely a lenticular akene . . . . POLYGONACEAE

Embryo curved or coiled; plants often mealy; perianth green or scarious when present, always in 1 series; stipules lacking; flowers not involucrate; fruit a utricle, pyxis, or rounded akene

Inflorescence without prickly bracts, or if prickly then not scarious . . . . CHENOPODIACEAE

Inflorescence with scarious, prickly bracts . . . . AMARANTHACEAE

Fruit an anthocarp, surrounded by the withered remains of the perianth tube; stamens alternate with the sepals . . . . . NYCTAGINACEAE

ORDER URTICALES

Herbs, shrubs, or trees with reduced flowers, which are small, greenish, apetalous or, rarely, naked, perfect to polygamous, monoecious, or dioecious, but not produced in catkins. Stamens variable, but mostly of the same number as the sepals and opposite them. Carpels 1–2, but the fruit with a single seed and becoming a samara, nutlet, akene, or drupe, or sometimes a multiple.

The order includes the following 4 families, of ancient origin and dubious affinities.

COMPARISON OF FAMILIES OF URTICALES

	Ulmaceae	*Urticaceae*	*Cannabinaceae*	*Moraceae*
Growth form	trees and shrubs	herbs, shrubs or trees in the tropics	herbs or vines, not woody	nearly all trees or shrubs, except *Dorstenia*
Plants with milky juice	no	no	no	yes
Plants aromatic	no	no	yes	no
Leaves alternate or opposite	alternate	alternate or opposite	alternate or opposite	alternate, rarely opposite
Flowers *pol*ygamous, *mon*oecious, *di*oecious, or *perf*ect	perf, pol, mon	mon or di, or sometimes perf	mostly di	mon or di
Fruit type	samara or drupaceous	akene, nutlet, or sometimes drupaceous	akene or nutlet	usually a multiple

Key to the Families of Urticales

Mostly trees or shrubs; fruit not a simple akene or nutlet

  Juice milky; fruit usually multiple . . . . MORACEAE

  Juice not milky; fruit a samara or a drupe . . . . ULMACEAE

Herbs or vines, woody only in the tropics; fruit an akene or nutlet, or sometimes drupaceous

  Plants aromatic, without stinging hairs . . . . CANNABINACEAE

  Plants not aromatic, often with stinging hairs . . . . URTICACEAE

ULMACEAE: Elm Family

Trees or shrubs, rarely herbs, with simple, alternate, stipulate leaves having blades that are often inequilateral at the base. Flowers perfect or polygamo-monoecious, the calyx with a very short or, often, a funnelform tube below the 4–9 lobes. Petals none. Stamens 4–6. Pistil 1, the ovary usually 1-celled and 1-seeded, the stigmas 2. Fruit a samara, drupe, or nutlet.

The family includes about 15 genera and 150 species, mostly in temperate regions. There are 5 genera in the United States, including some valuable timber and shade trees.

EXAMPLES: *Ulmus americana:* American, or White, Elm.
Celtis occidentalis: Hackberry, or Nettle Tree.
Planera aquatica: Planer Tree.

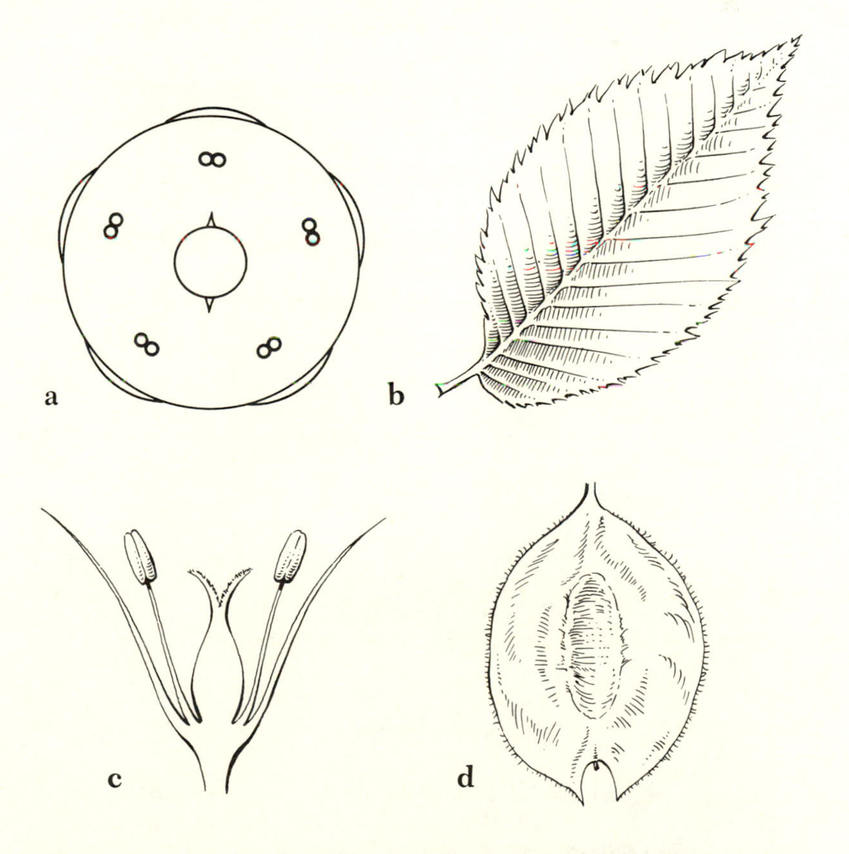

145 *Ulmus americana,* (**a**) floral diagram, (**b**) leaf, (**c**) longitudinal section of flower, enlarged, (**d**) fruit, enlarged.

URTICACEAE: Nettle Family

Annual or perennial herbs in ours, but arborescent in some genera in the tropics, with simple, alternate or (mostly) opposite, stipulate, toothed leaves and watery juice. Stems and leaves often armed with stinging hairs. Flowers small, greenish, variously clustered, often axillary, apetalous, monoecious, dioecious, or polygamous, usually 4–5-merous, the stamens as many as the sepals and opposite them. Pistil 1, with a 1-celled ovary and 1 ovule. Fruit an akene or sometimes drupaceous.

The family is mainly tropical; it includes about 42 genera and 600 species.

EXAMPLES: *Urtica:* Nettle, with about 50 species, widely distributed in both hemispheres.
Boehmeria nivea: Ramie, grown for its long fibers.
Laportea: includes Tree Nettles of the tropical Pacific.
Pilea: Clearweed, a large and mainly tropical genus.

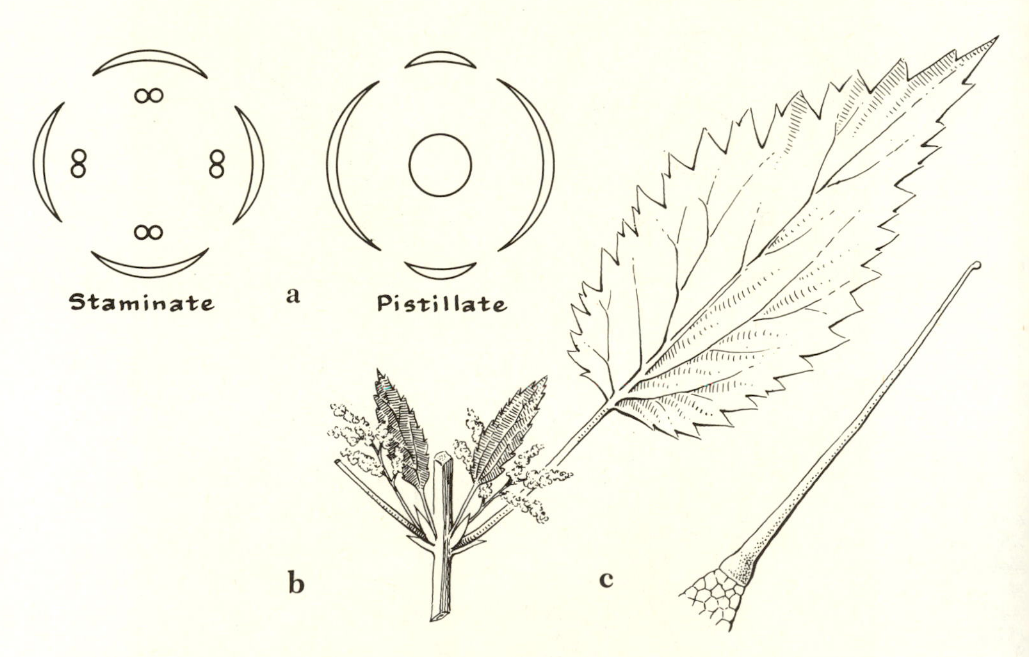

146 *Urtica dioica,* (**a**) floral diagrams of staminate and pistillate flowers, (**b**) portion of plant, (**c**) stinging hair, greatly magnified, showing the secretory cells at the base and a minute bulbous tip, which breaks off when touched to release the fluid within.

147 *Urtica dioica* L. var. *procera* (Muhl. ex Willd.) Wedd., a common Nettle, showing the upper part of the plant in bloom.

CANNABINACEAE: Hemp Family

Erect or climbing aromatic herbs with watery juice and alternate or often opposite leaves, which are simple and palmately veined or else palmately compound, the margins coarsely toothed. Stipules present and persistent. Flowers usually dioecious, apetalous, axillary; the staminate flowers in panicles, each with a usually 5-parted calyx and 5 stamens, the pistillate flowers in crowded bracteate spikes, each with a single sepal, which envelops the single 1-celled, 1-ovuled ovary. Fruit an akene, which is surrounded by the persistent perianth.

The family includes 2 genera and 3 species, widely distributed and mostly of considerable economic significance. Some authors merge this family with the *Moraceae;* others include it in the *Urticaceae.*

EXAMPLES: *Cannabis sativa:* Hemp. The plants are often cultivated for the strong fibers of the stem. It is also the source of marijuana and hashish, which are superior sexual intoxicants.

Humulus lupulus: Hop Vine, commonly used in the brewing industry. Numerous yellow glandular hairs on the bracteoles secrete lupulin, and to this the plants owe their economic value.

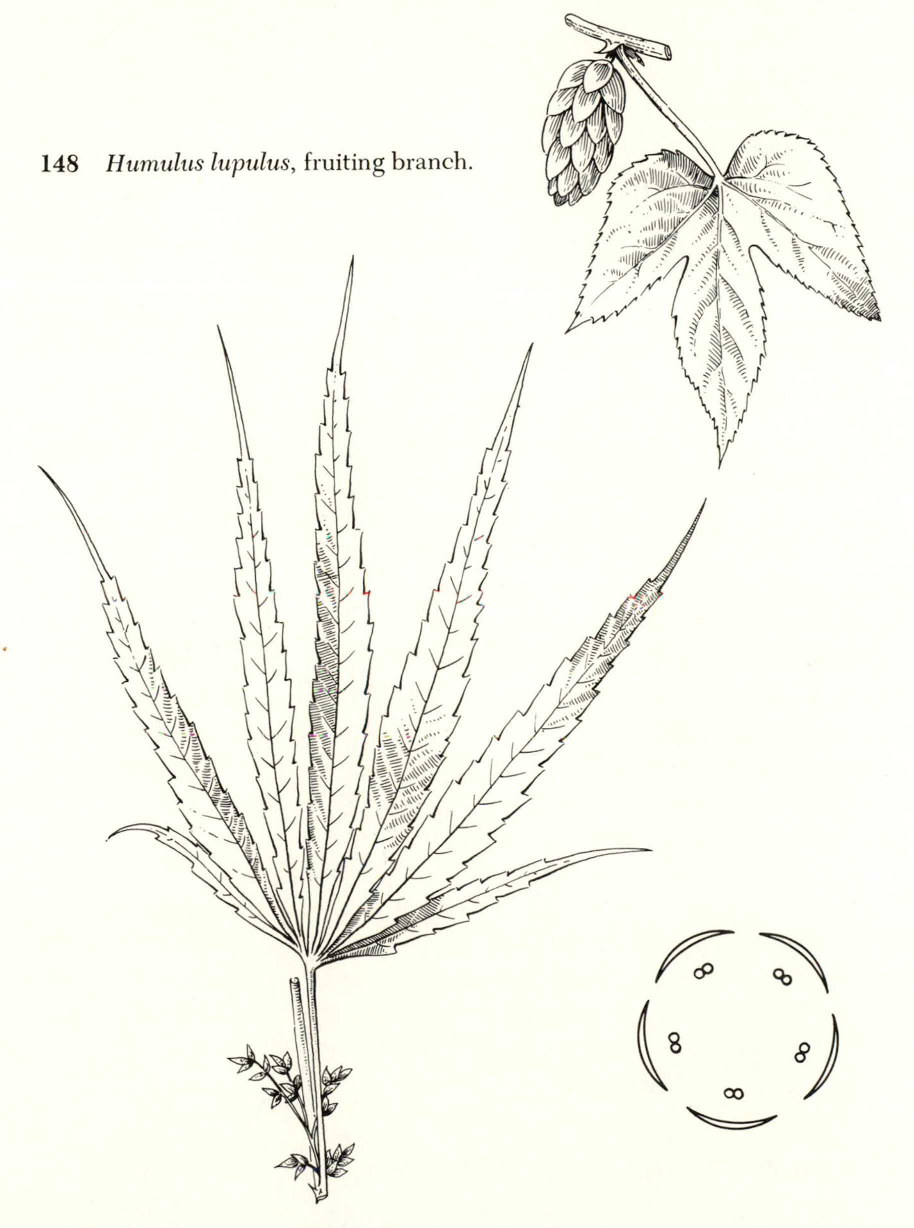

148 *Humulus lupulus*, fruiting branch.

149 *Cannabis sativa,* fruiting branch an floral diagram of staminate flowers.

MORACEAE: Mulberry Family

Trees or shrubs, rarely herbs, with milky sap, and alternate, simple, stipulate leaves. Flowers small, monoecious or dioecious, often in dense clusters, commonly 4-merous. Petals none. Pistil of 2 carpels, but one of these is often abortive and represented only by a style or rudiment. Fruit an akene or a drupe, invested by a fleshy perianth or axis and variously grown together to form a multiple fruit or a syconium (a hollow floral axis inside which the minute flowers are produced and which becomes fleshy at maturity as in a Fig).

A large family of some 55 genera and perhaps 2,300 species, mostly in tropical regions, but a few extend into the temperate zone.

EXAMPLES: *Morus:* Mulberry, with 10 species, some being used as food for silkworms, and the fruit sweet and edible.

Ficus: Fig, a genus of about 2,000 species,[1] including the well-known Banyan Tree (*F. benghalensis*) and the edible Fig (*F. carica*).

Artocarpus: a genus of some 40 species, including the Breadfruit (*A. incisa*).

Maclura pomifera: Osage Orange of North America.

[1] E. D. Merrill: "Some Economic Aspects of Taxonomy," *Torreya* 43:50–64 (1943). "Approaching 2,000 species even without splitting hairs on specific differences."

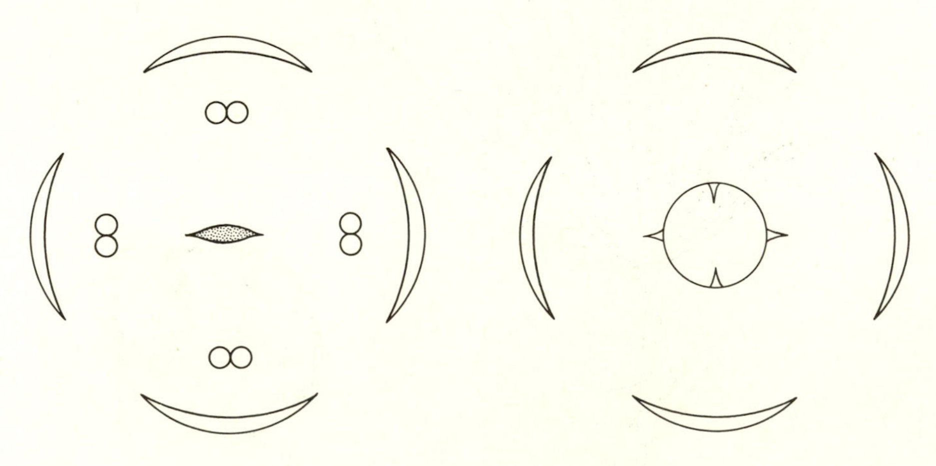

150 *Morus*, floral diagrams of staminate (left) and pistillate (right) flowers.

151 *Ficus benghalensis* L., Banyan Tree, of tropical Africa and India, rooting from the branches to form additional, accessory trunks. This is a small example.

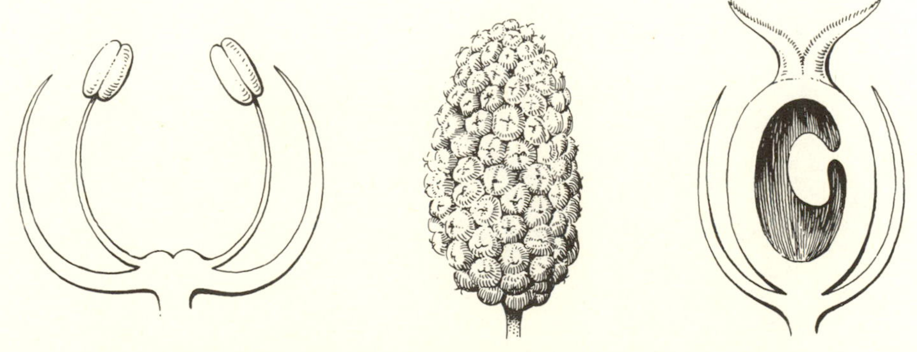

152 *Morus*, (left) longitudinal section of staminate flowers, (middle) multiple fruit, (right) longitudinal section of pistillate flower.

ORDER POLYGONALES

POLYGONACEAE: Buckwheat Family

Herbs, in ours, often with sour juice, and with mostly alternate, simple, entire leaves, with stipules forming sheaths (ocreae) around the stem at the joints (these lacking in *Eriogonum* and *Brunnichia*). Flowers small, often numerous, mostly perfect but sometimes polygamous or dioecious, the perianth often petaloid and in a single series, or sepaloid and in two series, consisting of 4–6 parts, sometimes persistent, and sometimes tubular or funnelform. Stamens 4–12, free or inserted at the base of the perianth tube. Pistil 1, the ovary 1-celled and 1-ovuled, the styles 2–3. Fruit a lenticular or triangular akene. Seed with a straight embryo.

The family includes about 32 genera and 700 species, widely distributed.

EXAMPLES: *Fagopyrum sagittatum:* Buckwheat.
Rheum rhaponticum: Rhubarb.
Rumex: Dock.
Eriogonum: False Buckwheat.

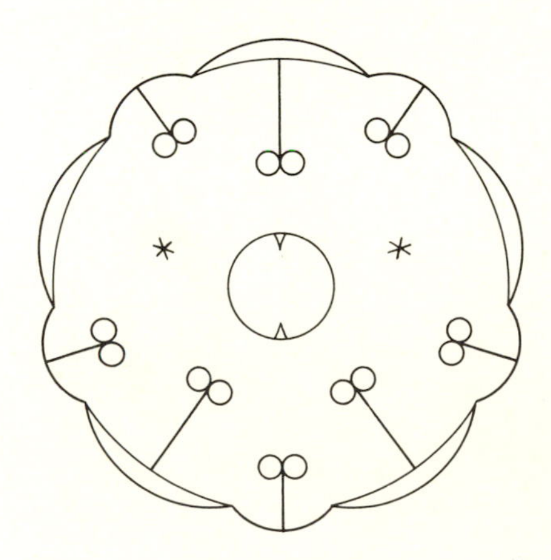

153 *Polygonum pensylvanicum*, floral diagram.

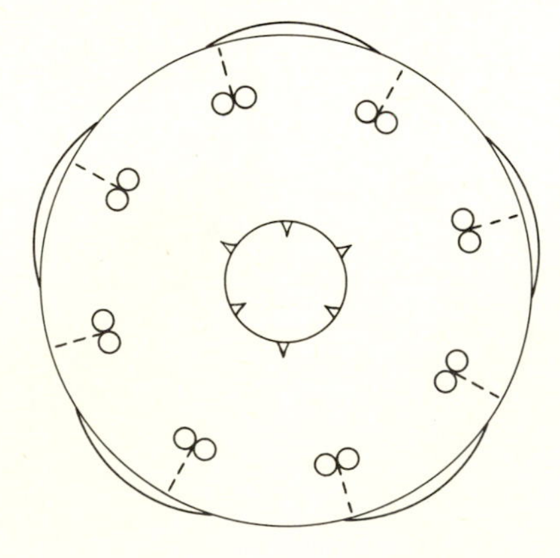

154 *Fagopyrum*, floral diagram.

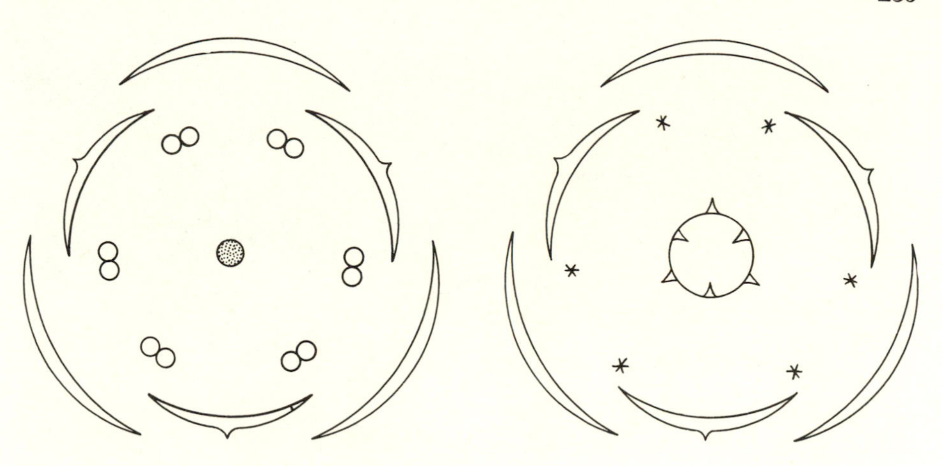

155 *Rumex acetosella*, floral diagrams of staminate (left) and pistillate (right) flowers.

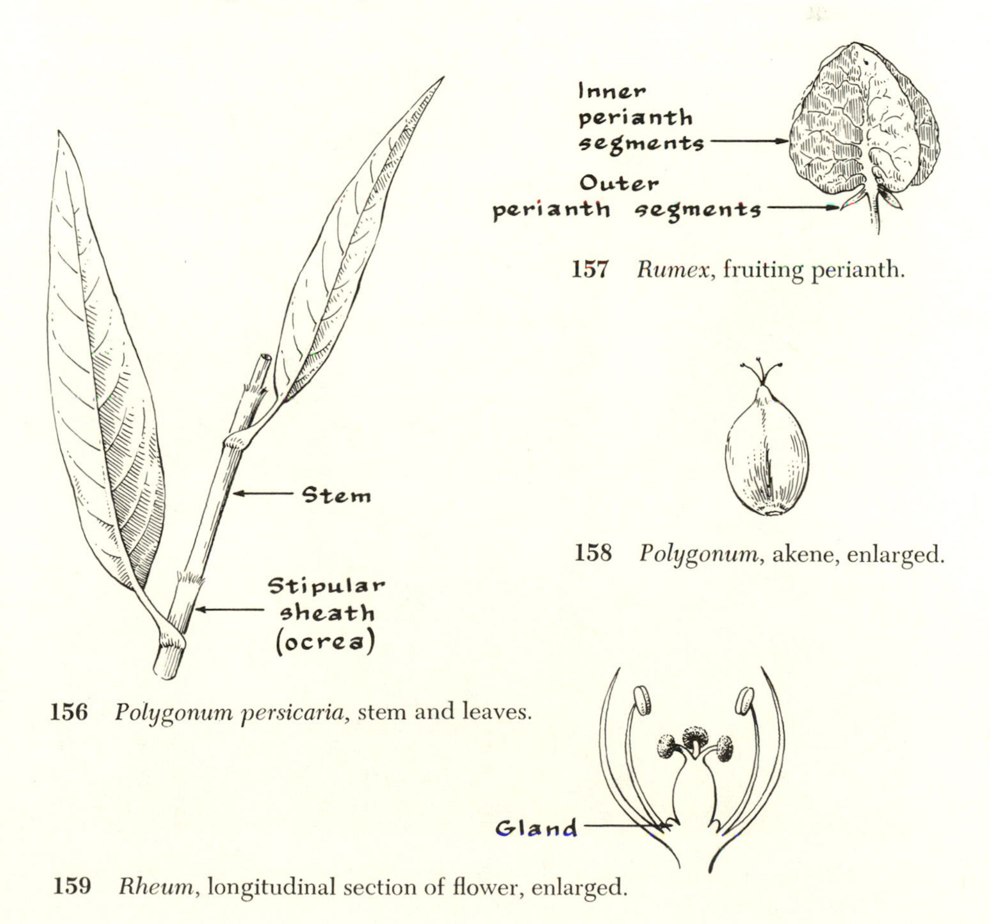

157 *Rumex*, fruiting perianth.

158 *Polygonum*, akene, enlarged.

156 *Polygonum persicaria*, stem and leaves.

159 *Rheum*, longitudinal section of flower, enlarged.

160 *Polygonum viviparum* L., a low, circumboreal species. The lower flowers in the spike are modified into bulblets.

161 *Rumex crispus* L., Sour Dock, a common weed. This shows the early fruit stage.

ORDER CHENOPODIALES

Mostly weedy herbs, sometimes shrubs, of dry regions, with alternate or opposite, simple, exstipulate leaves, and small, often densely clustered, greenish flowers without petals. Gynoecium composed of from 1 to several carpels, free or connate, with basal placentation, the seeds from 1 to several, with the embryo usually curved or coiled round the endosperm, rarely with a straight embryo.

The order includes 6 families as here defined: *Phytolaccaceae, Cynocrambaceae, Chenopodiaceae, Batidaceae, Amaranthaceae,* and *Basellaceae,* these being the apetalous members of the group of families often referred to as the *Centrospermae,* which also includes the related order *Caryophyllales* in the polypetalae. This relationship is best shown by the similarity of placentation and seed morphology.

CHENOPODIACEAE: Goosefoot Family

Annual or perennial, often halophytic herbs or shrubs, which are commonly mealy, with mostly alternate, simple, exstipulate leaves and small, greenish, perfect or sometimes unisexual flowers. Calyx usually of 1–5 sepals, which are often fleshy, sometimes lacking. Petals none. Stamens as many as the sepals and opposite them, or fewer. Pistil 1, the ovary 1-celled, the styles 2 or 3. Fruit 1-seeded and circumscissile (a utricle), or indehiscent (an akene). The embryo may be either curved or spirally coiled.

The family includes about 75 genera and 500 species, widely distributed.

EXAMPLES: *Spinacea oleracea:* Spinach.
Beta vulgaris: Beet.
Chenopodium: Goosefoot.
Eurotia lanata: Winterfat.
Atriplex: Saltbush.
Salsola: Russian Thistle.

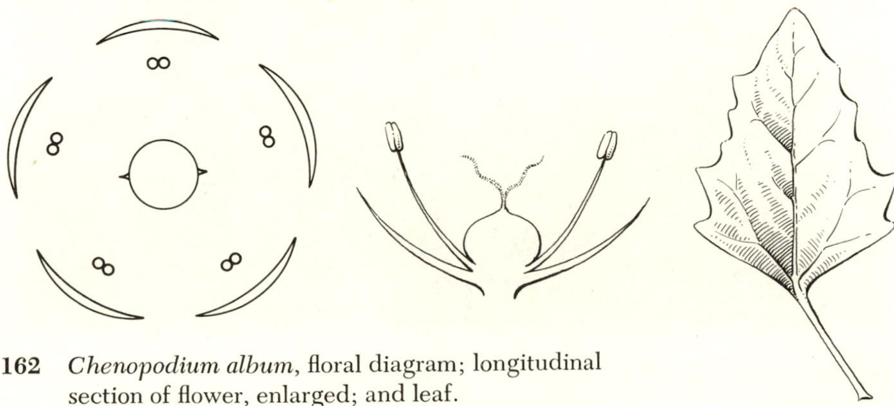

162 *Chenopodium album,* floral diagram; longitudinal section of flower, enlarged; and leaf.

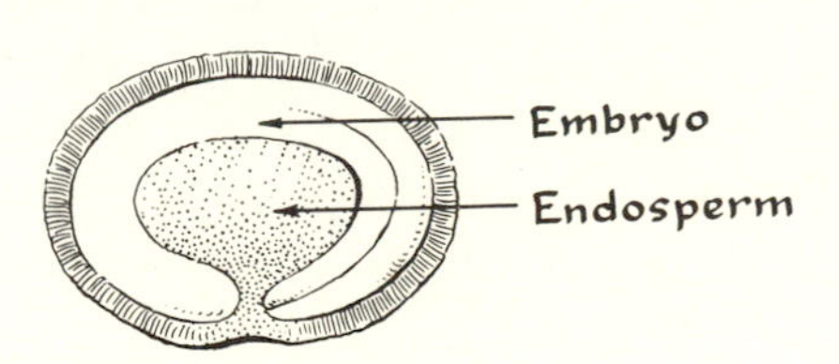

163 *Chenopodium*, section of seed, enlarged.

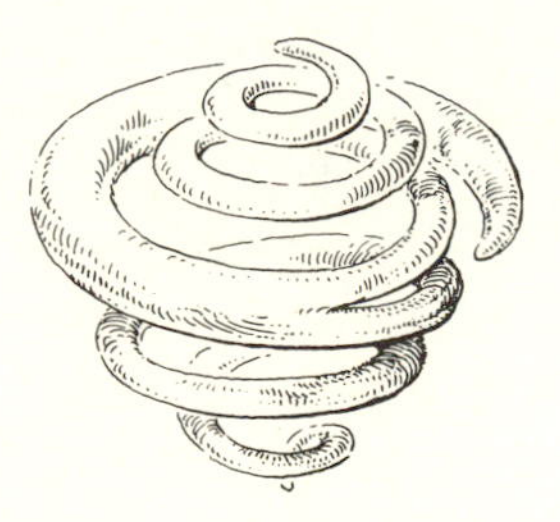

164 *Salsola*, embryo, enlarged.

165 *Chenopodium album* L., Lamb's Quarters, a common weed. Upper part of plants in flower.

AMARANTHACEAE: Pigweed Family

Annual or perennial herbs, rarely undershrubs or climbers, with opposite or alternate, simple, exstipulate leaves. Flowers small and usually crowded, often with scarious or prickly bracts and bracteoles, perfect or unisexual. Sepals usually 5, thin and dry. Petals none. Stamens of the same number as the sepals and opposite them, their filaments often united below into a short tube (monadelphous). Pistil 1, the ovary 1-celled and with 1 or more ovules from the base, the styles short or long, and the stigmas 1, 2, or 3. Fruit a utricle, akene, or pyxis.

The family includes about 50 genera and 500 species, in temperate and warm regions.

EXAMPLES: *Celosia cristata:* Cockscomb, a common ornamental.
Amaranthus retroflexus: Pigweed, a common weed.

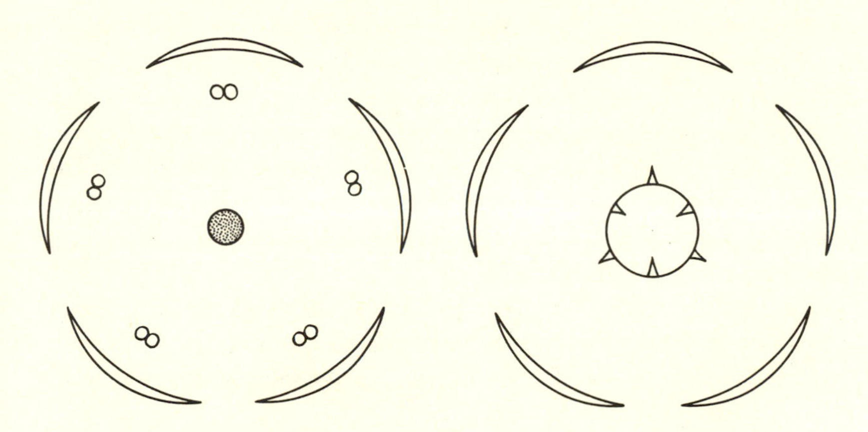

166 *Amaranthus retroflexus,* floral diagrams of staminate (left) and pistillate (right) flowers.

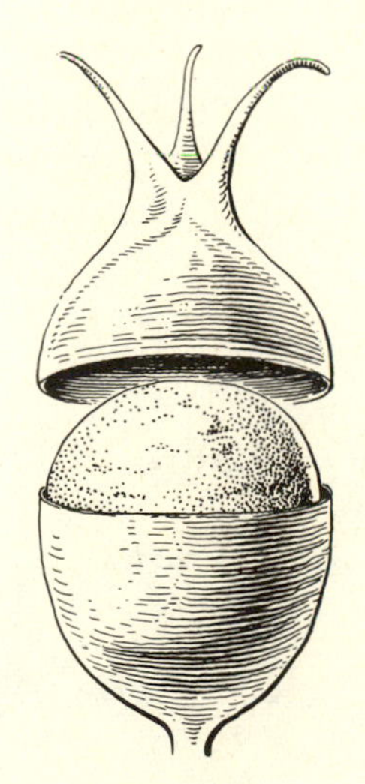

167 *Amaranthus,* fruit (utricle).

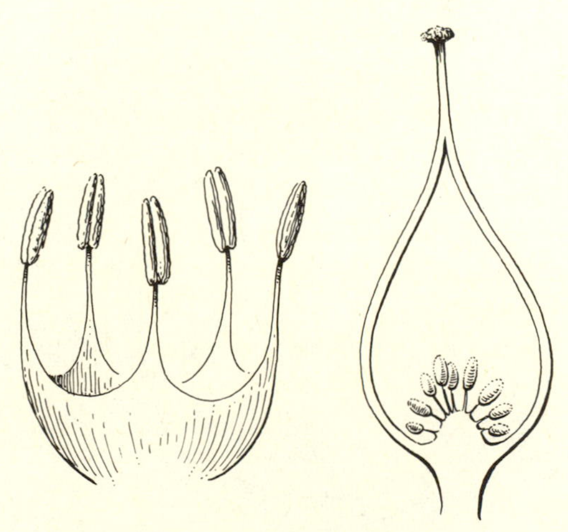

168 *Celosia,* monadelphous stamens and longitudinal section of pistil.

ORDER NYCTAGINALES

NYCTAGINACEAE: Four O'Clock Family

Herbs, shrubs, vines, or trees with simple, alternate or opposite, exstipulate leaves. Flowers perfect or sometimes unisexual, in cymes that are often dense and head-like, often subtended by prominent or colored bracts. Calyx tubular, salverform, or funnelform, 4–5-lobed, often petaloid. Petals none. Stamens 1 or more, free or connate at the base. Pistil 1, with a 1-celled ovary containing a single ovule, and a slender simple style. Fruit often enclosed in the persistent base of the calyx and indehiscent (an anthocarp), sometimes broadly winged at maturity.

The family includes about 20 genera and 160 species, mainly in tropical America but with some representatives in temperate regions.

EXAMPLES: *Mirabilis jalapa:* Four o'Clock, a commonly cultivated herb native to tropical America.

Bougainvillea glabra: Bougainvillea, a cultivated woody vine commonly growing over buildings in warm regions, having showy magenta or purple bracts, and native to Brazil.

Abronia: Sand Verbena, with several native North American species, usually in dry sandy soil.

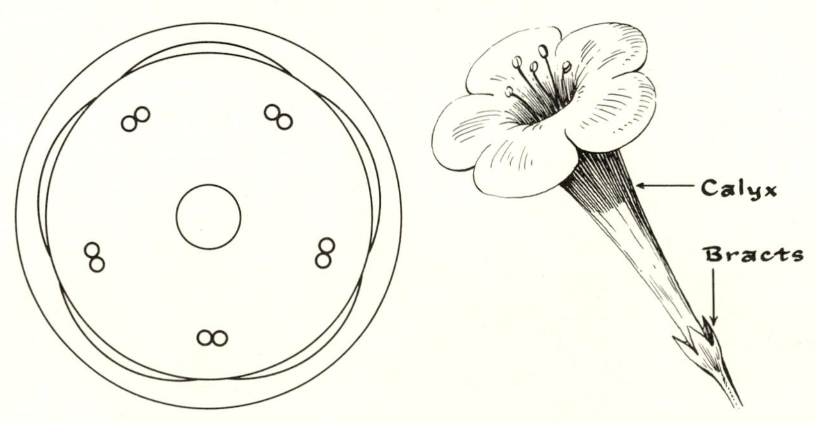

169 *Mirabilis*, floral diagram (note the abnormal position of the stamens) and flower.

170 *Abronia*, portion of plant, and flower, enlarged.

THE POLYPETALAE

Included in this group are the dicotyledons having flowers with separate petals. A few families contain genera or species that have apetalous flowers, but these are exceptional. Within the *Polypetalae* there is a general progression from hypogynous through perigynous to epigynous flowers, and in the hypogynous and perigynous members from apocarpous to syncarpous flowers. See Fig. 69 for major groupings.

ORDER CARYOPHYLLALES

Herbaceous or, rarely, shrubby, sometimes succulent plants, with mostly opposite or whorled leaves having stipules reduced or lacking. Flowers hypogynous (partly epigynous in *Portulaca*) and regular, usually with petals. Stamens mostly definite. Pistil 1, with free-central or basal placentation.

This order includes the polypetalous members of the *Centrospermae: Elatinaceae, Caryophyllaceae, Molluginaceae, Aizoaceae,* and *Portulacaceae.* Its connection with the apetalous *Centrospermae* is through the subfamily *Paronychioideae* of the family *Caryophyllaceae.*

CARYOPHYLLACEAE: Chickweed or Pink Family

Annual or perennial herbs with opposite, simple, and mostly exstipulate leaves (scarious stipules present in *Paronychia*). Flowers solitary or in cymes (dichasia), usually perfect, regular, hypogynous, with or sometimes without petals, 4–5-merous. Stamens 10 or fewer, distinct. Pistil 1, of 2–5 united carpels, the ovary 1-celled and with free-central placentation and 1–many ovules, the styles commonly distinct but sometimes connate. Fruit a utricle or, more commonly, a capsule opening by valves or apical teeth.

The family, named for the pre-Linnaean genus *Caryophyllus,* includes about 80 genera and 2,000 species, mostly in cool or north temperate regions. It includes three clearly defined subfamilies, sometimes treated as families.

Paronychioideae	*Alsinoideae*	*Silenoideae*
Petals lacking. Sepals distinct or united. Fruit a utricle.	Petals present. Sepals distinct. Fruit a capsule.	Petals present. Sepals united. Fruit a capsule.
EXAMPLES: *Paronychia:* Whitlow-wort. *Scleranthus:* Knawel.	EXAMPLES: *Cerastium:* Chickweed. *Arenaria:* Sandwort.	EXAMPLES: *Silene:* Catchfly. *Lychnis:* Campion. *Dianthus:* Carnation, or Pink.

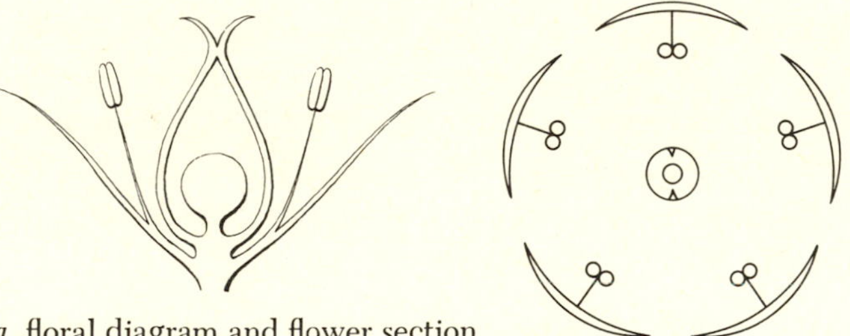

171 *Paronychia*, floral diagram and flower section. The stamens are inserted on the base of the sepals.

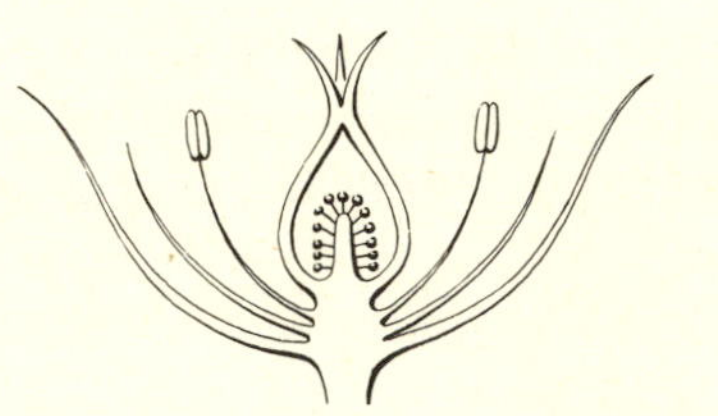

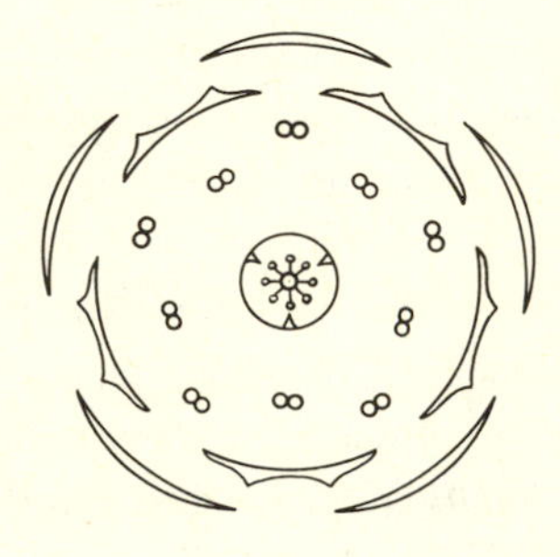

172 *Stellaria pubera*, floral diagram and flower section. The petals are bifid, deeply 2-lobed.

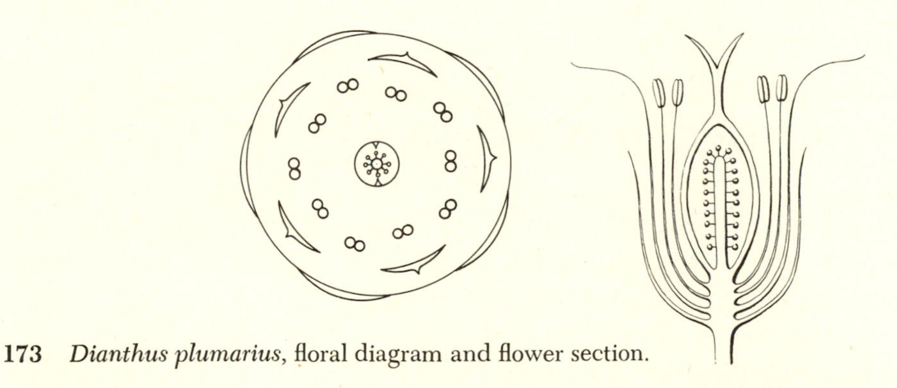

173 *Dianthus plumarius*, floral diagram and flower section.

174 *Cerastium arvense* L., Chickweed. This illustrates the characteristic dichasium inflorescence, the central flower in each group being the oldest, the corolla having withered, while on each side are younger flowers.

PORTULACACEAE: Purslane Family

Herbs or small shrubs with succulent, alternate or opposite leaves and scarious stipules (except *Claytonia*, which is exstipulate). Flowers solitary or in cymes or racemes, regular, perfect, and hypogynous or sometimes partly epigynous. Sepals usually 2, often persistent. Petals usually 4–6, often falling early. Stamens as many as the petals and opposite them, often epipetalous, and sometimes more numerous or fewer. Pistil 1, the ovary 1-celled with free-central placentation, the style variously divided but the branches often 3. Fruit a capsule, which is either circumscissile (a pyxis) or dehiscent by valves.

The family includes about 20 genera and 200 species, chiefly American in distribution.

EXAMPLES: *Portulaca grandiflora:* Portulaca, often grown for ornament.
Portulaca oleracea: Purslane, a succulent weed of cultivated areas.
Claytonia: Spring Beauty.
Lewisia rediviva: Bitterroot, named for Captain Meriwether Lewis, and the state flower of Montana.

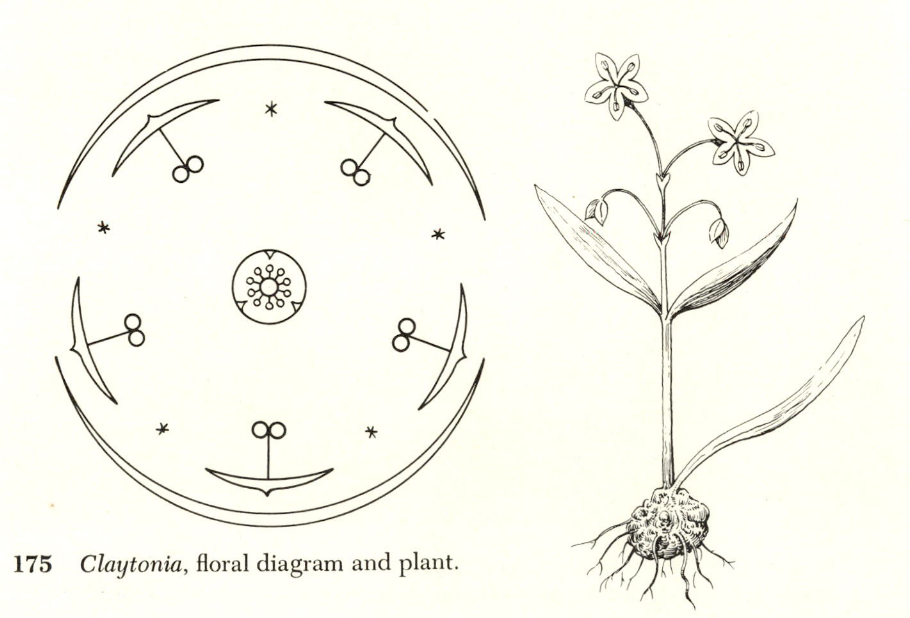

175 *Claytonia*, floral diagram and plant.

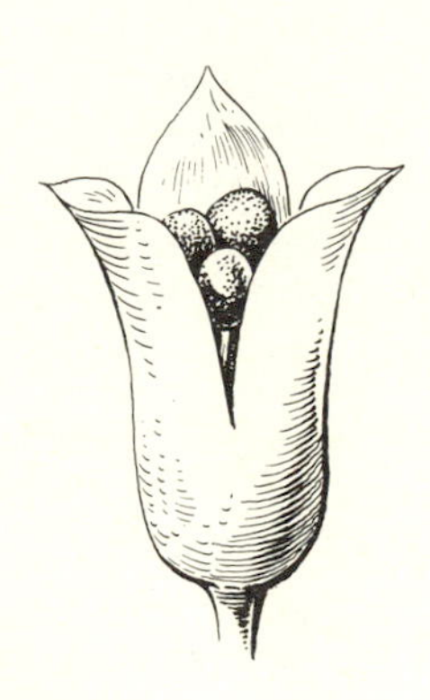

176 *Claytonia*, capsule, enlarged.

177 *Portulaca*, circumscissile capsule (pyxis), enlarged.

178 *Claytonia lanceolata* Pursh, Spring Beauty. The flowers are pink to nearly white, the leaves thick and succulent.

ORDER RANALES

Herbs or woody plants with spiral or cyclic, hypogynous, and regular to irregular flowers. Perianth consisting of sepals and petals, sometimes the petals lacking or reduced and the sepals petaloid. Stamens usually numerous. Carpels numerous and 1-seeded to few and several-seeded, sometimes single.

As treated here, the order includes 12 families according to Rendle[1] and 23 families according to Hutchinson,[2] who divides the group into 5 orders. This is generally regarded as the most primitive order of dicotyledons and is taken as the ancestral group that gave rise to the monocotyledons. The following families are well represented in the North American flora. They may be distinguished by the following key.

Key to the Families of Ranales

Stamens numerous; anthers not opening by terminal valves

Trees or shrubs . . . . MAGNOLIACEAE

Herbs, or sometimes woody vines

Plants aquatic; sepals or petals each in two or more series; flowers solitary . . . . NYMPHAEACEAE

Plants mostly terrestrial; sepals and petals each in one series; flowers often in several-flowered inflorescences . . . . RANUNCULACEAE

Stamens definite (10 or less); anthers opening by terminal valves

Plants aromatic; leaves never spiny . . . . LAURACEAE

Plants not aromatic; leaves usually spiny . . . . BERBERIDACEAE

[1] A. B. Rendle: *The Classification of Flowering Plants* (1925), Vol. II, p. 124.
[2] J. Hutchinson: *The Families of Flowering Plants* (1926), Vol. I, pp. 10–11.

MAGNOLIACEAE: Magnolia Family

Shrubs or trees with mostly alternate simple leaves, with or without stipules. Flowers usually solitary and showy, perfect, regular, and hypogynous, the floral axis usually elongated and the floral parts spirally arranged on it. Perianth usually but not always clearly differentiated into calyx and corolla, the sepals or the petals in two or more series. Stamens numerous. Carpels (pistils) usually numerous and distinct, becoming follicles, samaras, or berries in fruit, or sometimes aggregates.

The family includes 10 genera and about 100 species in tropical Asia and tropical America, extending northward into the temperate zone, but with a much wider distribution in the upper Cretaceous and Tertiary periods.

EXAMPLES: *Magnolia:* Magnolia, ornamental small trees and shrubs.
Liriodendron tulipifera: Tulip-tree, a valuable timber tree, often reaching immense size. (See the frontispiece.)

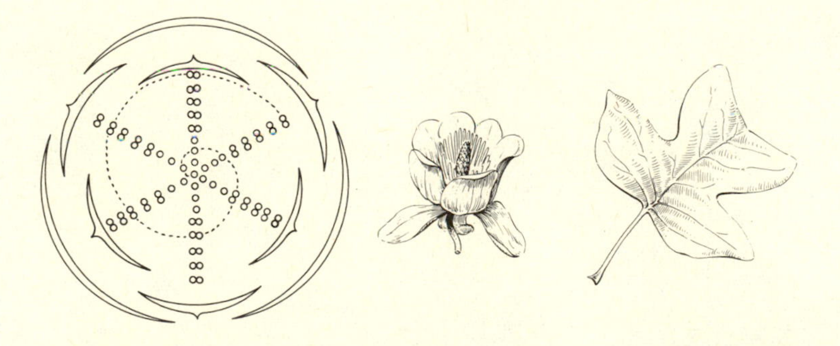

179 *Liriodendron tulipifera*, floral diagram (the broken line indicates that the parts are spirally arranged), flower, and leaf.

180 *Magnolia graniflora* L., a handsome tree of the southeastern United States. The stamens in the flower have fallen off, and the pistils are beginning to mature into fruits.

NYMPHAEACEAE: Water Lily Family

Perennial aquatic herbs with submerged, floating, or sometimes emersed and often large leaves. Flowers usually solitary and showy, hypogynous or more or less epigynous, the perianth composed of sepals and petals that are often indefinite in number and in two or more series, the calyx and corolla poorly differentiated. Stamens few or usually many. Carpels 2 to many, free or united. Fruit nut-like or berry-like.

As treated here, the family includes 8 genera and about 50 species, in tropical and north temperate regions. Because of the varied floral structure found in this group, some authors have proposed that several families be recognized [1] instead of the one given here.

EXAMPLES: *Nymphaea:* Water Lily, the largest genus, with about 32 species.

Nuphar: Yellow Water Lily, or Spatterdock.

Nelumbo: Lotus Lily.

Victoria amazonica: Giant Water Lily of the Amazon region, with leaves 6–7 feet in diameter, often cultivated.

[1] Five families are recognized by Hui-Lin Li. See Am. Midl. Nat. 54: 33–41 (1955).

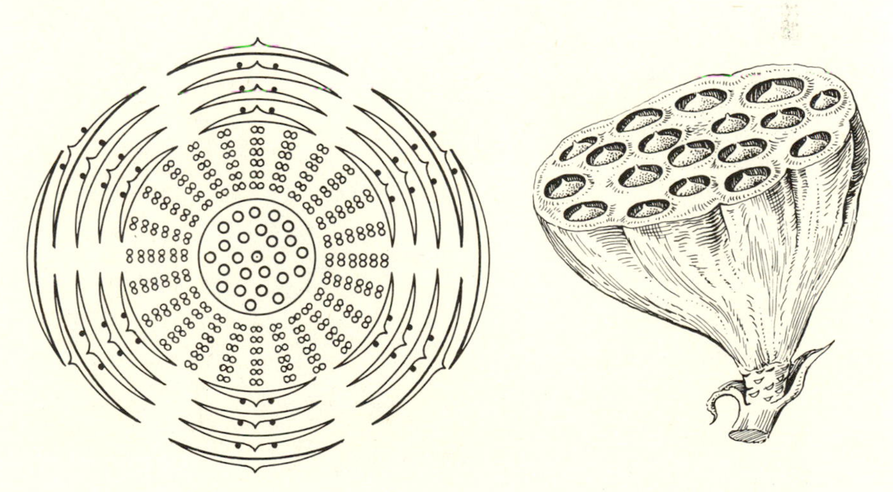

181 *Nelumbo,* floral diagram and gynoecium axis and pistils.

182 *Nymphaea*, flower and leaf.

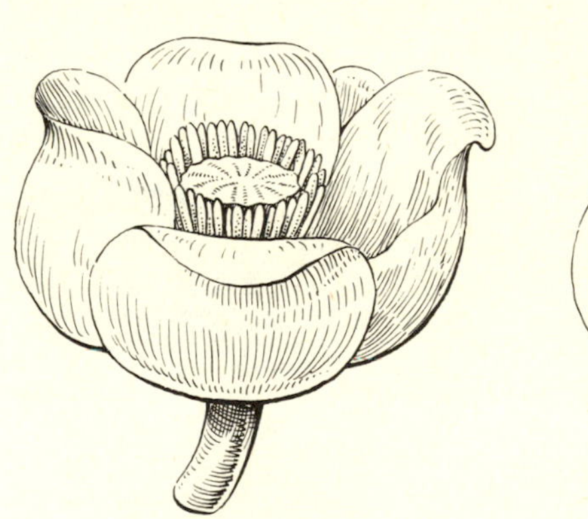

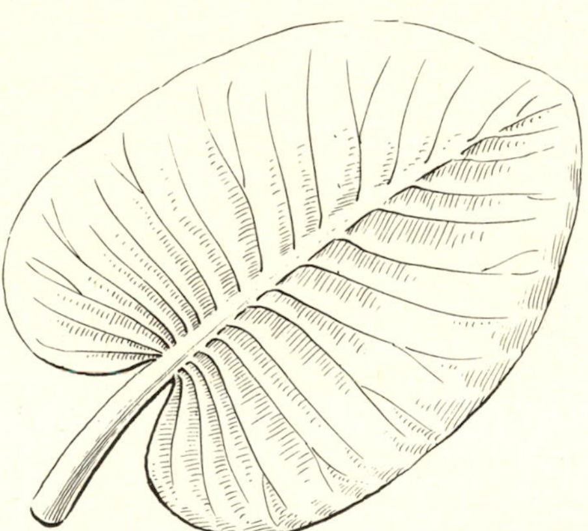

183 *Nuphar*, flower and leaf.

184 *Nuphar polysepalum* Engelm., the western Yellow Water Lily, or Spatterdock, shown growing in a lake in about five feet of water, rooted in the mud on the bottom and with leaves and flower floating on the surface of the water.

RANUNCULACEAE: Buttercup Family

Mostly perennial herbs with alternate or basal and often divided leaves without stipules, or, in *Clematis,* sometimes soft-woody climbers with opposite leaves. Flowers solitary or in racemes or panicles, hypogynous, usually perfect, quite variable, regular to irregular, sometimes spurred. Sepals 3 or more, sometimes petaloid. Petals none to indefinite, often with a nectariferous claw, the sepals and petals each usually in one series. Stamens usually numerous and spirally arranged. Pistils simple (flowers apocarpous), numerous, spirally arranged, and forming akenes in fruit, or few and cyclic and forming follicles or berries in fruit.

The family includes about 40 genera and 1,500 species, in temperate and cold regions, often circumboreal.

EXAMPLES: *Ranunculus:* Buttercup, with about 300 species, nearly 100 of these in North America.

Delphinium: Larkspur, often cultivated, and some species of western ranges poisonous to livestock.

Aquilegia: Columbine, with both native and cultivated species.

Paeonia: Peony, often cultivated for its large flowers.

Anemone: Anemone, or Windflower, with about 100 species.

Caltha: Marsh Marigold, with about 20 species in cold and north temperate regions.

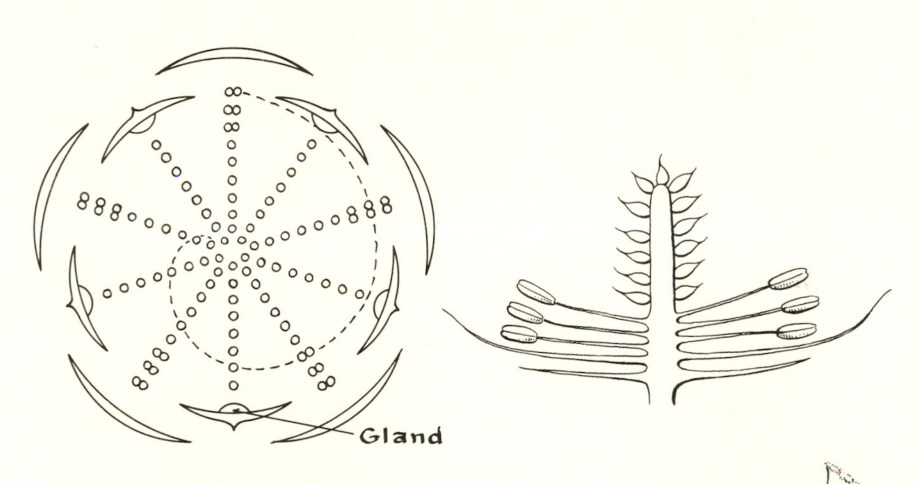

185 *Ranunculus*, floral diagram and diagrammatic longitudinal section of flower.

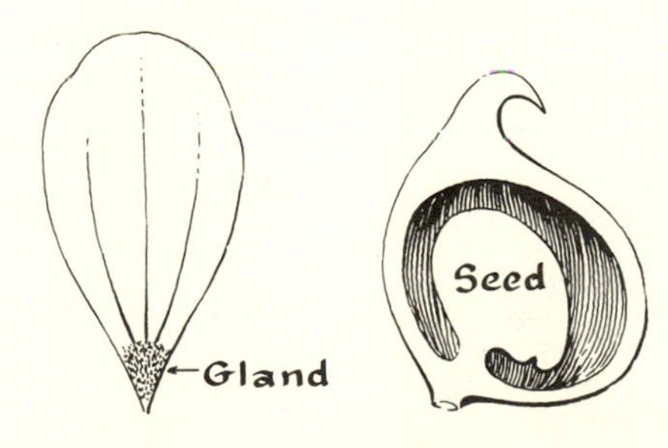

186 *Ranunculus*, petal and fruit (akene) cut lengthwise.

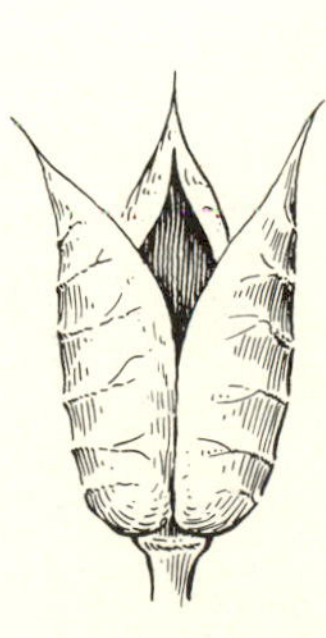

187 (Left) *Delphinium*, fruits (follicles). (Right) *Caltha*, follicle cut lengthwise.

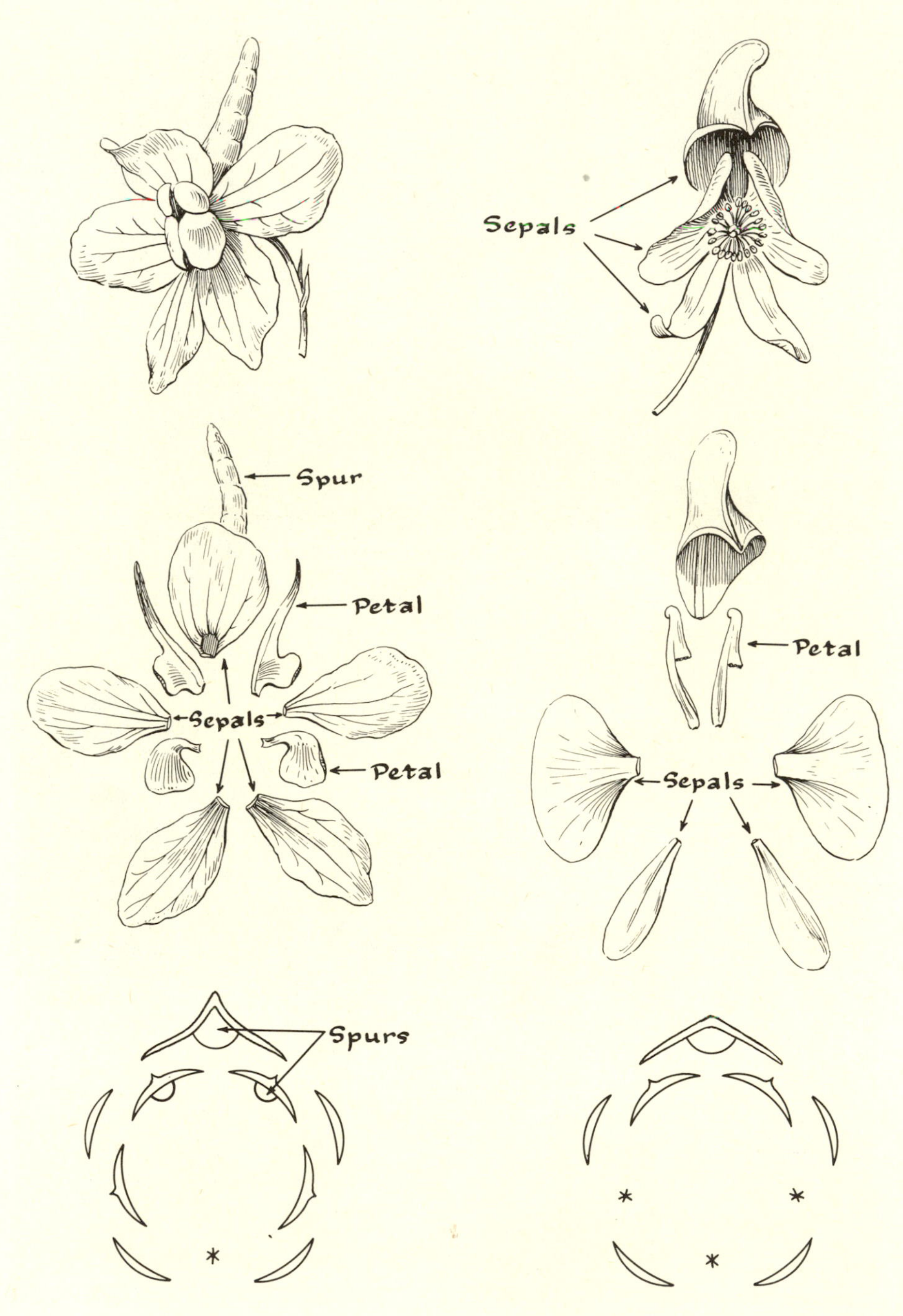

188 *Delphinium*, flower, perianth, and diagram of perianth.

189 *Aconitum*, flower, perianth, and diagram of perianth.

190 *Caltha leptosepala* DC., a species of Marsh Marigold found in cool, moist situations in western North America. It has white flowers.

191 *Ranunculus glaberrimus* Hook., a common early Buttercup of the western United States.

192 *Anemone patens* L., Pasque Flower, a circumboreal species with several races. The large sepals are blue and resemble petals, but the flowers have no petals. The leaf-like whorl below the flower is an involucre.

193 *Aquilegia caerulea* James, Columbine, of the Rocky Mountains, and the State Flower of Colorado. The petals have long spurs projecting backward from the flower.

LAURACEAE: Laurel Family

Mainly aromatic woody plants with alternate, simple, entire or lobed, and exstipulate leaves. Flowers numerous, in mixed inflorescences, perfect to unisexual, regular, 3-merous, the perianth of similar parts (probably representing sepals and petals). Stamens definite, the inner often reduced to staminodia, the fertile anthers opening by valves. Pistil 1, with a single style and often 3 stigmas (probably representing 3 carpels but only 1 developing), the ovary with a single ovule. Fruit a drupe.

The family includes about 40 genera and 1,000 species, mainly tropical but with a few representatives in North America.

EXAMPLES: *Laurus nobilis:* Laurel, or Bay Tree, of Europe, often cultivated.

Sassafras: a genus of 3 species, one in the eastern United States, one in China, and one in Taiwan (Formosa).

Cinnamomum zeylandicum: Cinnamon, of southern Asia.

Cinnamomum camphora: Camphor tree, of southern Asia.

Persea gratissima: Avocado, native of tropical America.

Umbellularia californica: Oregon Myrtle, or California Laurel.

BERBERIDACEAE: Barberry Family

Perennial herbs or shrubs with simple or compound, often spiny-toothed, mostly exstipulate leaves. Flowers usually small, solitary or in mixed inflorescences, perfect, hypogynous, 2–3-merous, the sepals and petals often similar and the petals sometimes lacking. Stamens 4–9, opposite the petals, their anthers opening by valves. Pistil 1, of 1 carpel. Fruit usually a berry.

The family includes 9 genera and about 200 species, mainly in north temperate regions.

EXAMPLES: *Berberis:* Barberry, the largest genus, with nearly 200 species. *B. vulgaris* is the alternate host of Wheat Rust and has been largely exterminated in the United States. The Japanese Barberry, *B. thunbergii,* is often cultivated.
Podophyllum peltatum: May Apple.
Mahonia: Oregon Grape.

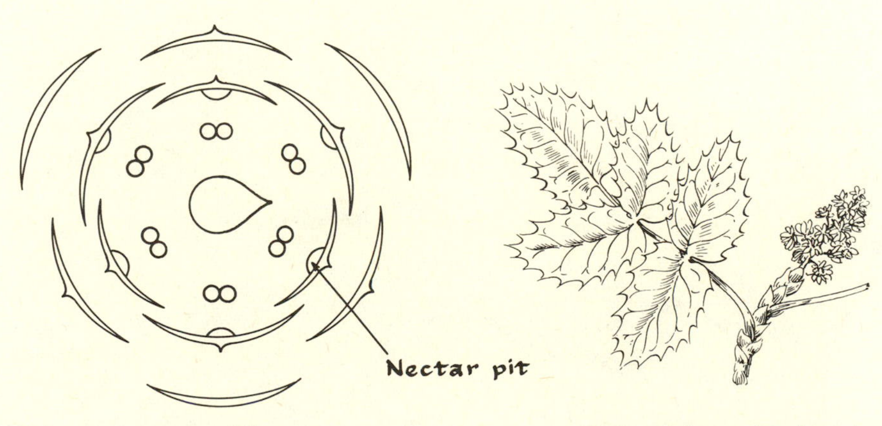

194 *Berberis,* floral diagram.

195 *Mahonia* (*Berberis*), flowering branch.

ORDER PAPAVERALES (RHOEDALES)

Mostly herbaceous plants with alternate exstipulate leaves. Flowers hypogynous, perfect, regular to irregular, with separate petals, mostly 4-merous, the stamens from numerous to 2, the pistil syncarpous, with or without septation in the ovary, the placentation parietal. Fruit a capsule or silique.

The order represents an early modification of the *Ranales* by reduction and specialization of flower parts, and as here treated it includes the families *Papaveraceae, Fumariaceae, Brassicaceae* (*Cruciferae*), and *Capparaceae.* These may be distinguished by the following key.

Key to the Families of Papaverales

Juice milky or colored; stamens numerous . . . . PAPAVERACEAE

Juice watery; stamens mostly definite

Flowers irregular, one or both of the outer petals spurred or saccate at the base . . . . FUMARIACEAE

Flowers regular, the petals not spurred

Ovary 1-celled, without a partition; fruit a capsule; stamens often exserted . . . . CAPPARACEAE

Ovary 2-celled, with a thin partition; fruit a silique or silicle; stamens included . . . . BRASSICACEAE

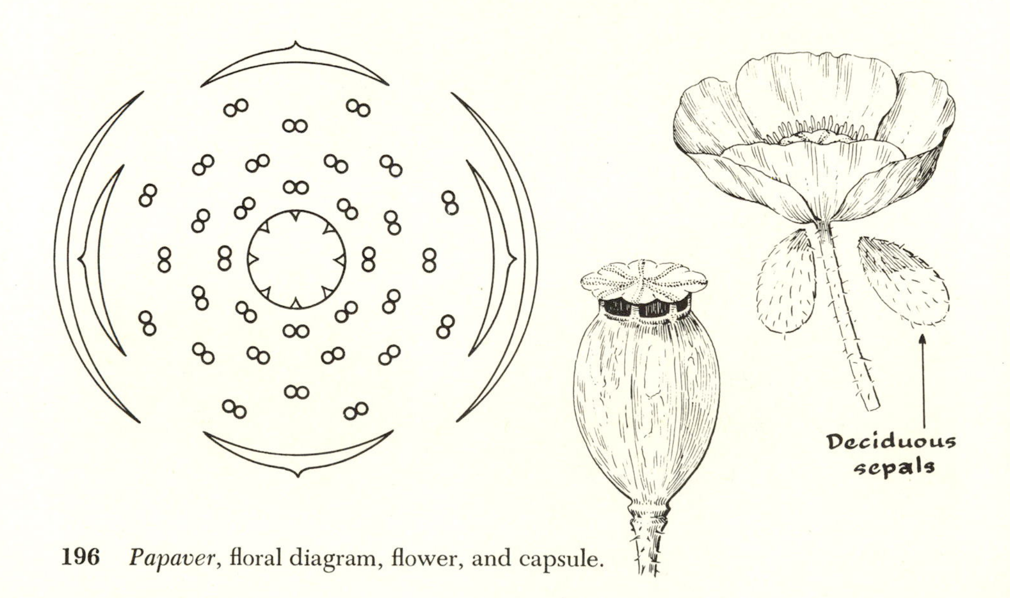

196 *Papaver*, floral diagram, flower, and capsule.

PAPAVERACEAE: Poppy Family

Annual or perennial herbs, rarely shrubs or trees, with milky or colored juice and exstipulate leaves. Flowers usually solitary, often showy, regular, perfect, and hypogynous, or rarely somewhat perigynous. Sepals 2–3, falling as the flowers open. Petals 4–12, in two series, crumpled in the bud. Stamens numerous and distinct. Pistil 1, of 2 or more united carpels; the ovary 1-celled, with parietal placentation, or sometimes becoming several-celled by intrusion of the placentae; the stigmas as many as the placentae, often forming a lobed disk on top of the ovary or on a short style. Fruit a many-seeded capsule opening by a ring of pores under the stigmas, or opening by valves.

The family includes about 24 genera and 450 species, chiefly in north temperate and subtropical regions.

EXAMPLES: *Papaver:* Poppy, a genus of about 100 species, including *P. somniferum,* which yields opium from the latex of the green pods; *P. nudicaule,* the Iceland Poppy; and *P. orientale,* the Oriental Poppy of our gardens.

Argemone: Prickly Poppy, common in the western United States.

Sanguinaria canadensis: Bloodroot, a common spring flower of the wooded areas of the eastern United States.

197 *Argemone hispida* Gray, Prickly Poppy. The petals are white.

FUMARIACEAE: Fumitory Family

Delicate smooth herbs or climbers with watery juice and alternate compound and dissected exstipulate leaves. Flowers usually irregular, perfect, hypogynous. Sepals 2, small and deciduous. Corolla closed, the petals 4, in two pairs, somewhat connivent, one or both of the outer petals spurred or saccate at the base, the two inner ones smaller and sometimes coherent. Stamens 4, free and opposite the petals, or 6 and united by their filaments in two groups (diadelphous). Pistil 1, of 2 united carpels, the ovary 1-celled with 2 parietal placentae and 2 to many ovules, the style single, and the stigma capitate or lobed. Fruit a 2-valved capsule or rarely (in *Fumaria*) indehiscent and nut-like.

The family includes 5 genera and about 150 species, found chiefly in north temperate regions, especially in the Mediterranean basin. It is included with the *Papaveraceae* by some authors.

EXAMPLES: *Dicentra spectabilis:* Bleeding-heart.
Dicentra cucullaria: Dutchman's-breeches.
Corydalis aurea: Golden Corydalis.

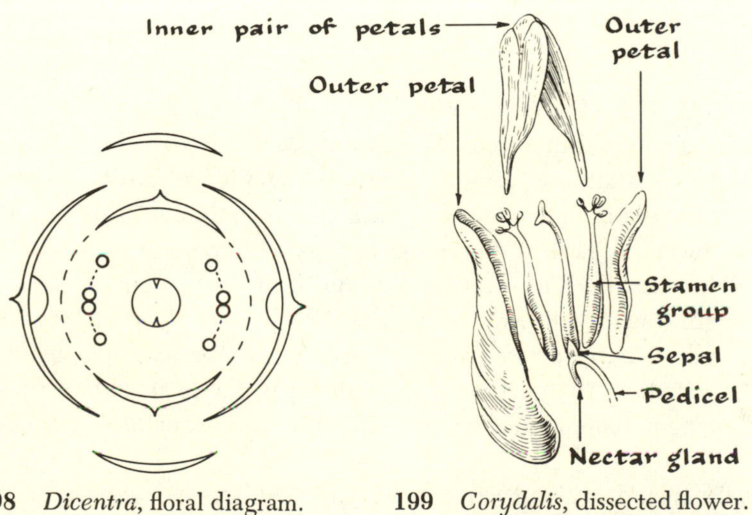

198 *Dicentra*, floral diagram.

199 *Corydalis*, dissected flower. [After Gray, *Genera Florae Americae Boreali-Orientalis Illustrata*, **1**: pl 52 (1848).]

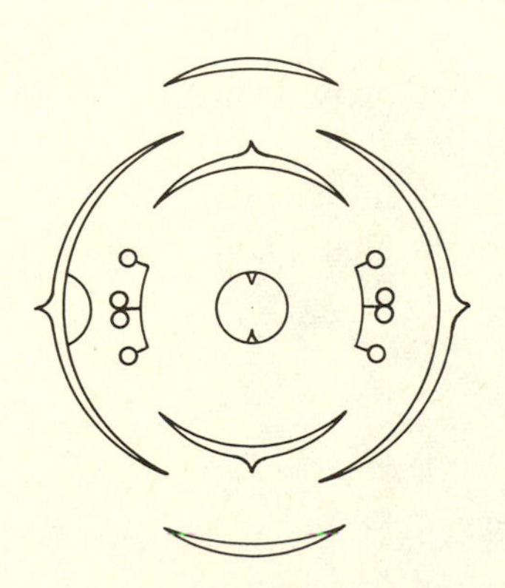

200 *Corydalis*, floral diagram.

CAPPARACEAE: Caper Family

Mostly herbs, in ours, but often arborescent in the tropics, with watery juice, with leaves that are usually alternate, simple or often digitately 3–7-foliolate, and exstipulate or with minute or spiny stipules. Flowers often in bracteate racemes, usually 4-merous, hypogynous, regular or somewhat irregular. Petals usually long-clawed. Stamens 4 or more, usually 6, about equal in length and usually much exserted. Pistil 1, of 2 or more united carpels (usually 2), the ovary usually 1-celled with parietal placentation and often elevated on a long stipe or gynophore, the style short and simple or none, the stigma commonly capitate. Fruit a capsule or less commonly a berry.

The family includes about 40 genera and 450 species, mostly in the tropics. It is generally regarded as the ancestral group to the *Brassicaceae* (*Cruciferae*).

EXAMPLES: *Capparis spinosa:* Capers, the condiment being the flower buds; a Mediterranean shrub.

Cleome spinosa: Spider Plant, often cultivated.

Cleome serrulata: Rocky Mountain Bee Plant, a common roadside weed.

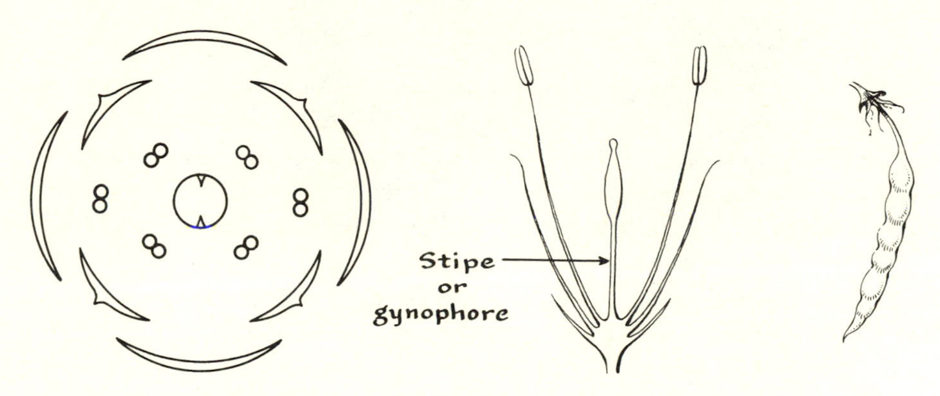

201 *Cleome,* floral diagram, flower section, and fruit.

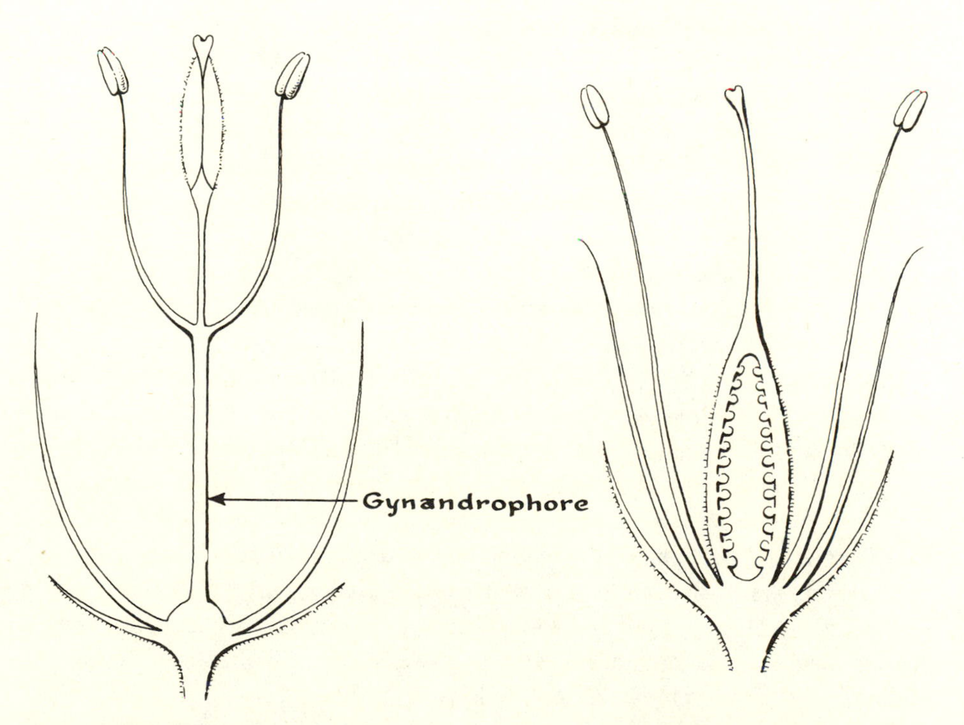

202 *Gynandropis,* flower section.

203 *Polanisia,* flower section.

BRASSICACEAE (CRUCIFERAE): Mustard Family

Mostly herbs with pungent watery juice and alternate exstipulate leaves. Flowers in simple or sometimes branched ebracteate racemes, 4-merous, hypogynous and regular. Petals usually long-clawed, rarely absent. Stamens usually 6, the two outer ones short and the four inner ones long, or sometimes reduced to 2, included. Pistil 1, of 2 united carpels, the ovary 2-celled with 1 or more ovules in each cell, the style simple or none, the stigma capitate or 2-lobed. Fruit a silique or silicle, the valves falling away from the persistent partition at maturity, or, rarely, indehiscent.

The family includes about 200 genera and 1,800 species, mostly in temperate and cold regions, including numerous alpine species. Some are noxious weeds, some are ornamentals, and some are common garden vegetables.

EXAMPLES: *Brassica oleracea:* Cabbage, with var. *gemmifera,* Brussels Sprouts; var. *botrytis,* Cauliflower; var. *caulorapa,* Kohlrabi; and var. *italica,* Broccoli.
Brassica rapa: Turnip.
Brassica alba: White Mustard.
Raphanus sativa: Radish.
Cheiranthus cheiri: Wallflower.
Iberis amara: Candytuft.
Sisymbrium altissimum: Tumble Mustard.
Cardaria draba: White-top.

In most members of the family the ovary is sessile on the receptacle, but in some, such as *Stanleya,* there is a *stipe* between the ovary and the receptacle, and this may elongate in fruit (Fig. 207). When the fruit is strongly flattened, the flattening may be parallel to a broad septum or partition; or contrary to a narrow septum or partition (Fig. 208). Except for the genus *Leavenworthia,* which has a straight embryo, the embryos of the members of the family are bent in such a way as to bring the cotyledons and radicle together in various ways, as shown in Figure 209. These arrangements are sometimes used as a basis for classification. In some manuals the arrangement of cotyledons is indicated by a symbol, also shown, which indicates the appearance of the embryo in cross section.

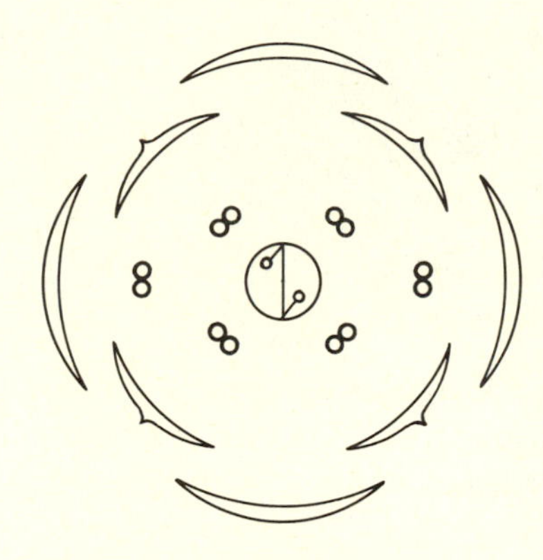

204 *Brassica,* floral diagram.

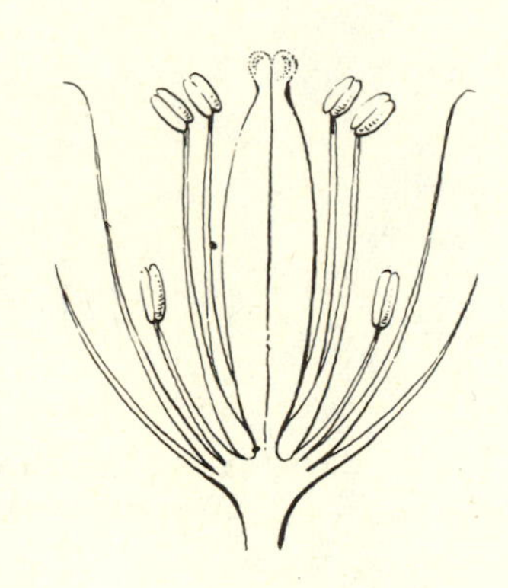

205 *Brassicaceae,* representative flower section showing all the stamens.

206 *Capsella,* plant.

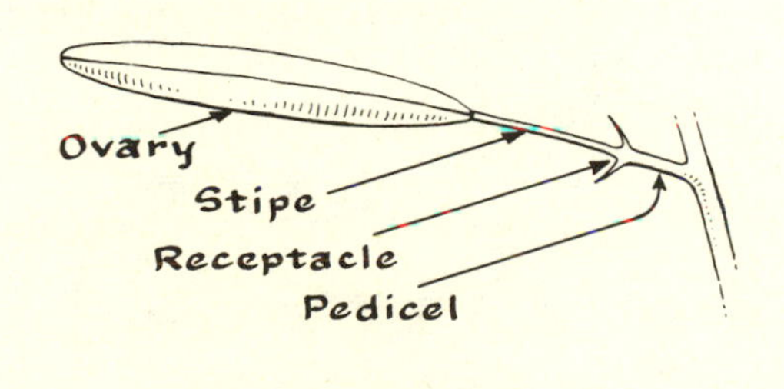

207 *Stanleya,* showing the stipitate pod.

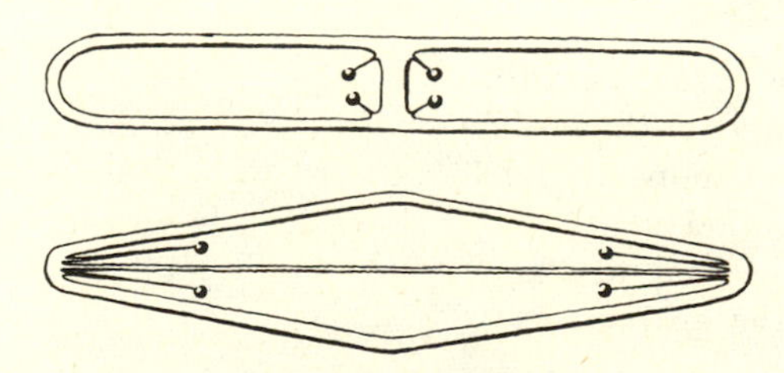

208 Two types of flattening of fruits of the Mustard Family, one flattened contrary to the septum, the other flattened parallel to the septum.

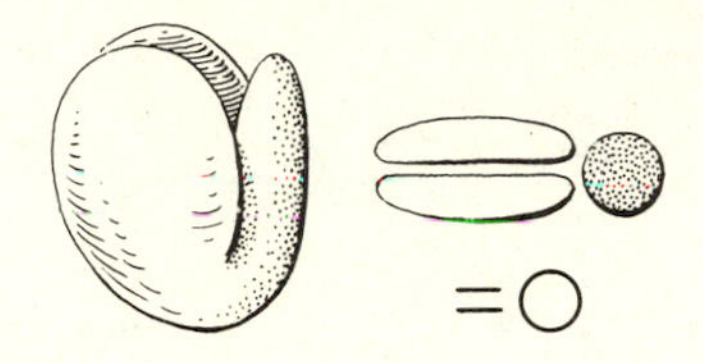

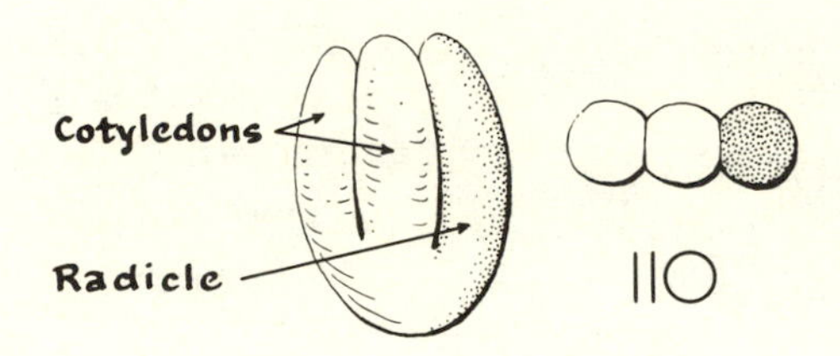

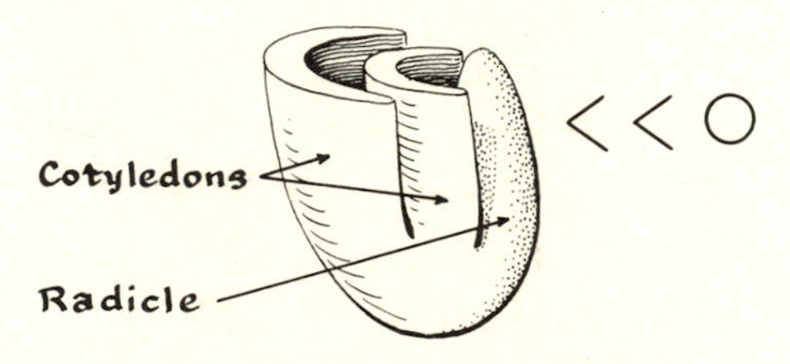

209 Types of embryos: *Barbarea*, with cotyledons accumbent; *Sisymbrium*, with cotyledons incumbent; and *Brassica*, with cotyledons conduplicate. The small figures are diagrams of these structures.

210 *Stanleya pinnata* (Pursh) Britt., var. *integrifolia* (James) Rollins. Plants of this genus indicate seleniferous soils, which often produce plants poisonous to livestock. The flowers are yellow.

ORDER SARRACENIALES

Herbaceous plants, often in bogs, having their leaves variously modified to capture insects. Flowers perfect, regular, hypogynous or somewhat perigynous, polypetalous, the stamens numerous to 4 or 5, the pistil 1, of 3–5 united carpels, the placentation parietal or axillary. Fruit a loculicidal capsule.

The order, as interpreted here, includes the families *Droseraceae* and *Sarraceniaceae,* but not the *Nepenthaceae* or Tropical Pitcher-plant Family of Indo-Malaya and Madagascar, which is sometimes placed in this order. The group represents an early modification, and a terminal one, from the *Ranales* by specialization of the leaves and the adoption of a specialized habitat.

DROSERACEAE: Sundew Family

Mostly bog herbs with a basal rosette of leaves, which are modified to capture insects and are sensitive to touch, the blades commonly covered with large stalked glands (*Drosera*) or provided with a marginal fringe of stiff bristles, which intermesh when the blade folds upward to trap an insect (*Dionaea*). Flowers in racemes or cymes, scapose, hypogynous to somewhat perigynous, mostly 5-merous, regular and perfect. Stamens commonly 5, but sometimes more, the pollen in tetrads. Pistil 1, of mostly 3–5 united carpels, the ovary usually 1-celled, with parietal placentation, the styles commonly free and often divided so as to appear to be twice as many as the carpels. Fruit a loculicidal capsule.

The family includes about 6 genera and nearly 100 species, widely distributed.

EXAMPLES: *Drosera:* Sundew, with about 90 species, temperate and tropical.

Dionaea: Venus'-flytrap, with a single species along the southeastern seacoast of the United States.

Drosophyllum: with a single species in Spain, Portugal, and Morocco.

Aldrovanda: with a single species in Australia, Bengal, northeast Asia, and Europe.

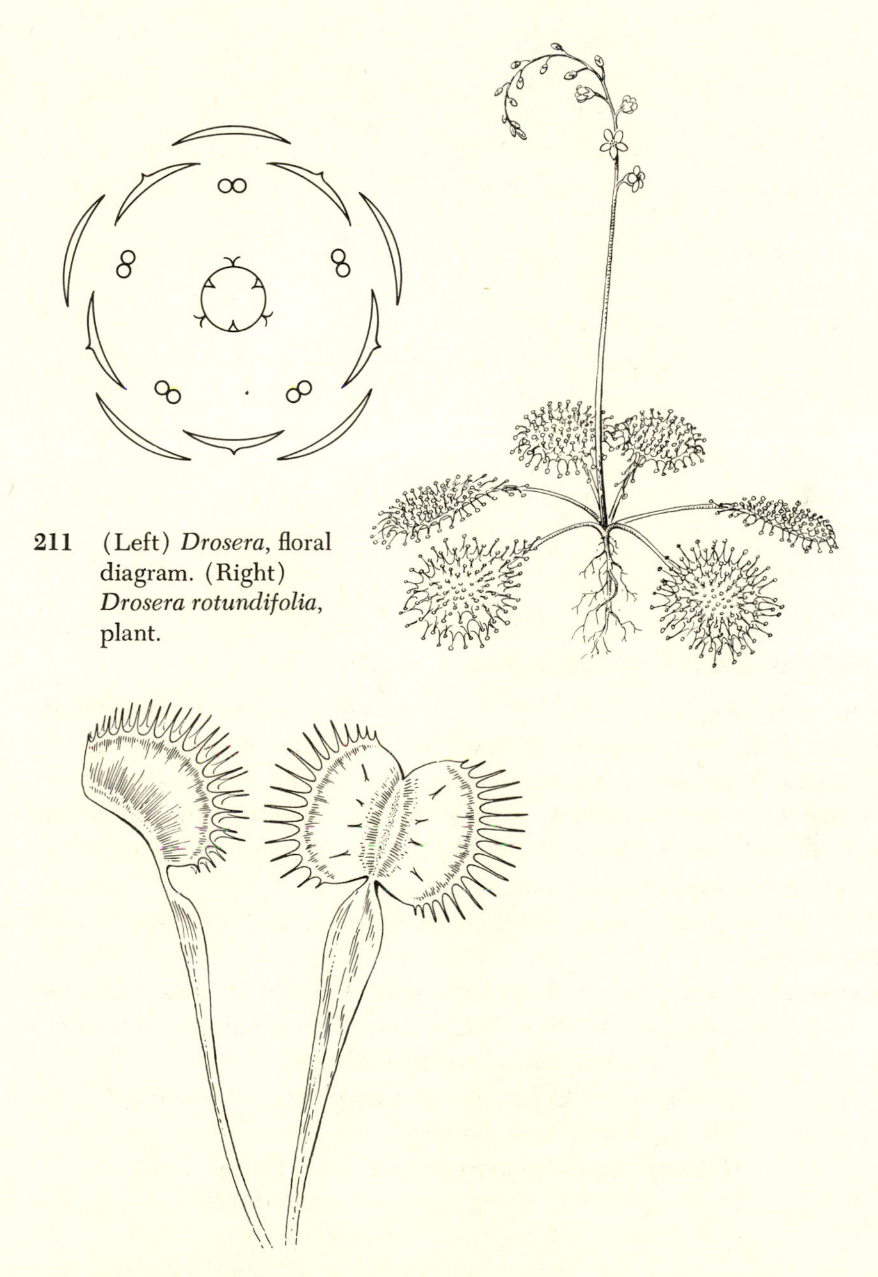

211 (Left) *Drosera*, floral diagram. (Right) *Drosera rotundifolia*, plant.

212 *Dionaea*, leaves. The bristle-like hairs near the middle of the upper part of the blades are sensitive to touch and cause the blades to snap shut.

SARRACENIACEAE: Pitcher-plant Family

Scapose perennial bog herbs with tubular leaves having a small blade. Flowers solitary (in ours) or in few-flowered racemes, perfect, regular, hypogynous, 5-merous, polypetalous (the petals lacking in *Heliamphora*), the stamens numerous. Pistil 1, of 3 or 5 united carpels; the style often greatly dilated and umbrella-like, with 5 small stigmas; the placentation axillary. Fruit a loculicidal capsule.

The family includes 3 genera and about 8 species, in North America and British Guiana.

EXAMPLES: *Sarracenia:* Pitcher-plant, with about 6 species in eastern and northeastern North America, extending westward to Texas and to the Canadian Rockies.

Darlingtonia: California Pitcher-plant, a monotypic genus of California and Oregon.

Heliamphora: a monotypic genus of British Guiana.

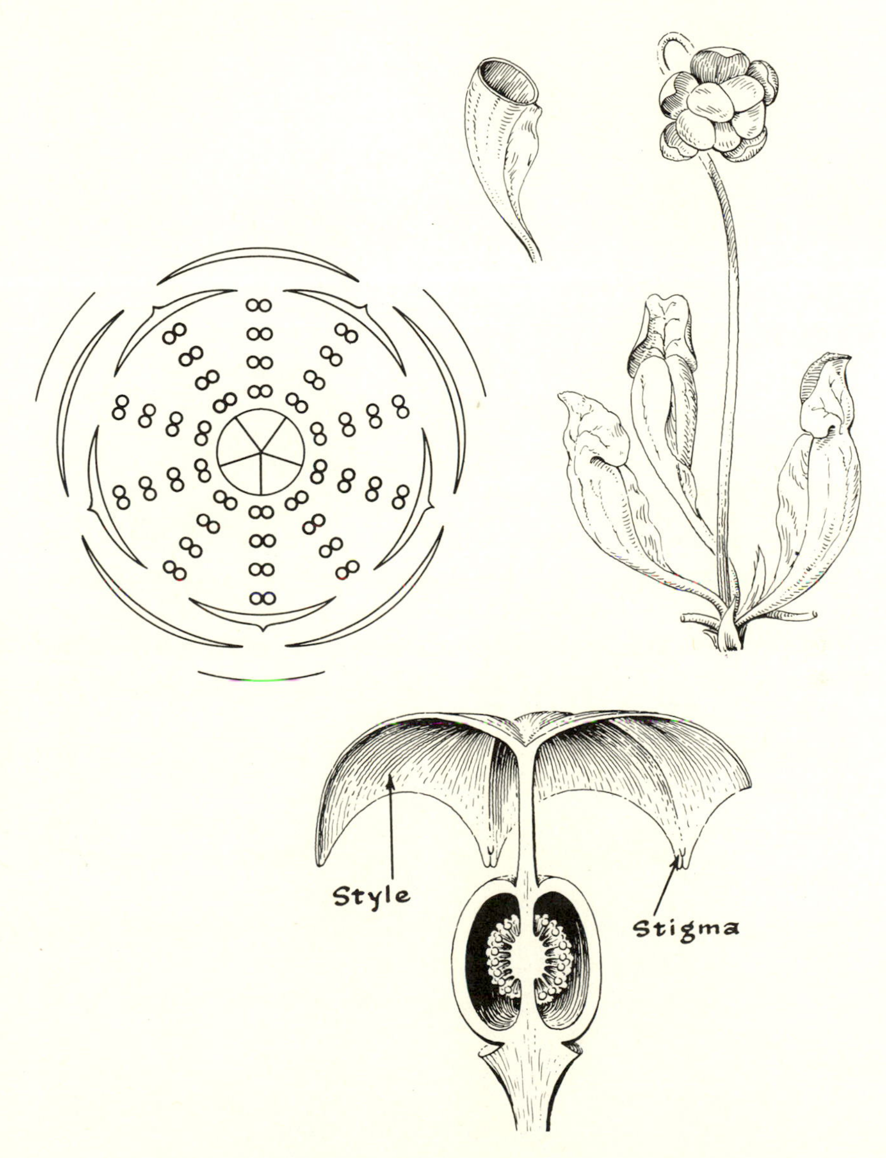

213 *Sarracenia,* floral diagram, plant, portion of leaf cut transversely, and pistil cut lengthwise.

ORDER ROSALES

Herbs, shrubs, or trees with simple or compound leaves, which frequently have conspicuous stipules but may be exstipulate. Flowers mostly perfect, polypetalous, 5-merous, regular to irregular, sometimes hypogynous but characteristically perigynous. Stamens numerous to definite, distinct, monadelphous or diadelphous. Carpels numerous to 1, free from each other or united. Fruit various.

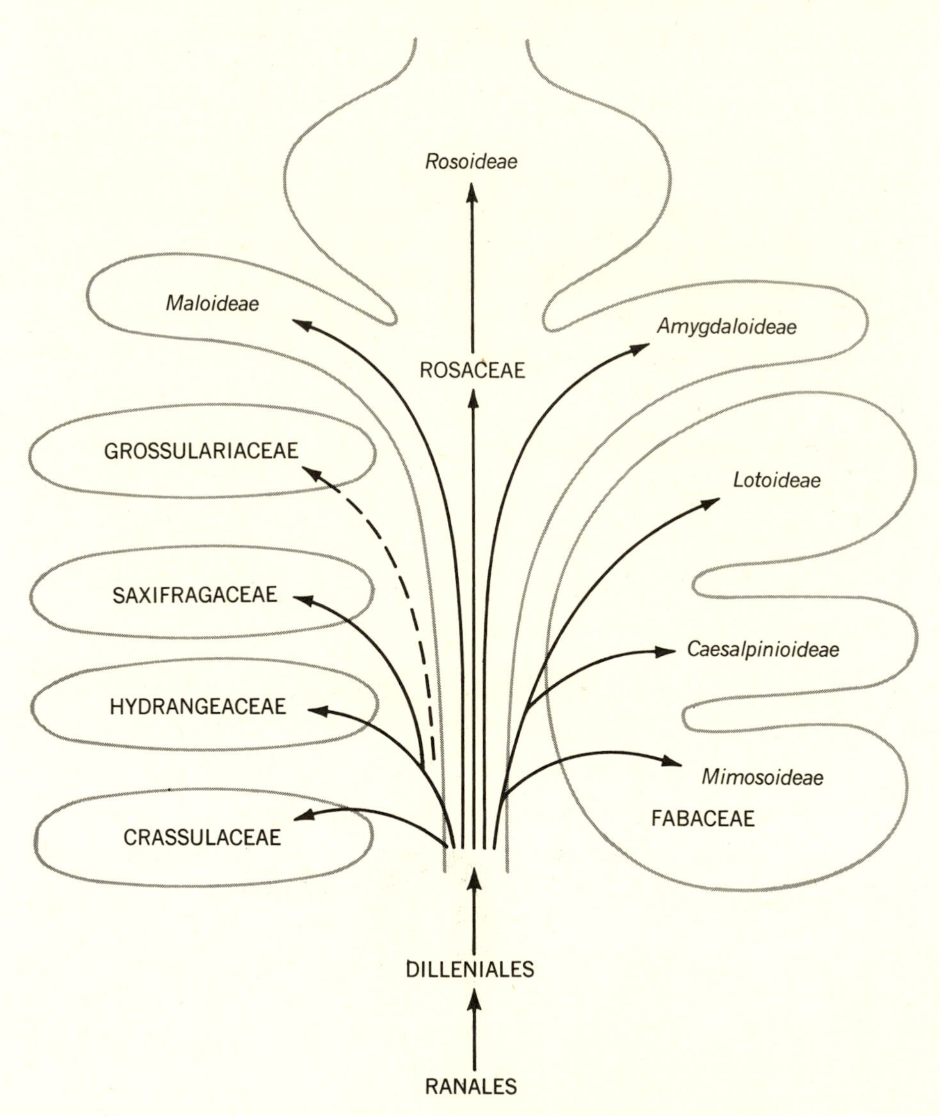

214 Diagram of possible relationships in the *Rosales*. The *Grossulaiaceae*, with epigynous flowers, are probably misplaced here.

This is one of the largest orders of flowering plants, including the families *Rosaceae, Saxifragaceae, Hydrangeaceae, Grossulariaceae,*[1] *Crassulaceae,*[1] and *Fabaceae* (*Leguminosae*), with a total of at least 600 genera and 17,000 species. It represents an early and eminently successful modification from the *Ranales* by the frequent development of a hypanthium, which, together with specialization of the corolla, stamens, and gynoecium axis in many groups, has produced the variety of morphological flower types characteristic of the order, and often misinterpreted. The order *Dilleniales,* a tropical group having flowers with connate sepals and thus perhaps the beginnings of a hypanthium, seems to be intermediate between the *Ranales* and the *Rosales.* In any case there is certainly a more than accidental similarity between the flowers of a Buttercup and those of *Potentilla* (*Cinquefoil*), the essential difference being the presence of a hypanthium in *Potentilla.* Figure 214 shows possible relations and, in a general way, the relative sizes of the families and subfamilies treated here.

The following key will aid in the recognition and distinction of the above families.

Key to the Families of Rosales

Fruit a legume; flowers often irregular . . . . FABACEAE

Fruit otherwise; flowers regular or nearly so

  Plants succulent; flowers hypogynous . . . . CRASSULACEAE

  Plants not succulent; flowers perigynous or epigynous

    Leaves opposite; stems woody . . . . HYDRANGEACEAE

    Leaves alternate; stems woody or herbaceous

      Stamens definite, mostly 5 or 10

        Flowers perigynous; plants herbaceous, often scapose . . SAXIFRAGACEAE

        Flowers epigynous; plants shrubby . . . . GROSSULARIACEAE

      Stamens numerous . . . . ROSACEAE

[1] The flowers of the *Grossulariaceae* are epigynous and exceptional in the order. The family is probably misplaced here. The flowers of the *Crassulaceae* are hypogynous, and so exceptional also, but vegetative parts are specialized.

ROSACEAE: Rose Family

Trees, shrubs, and herbs, with mostly alternate, simple or compound leaves, which often have prominent stipules. Flowers perfect, regular, 5-merous, and perigynous, the hypanthium saucer-shaped, cup-like, or urn-like, free or adnate to the ovary. Stamens numerous and free from each other. Carpels 1–many, separate or united, the styles commonly as many as the carpels when the latter are united or sometimes joined. Fruit a group of akenes or follicles, a pome, or a drupe, or an aggregate of akenes or drupelets.

The family, as treated here, includes about 100 genera and 3,000 species, cosmopolitan in distribution. There are three subfamilies, well represented in North America, which are treated as families by some.

Key to the Subfamilies of Rosaceae

Hypanthium free from the gynoecium; fruit not a pome

Woody or mostly herbaceous plants, often with persistent stipules; fruit a group of akenes or follicles, or aggregates of akenes or drupelets . . . . 1. ROSOIDEAE

Woody plants with deciduous stipules; fruit usually a single drupe . . . . 2. AMYGDALOIDEAE

Hypanthium adnate to the gynoecium, the flowers thus appearing as if epigynous; fruit a pome . . . . 3. MALOIDEAE

SUBFAMILY 1. ROSOIDEAE: *Rose Subfamily*

Mostly herbs, but including some woody plants, the leaves often with persistent stipules. Hypanthium free from the gynoecium. Carpels usually separate and often numerous. Fruit a group of follicles or akenes, a single akene, or an aggregate of akenes or drupelets.

The subfamily includes about 75 genera and 1,200 species, widely distributed.

EXAMPLES: *Rosa:* Roses, a taxonomically difficult genus.

Potentilla: Cinquefoil, with about 300 species.

Rubus: Raspberries and Blackberries, with perhaps 200 species.

Fragaria: Strawberry.

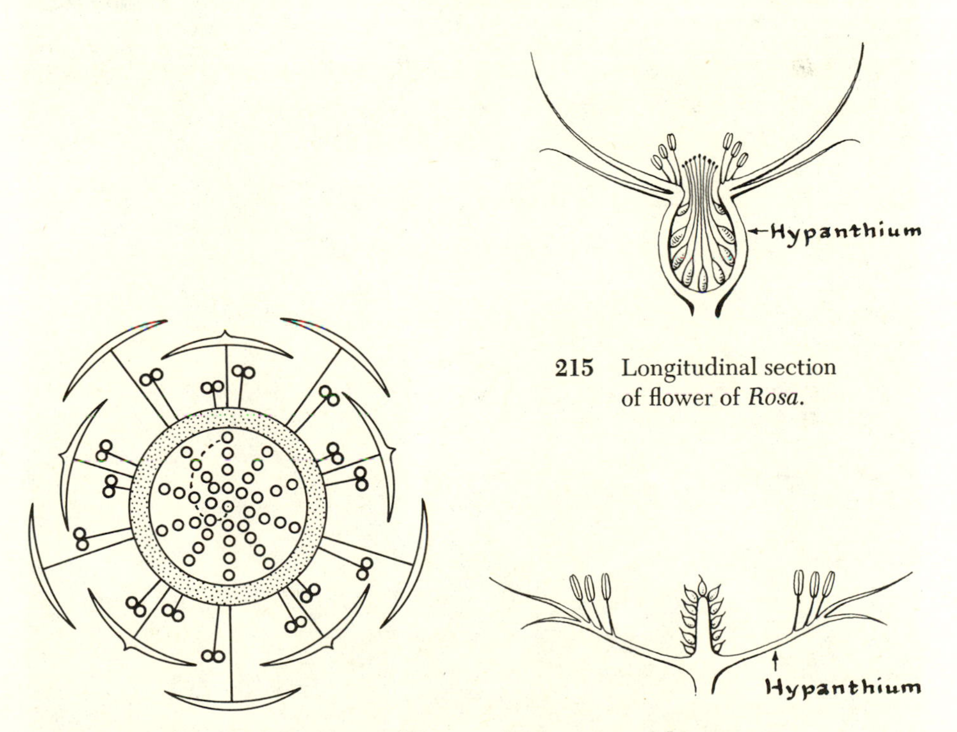

215 Longitudinal section of flower of *Rosa*.

216 *Potentilla*, floral diagram and longitudinal section of flower.

217 *Potentilla fissa* Nutt., a species of Cinquefoil (subfamily *Rosoideae*).

218 *Rosa woodsii* Lindl., a common western American species with pink flowers.

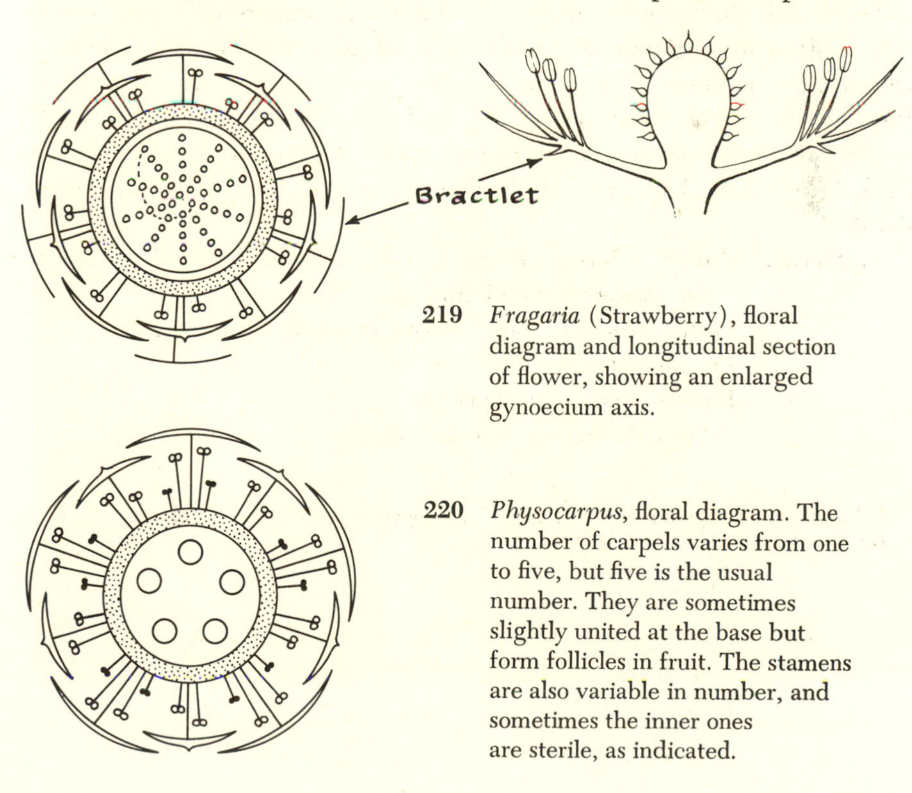

219 *Fragaria* (Strawberry), floral diagram and longitudinal section of flower, showing an enlarged gynoecium axis.

220 *Physocarpus*, floral diagram. The number of carpels varies from one to five, but five is the usual number. They are sometimes slightly united at the base but form follicles in fruit. The stamens are also variable in number, and sometimes the inner ones are sterile, as indicated.

SUBFAMILY 2. AMYGDALOIDEAE: *Peach Subfamily*

Trees and shrubs with alternate simple and serrate or entire leaves, which often have glandular petioles and stipules that are small and deciduous. Hypanthium usually cup-shaped, free from the ovary, deciduous. Pistil usually single, the ovary 1-celled and maturing into a drupe in fruit.[1]

The subfamily includes about 6 genera and 120 species, widely distributed. The leaves, bark, and seeds contain prussic acid and may poison livestock.

EXAMPLES: *Prunus serotina:* Black Cherry, a valuable timber tree of the eastern United States.

Prunus cerasus: Sour Cherry, cultivated for its fruit.

Prunus persica: Peach.

Prunus armeniaca: Apricot.

Prunus domestica: Garden Plum.

[1] The genus *Osmaronia* of the Pacific Northwest is anomalous in having mostly 5 pistils, which mature into 1–5 drupes in fruit.

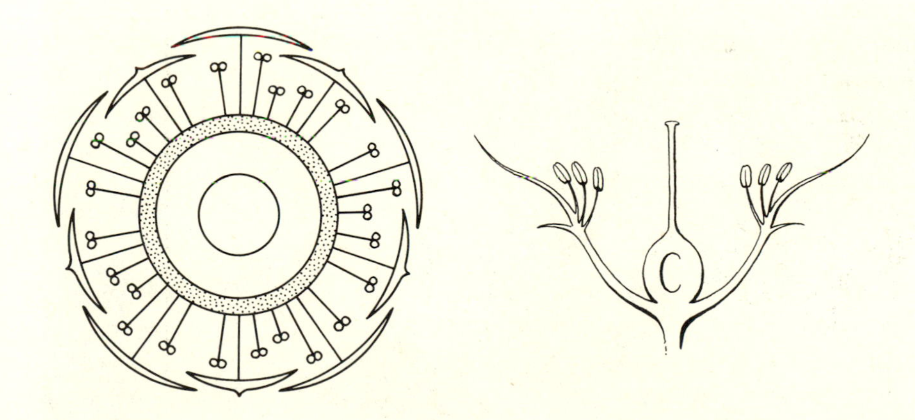

221 *Prunus*, floral diagram and longitudinal section of flower.

222 *Prunus virginiana* L., var. *melanocarpa* (A. Nels.) Sargent, the western American dark-fruited Choke Cherry (subfamily Amygdaloideae).

223 *Prunus americana* Marsh., Wild Plum, a shrub or small tree with white flowers (subfamily *Amygdaloideae*).

SUBFAMILY 3. MALOIDEAE: *Apple Subfamily*

Trees and shrubs with alternate, simple or pinnately compound leaves having small deciduous stipules. Hypanthium adnate to the ovary throughout. Pistil 1, of 1–5 united carpels and maturing into a more or less fleshy pome, the flesh being derived from the enlarged hypanthium, which encloses the bony, leathery, or papery carpels.

The subfamily includes about 20 genera and 500 species, widely distributed.

EXAMPLES: *Malus pumila:* Common Apple, with many varieties.
Pyrus communis: Common Pear.
Cydonia oblonga: Quince.
Crataegus: Hawthorn, or Thorn-apple, a taxonomically difficult genus of at least 300 species.
Sorbus: Mountain Ash, plants with pinnately compound leaves.

CRASSULACEAE: Stonecrop Family

Mostly succulent herbs or small shrubs with thick, alternate or opposite, usually simple, exstipulate leaves. Flowers in cymes, usually perfect, regular, commonly 5-merous, hypogynous. Sepals free or united into a tube. Petals free or somewhat united. Stamens definite and, in gamopetalous flowers, inserted on the corolla tube. Carpels usually free and 5 in number, sometimes united at the base, each carpel usually associated with a glandular appendage perhaps representing additional carpels. Fruit usually a group of follicles, rarely a capsule.

The family includes about 20 genera and 900 species, widely distributed and especially common in South Africa, rare in Australia and the southwest Pacific.

EXAMPLES: *Crassula portulacea:* Japanese Rubber-plant, a common house plant.
Sempervivum tectorum: Houseleek or Hen-and-chickens.
Bryophyllum calycinum: Air-plant, often propagated by leaves.
Sedum: Stonecrop, the largest genus, with about 140 species.

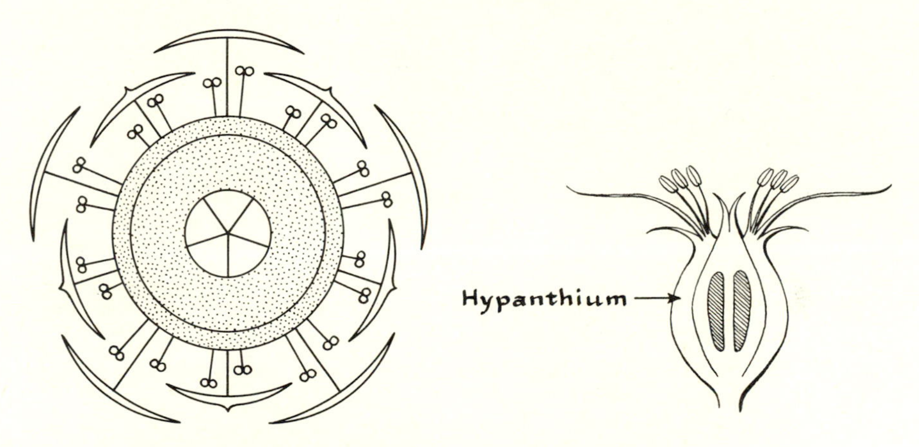

224 *Malus*, floral diagram and longitudinal section of flower. The stippling between the hypanthium and the ovary in the floral diagram indicates adnation and subsequent enlargement.

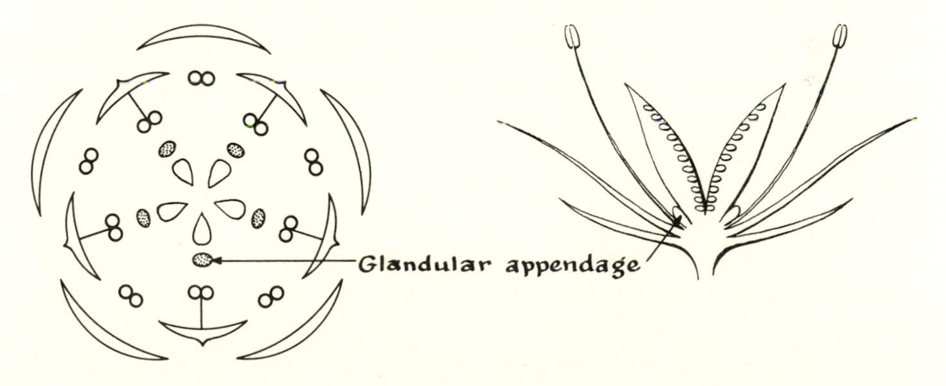

225 *Sedum*, floral diagram and longitudinal section of flower.

HYDRANGEACEAE: Hydrangea Family

Shrubs, vines, or small trees with simple, opposite, exstipulate leaves. Flowers sometimes showy and solitary, but usually small and in corymbose or paniculate cymes, usually perfect and regular (sometimes the marginal flowers sterile and with enlarged sepals), 4–5-merous, perigynous, but the hypanthium somewhat adnate to the lower part of the ovary. Stamens mostly definite. Pistil 1, of 2–10 carpels, which are often free at the apex. Fruit usually a many-seeded capsule.

The family includes about 16 genera and 80 species, mainly in North America and eastern Asia, some of the members being handsome ornamentals.

EXAMPLES: *Philadelphus coronarius:* Syringa, or Mock-orange.
Hydrangea paniculata and *H. arborescens:* Hydrangea.

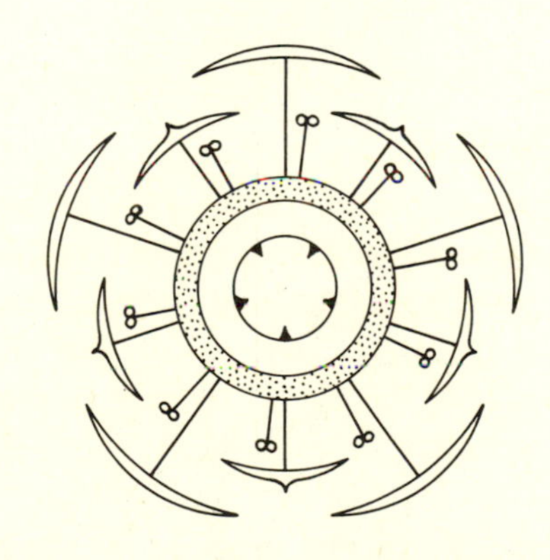

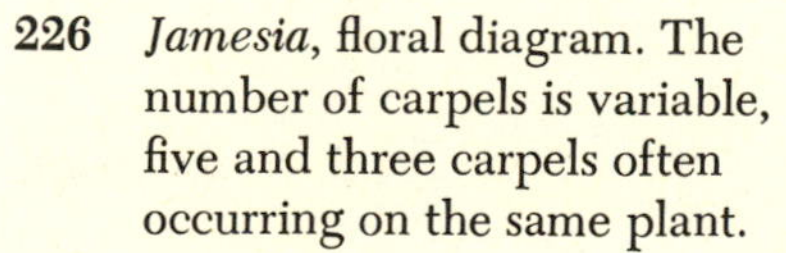

226 *Jamesia*, floral diagram. The number of carpels is variable, five and three carpels often occurring on the same plant.

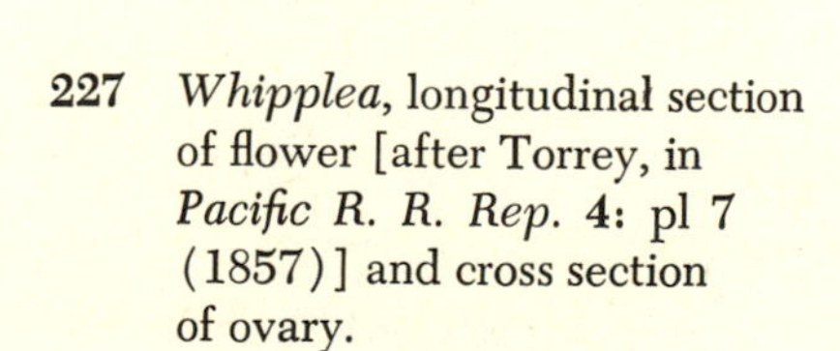

227 *Whipplea*, longitudinal section of flower [after Torrey, in *Pacific R. R. Rep.* **4**: pl 7 (1857)] and cross section of ovary.

SAXIFRAGACEAE: Saxifrage Family

Herbs with mostly alternate or basal exstipulate leaves, which are often palmately veined, are variously toothed or lobed, and are frequently hirsute. Flowers often scapose, usually racemose or paniculate, perigynous, the hypanthium free or rarely adnate to the ovary, regular, 5-merous, the stamens definite and usually 5 or 10. Pistil 1, of 2 (rarely 3 or 4) united carpels, commonly 2-horned at the apex, with 2 styles or stigmas. Ovary 1-celled with 2–3 parietal placentae, or 2–3-celled with axillary placentation. Fruit a many-seeded capsule.

The family includes about 36 genera and 500 species, in temperate, boreal, arctic, and alpine regions.

EXAMPLES: *Saxifraga:* Saxifrage, a genus of about 300 species, some cultivated in rock gardens.

Parnassia: Grass of Parnassus, with about 40 species.

Heuchera: Alumroot, an American genus of about 50 species.

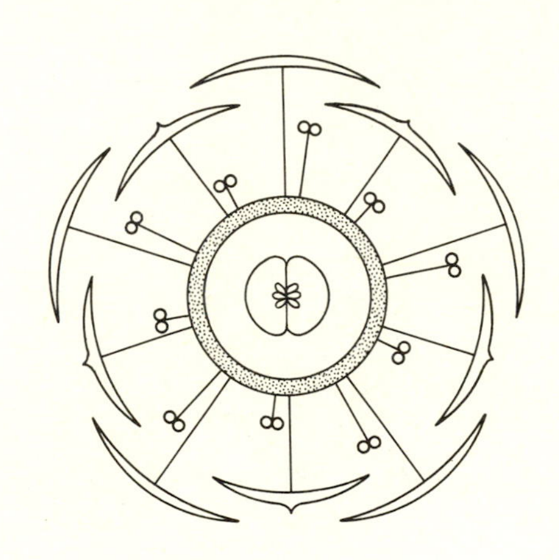

228 *Saxifraga*, floral diagram.

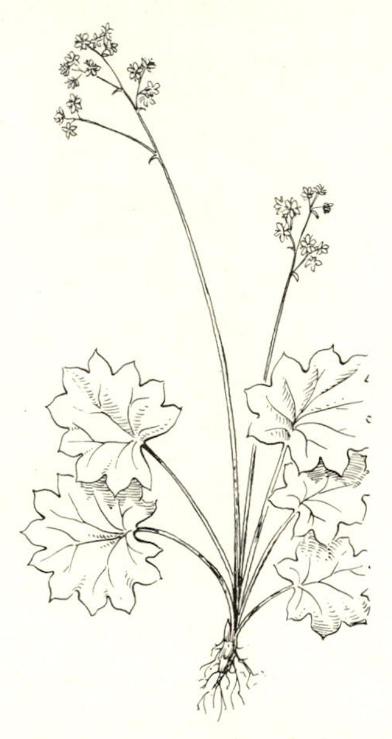

229 *Saxifraga aestivalis*, plant.

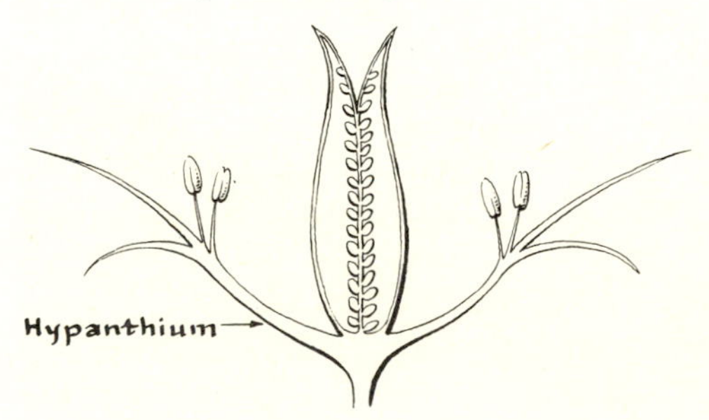

230 *Saxifraga*, longitudinal section of flower, enlarged.

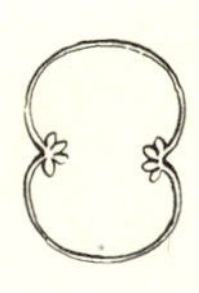

231 *Tiarella*, cross section of ovary, enlarged.

232 *Saxifraga rhomboidea* Greene, a species of Saxifrage that ranges from Montana to New Mexico. The scapes are mostly1–3 decimeters high.

GROSSULARIACEAE: Gooseberry Family

Shrubs, often spiny, with simple alternate leaves, which are palmately veined and often lobed, with or without stipules. Flowers commonly in racemes, usually perfect, 5-merous, regular, with a short or long tubular epigynous hypanthium, this as well as the sepals usually petaloid. Petals smaller than the sepals. Stamens 5, alternate with the petals, inserted on the hypanthium. Pistil 1, of 2 united carpels, the ovary inferior and 1-celled with 2 parietal placentae, the styles and stigmas 1 or 2. Fruit a several-seeded berry.

The family is usually regarded as including the single genus *Ribes,* the Gooseberries and Currants, with some 130 species, in temperate and alpine regions. Affinities seem to lie with the *Rosales,* but the clearly epigynous flowers set the family apart in an anomalous position.

EXAMPLES: *Ribes aureum:* Golden Currant.
Ribes nigrum: Black Currant.
Ribes rubrum: Red Currant.

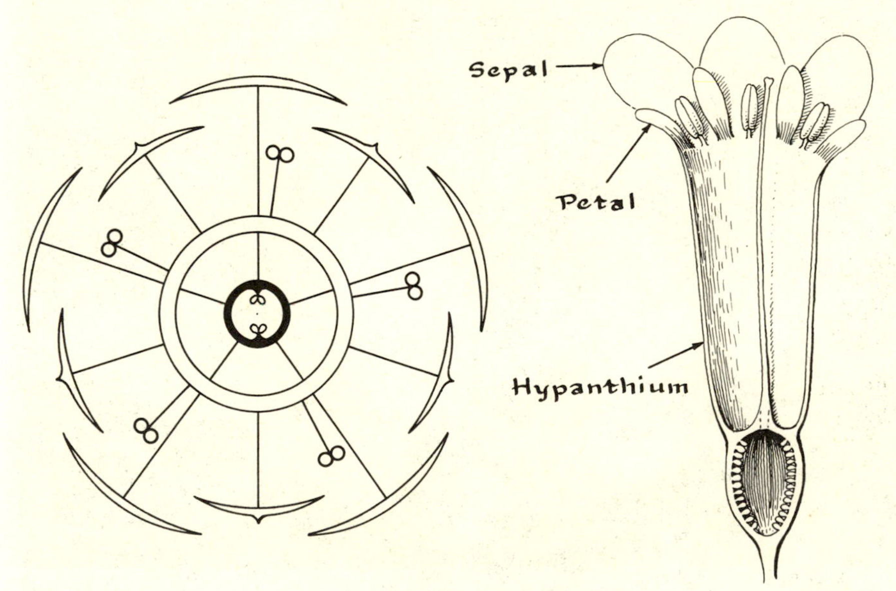

233 *Ribes aureum,* floral diagram and longitudinal section of flower, enlarged.

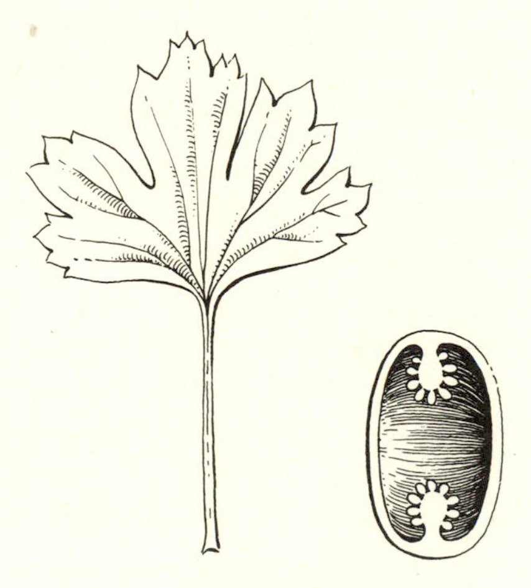

234 *Ribes aureum,* leaf, and cross section of ovary, enlarged.

235 *Ribes cereum* Dougl., Squaw Currant, a shrub with reddish berries. This shows the fruiting stage, with the withered remains of the hypanthium adhering to the top of the ovary.

FABACEAE (LEGUMINOSAE): Pea Family

Herbs, shrubs, or trees with mostly alternate compound leaves, with or without stipules. Flowers usually perfect, regular or irregular, usually perigynous but the hypanthium often very short, mostly 5-merous, the stamens numerous to 10 or less and often monadelphous or diadelphous. Pistil 1, of 1 carpel, forming a legume in fruit.

The family includes about 500 genera and 14,000 species, world-wide in distribution, and including many valuable food and forage crops, ornamentals, and some poisonous plants.

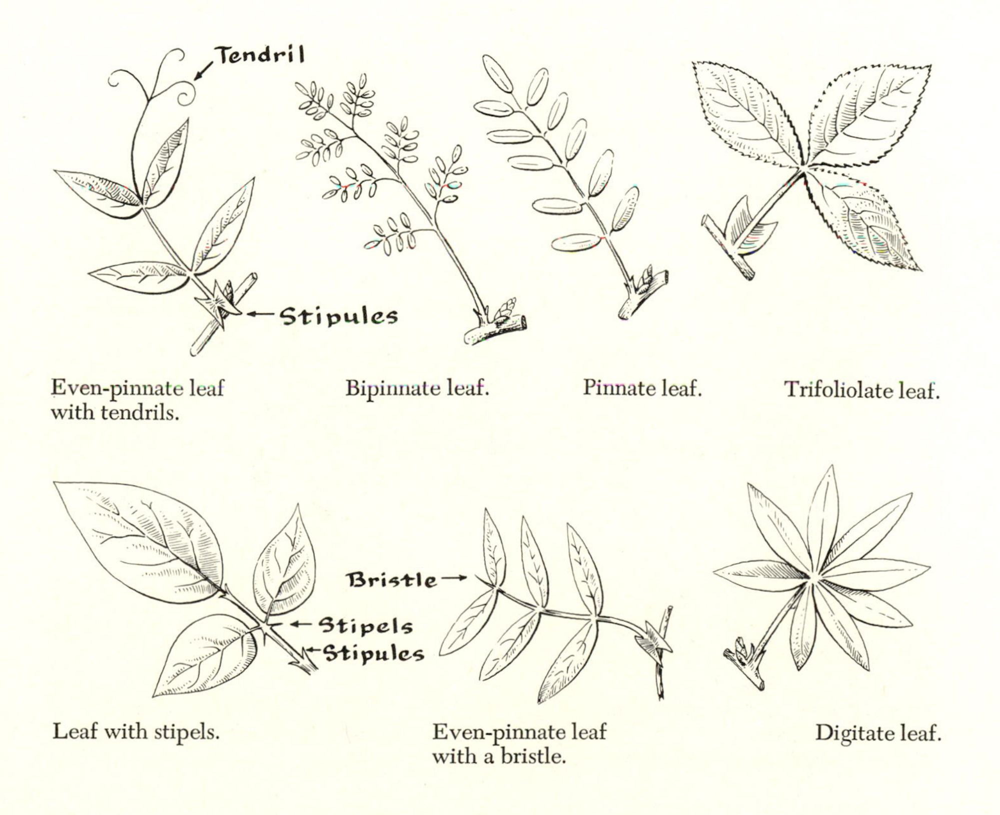

236 Some terms used in descriptions of leguminous plants.

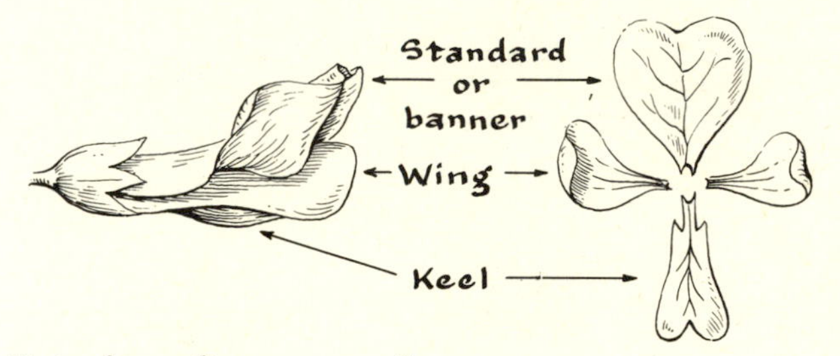

Parts of a papilionaceous corolla.

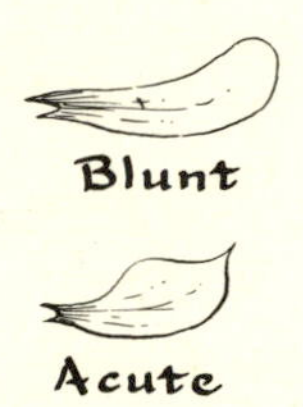

Types of keel petals.

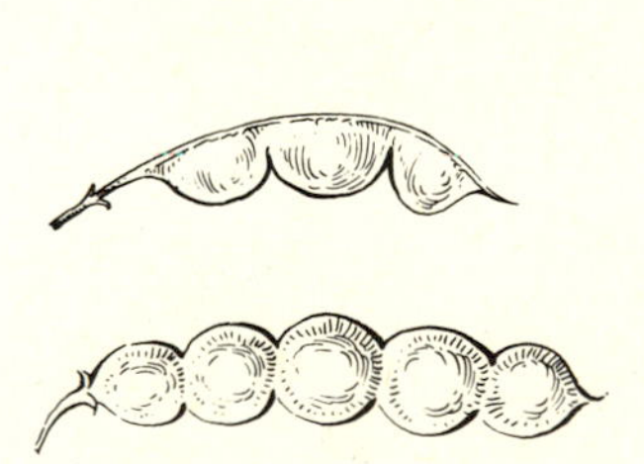

Types of loments, the dehiscence between the seeds.
This is a type of modified legume.

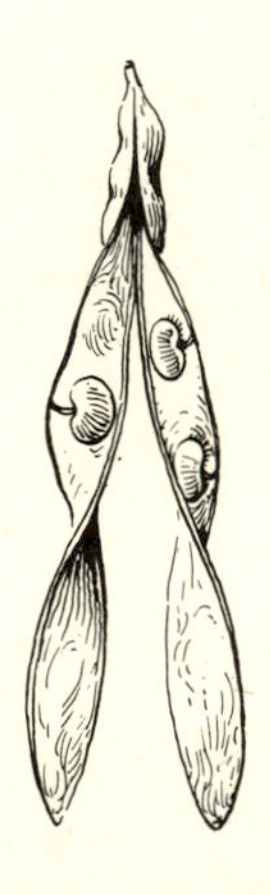

The ordinary type of legume with longitudinal dehiscence.

237 Some terms used in descriptions of leguminous plants.

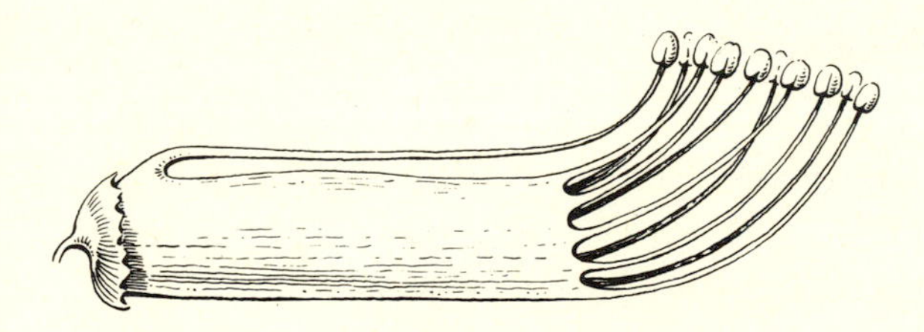

Diadelphous stamens (9+1).

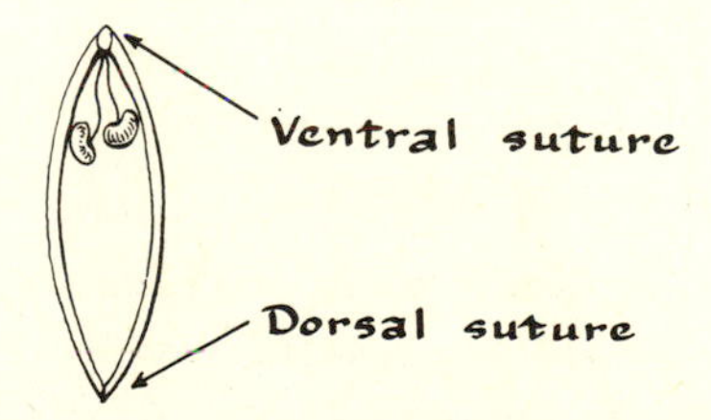

A laterally flattened pod in cross section.

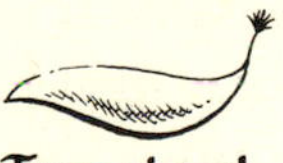

Types of stigmas, lateral and terminal.

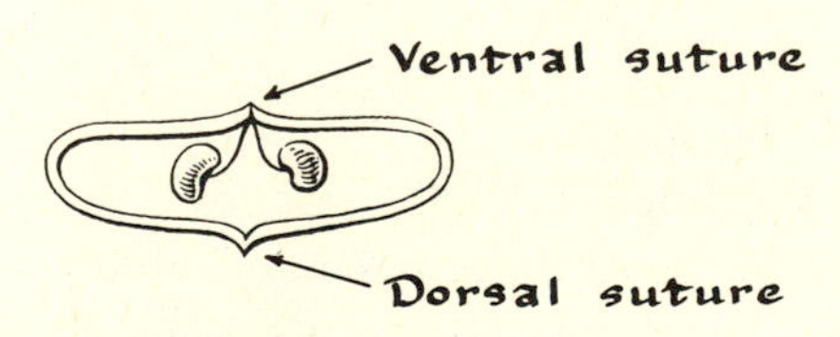

A dorsally flattened pod in cross section.

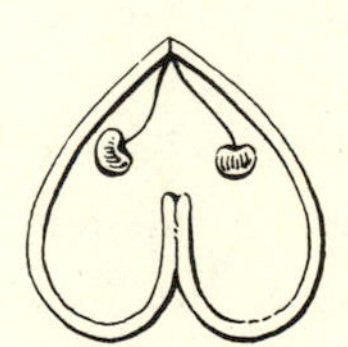

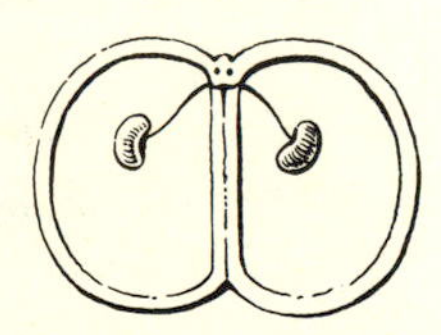

Partly and completely two-celled pods in cross section.

Key to the Subfamilies of Fabaceae

Flowers regular; stamens 10 or more; leaves bipinnate .... 1. MIMOSOIDEAE

Flowers irregular (papilionaceous); stamens 10 or less; leaves mostly pinnate or digitate, rarely simple

Lateral petals (wings) covering the upper one (standard) in the bud .... 2. CAESALPINOIDEAE

Lateral petals (wings) enclosed by the standard in the bud ..... 3. LOTOIDEAE

SUBFAMILY 1. MIMOSOIDEAE: *Mimosa Subfamily*

Trees or shrubs, rarely herbs. Leaves mostly bipinnate, with small leaflets. Flowers perfect, small, spicate, racemose, or capitate, regular, usually 5-merous. Stamens distinct or monadelphous, mostly numerous.

The subfamily includes about 35 genera and 2,000 species, mainly in dry, tropical or subtropical regions.

EXAMPLES: *Acacia:* a genus of about 450 species. *A. senegal* is the source of gum arabic, from tropical Africa.
Mimosa: a genus of about 300 species.
Schrankia uncinata: Sensitive Briar of North America.
Prosopis: the Mesquites of the southwestern United States.

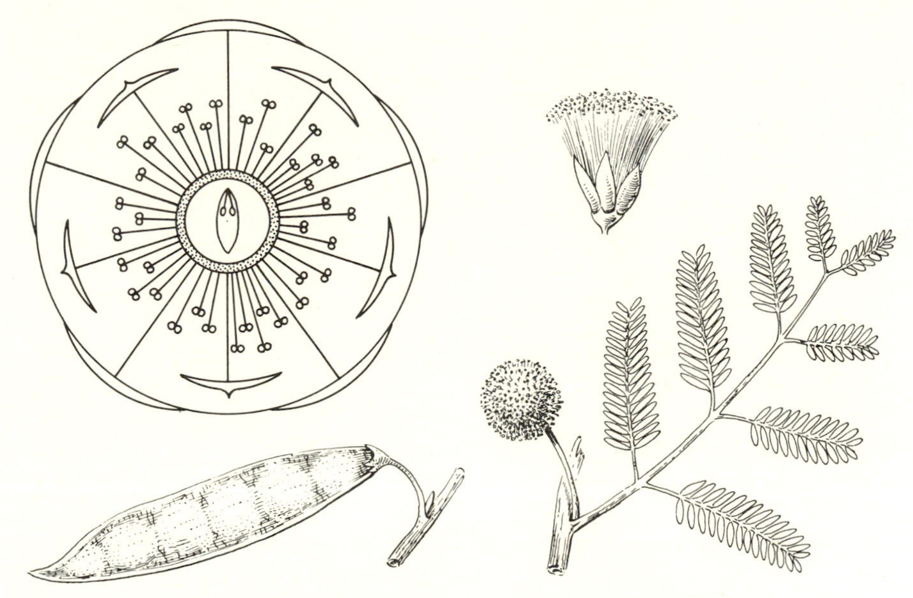

238 *Acacia*, floral diagram, portion of plant, flower, enlarged, and fruit.

239 *Calliandra,* or Powder-puff, a shrub of the subfamily Mimosoideae, with showy clusters of rose-colored stamens from each of the small flowers in the cluster. It is sometimes cultivated in warm regions.

SUBFAMILY 2. CAESALPINIOIDEAE: *Senna Subfamily*

Trees, shrubs, or, rarely, herbs. Leaves usually pinnate or bipinnate, with or without stipules. Flowers showy, racemose or spicate, rarely cymose, somewhat irregular, usually 5-merous, the stamens mostly 10 or fewer, distinct or monadelphous.

The subfamily includes about 60 genera and 2,200 species, some of them valuable timber trees. It is mainly tropical.

EXAMPLES: *Cercis:* Redbud of North America.
Cassia: Senna, species used for the production of dyes.
Gleditsia: Honey Locust.
Gymnocladus: Kentucky Coffee-tree.
Copaifera pubiflora: Purple-heart Wood of British Guiana.
Haematoxylon: Logwood, of tropical America, the source of a dye often used in microscopy.

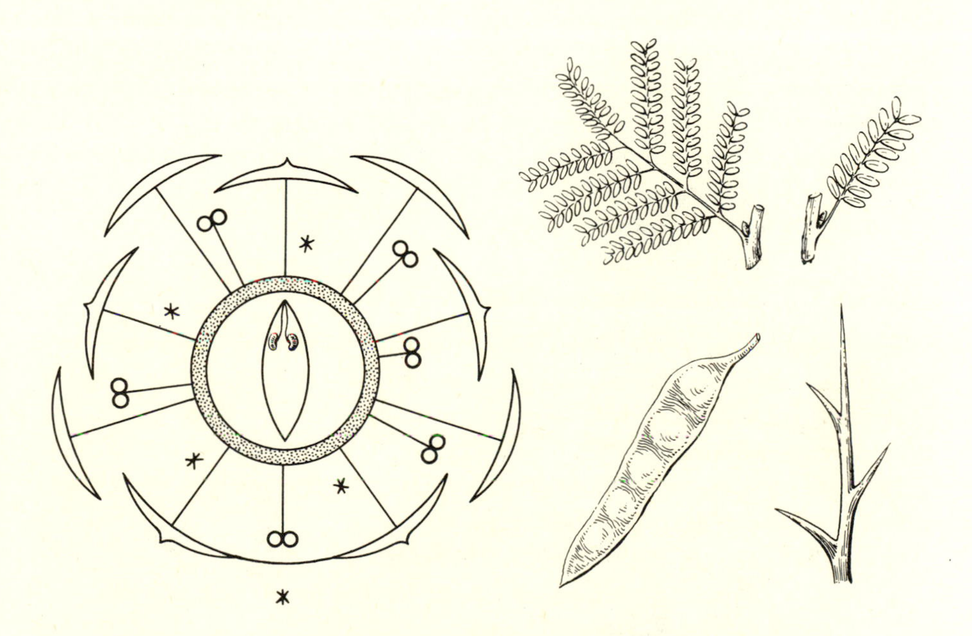

240 *Gleditsia.* (Left) Floral diagram. The flowers are polygamodioecious, and the stamens, sepals, and petals are variable in number. The two lower petals are sometimes united. (Right) Pinnate and bipinnate leaves, fruit, and branched spine from the trunk.

SUBFAMILY 3. LOTOIDEAE: *Pea Subfamily*

Mostly herbs, but some shrubs and trees. Leaves usually pinnate or digitate, less commonly simple, mostly with prominent and persistent stipules, and occasionally with tendrils. Flowers strongly irregular (papilionaceous), perfect, 5-merous, the stamens usually diadelphous (9 united and 1 free) but sometimes monadelphous or all distinct. Ovary and fruit normally 1-celled, but sometimes partly or completely 2-celled by the intrusion of one or both of the sutures.

The subfamily includes about 400 genera and at least 10,000 species, many of which are valuable forage plants, some of them food plants, and several of them commonly cultivated ornamentals. Some members, such as the genus *Oxytropis*, are poisonous to livestock because of their alkaloidal content; others, such as species of *Astragalus*, are selenium absorbers and poisonous because of the accumulation of that element.

EXAMPLES: *Pisum sativum:* Peas.
Lathyrus odoratus: Sweet Pea.
Glycine hispida: Soybean.
Trifolium: Clover.
Medicago sativa: Alfalfa.
Dalbergia: East Indian Rosewood.

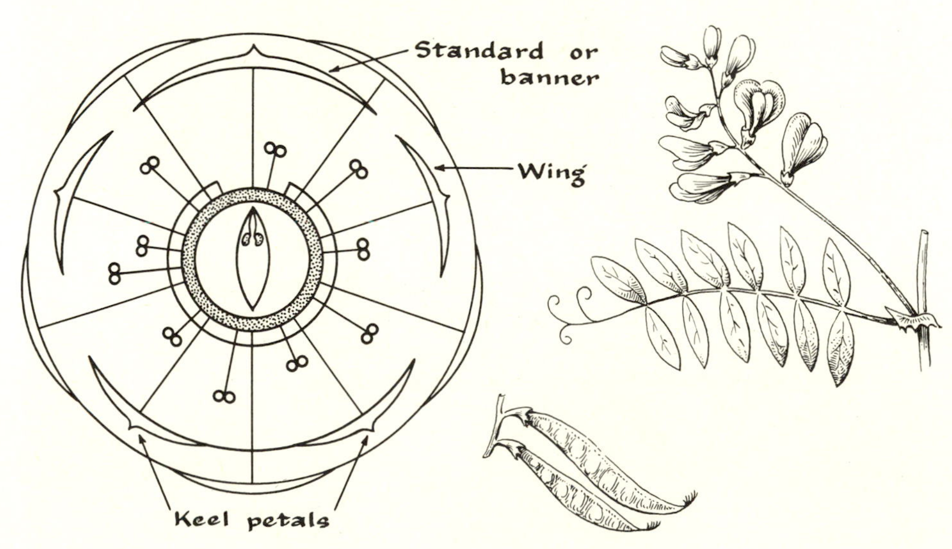

241 *Lathyrus,* floral diagram (stamens diadelphous), portion of plant, and fruit.

242 *Lupinous plattensis* S. Wats., a species of Lupine.

Key to the Tribes of Lotoideae[1]

Stamens distinct, not monadelphous or diadelphous

Leaves pinnate or with a single leaflet; trees or shrubs . . . . SOPHOREAE

Leaves digitate, often 3-foliolate, or with a single leaflet; shrubs or herbs . . . . PODALYRIEAE

Stamens monadelphous or diadelphous

Leaflets not stipellate

Leaves odd-pinnate, without tendrils

Pod not a loment

Leaves glandular-dotted . . . . PSORALEAE

Leaves glandless (except *Glycyrrhiza*, which has pods with hooked prickles and small white flowers)

Stamens mostly monadelphous; anthers of 2 kinds, some large and some small . . . . GENISTEAE

Stamens mostly diadelphous; anthers all alike

Leaves mostly 3–5-foliolate

Leaflets denticulate . . . . TRIFOLIEAE

Leaflets entire . . . . LOTEAE

Leaves mostly pinnate

Pod dehiscent; plants chiefly herbaceous . . . . GALEGEAE

Pod indehiscent; plants woody . . . . DALBERGIEAE

Pod a loment, dividing transversely into 1-seeded joints or articles . . . . HEDYSAREAE

Leaves mostly with tendrils or terminated by a bristle, mostly even-pinnate; shrubs or mostly herbs or herbaceous vines . . . . VICIEAE

Leaflets stipellate; woody or herbaceous vines, or herbs, often twining . . . . PHASEOLEAE

[1] This simplified key will usually produce the correct answer, but it should be remembered that exceptions will occur occasionally.

ORDER GERANIALES

Herbs, undershrubs, shrubs, or trees with hypogynous, perfect, regular to irregular flowers. Corolla of separate and often clawed petals. Stamens typically in two sets, the outer sometimes opposite the petals, or reduced to one set, and sometimes with sterile stamens present. Pistil syncarpous, with axillary placentation, the ovules few or solitary in each carpel.

The order is a large one, of perhaps 20 families, and is treated in different ways by various authors. It may represent an early modification from the *Malvales* by a reduction in stamen number and often by the development of variously modified fruits. The following families are representative.

Key to the Families of Geraniales

Herbage glandular-dotted; plants aromatic; fruit often a hesperidium . . . RUTACEAE

Herbage not glandular-dotted; plants sometimes aromatic; fruit not a hesperidium

  Flowers regular, never spurred

    Herbs with sour juice and 3-foliolate leaves . . . . OXALIDACEAE

    Herbs or shrubs, the leaves not 3-foliolate; juice not sour

      Fruit an elastic capsule with a long beak . . . . GERANIACEAE

      Fruit sometimes elastic but not beaked

        Leaves pinnately compound; stamens distinct; mostly shrubs of the southwestern desert regions, but some annual herbs . . . ZYGOPHYLLACEAE

        Leaves simple and entire; stamens monadelphous; mostly herbs of northern and temperate regions . . . . LINACEAE

  Flowers irregular, with or without a spur

    Flowers not spurred . . . . POLYGALACEAE

    Flowers spurred

      Fruit an elastic capsule with a long beak (the African genus *Pelargonium*) . . . . GERANIACEAE

      Fruit often elastic but not beaked

        Leaves peltate; stamens 8; spur straight . . . . TROPAEOLACEAE

        Leaves not peltate; stamens 5; spur curved . . . . BALSAMINACEAE

RUTACEAE: Rue Family

Aromatic trees or shrubs, rarely herbs, with simple or compound leaves, which are exstipulate, glabrous, often leathery, and usually glandular-dotted. Flowers perfect or rarely unisexual, usually regular, typically 5-merous, with a fleshy disk between the stamens and the ovary. Stamens numerous and in several bundles (polyadelphous) or 5 or 10, distinct, the outer stamens often opposite the petals. Pistil 1, usually of 4–5 united carpels, the ovary often deeply lobed. Fruit a berry with a leathery pericarp (hesperidium), a drupe, or rarely a capsule or samara.

The family includes 5 subfamilies, about 140 genera, and 1,500 species, widely distributed, mostly in tropical and subtropical regions, and best developed in South Africa and Australia.

EXAMPLES:

Citrus sinensis: Orange
Citrus limonia: Lemon
Citrus maxima: Grapefruit
Citrus media: Citron
} citrus fruits, natives of subtropical Asia.

Xanthoxylum americanum: Prickly Ash of North America.
Ptelea trifoliata: Wafer-ash of North America.
Ruta graveolens: Rue, a native of Europe and cultivated.

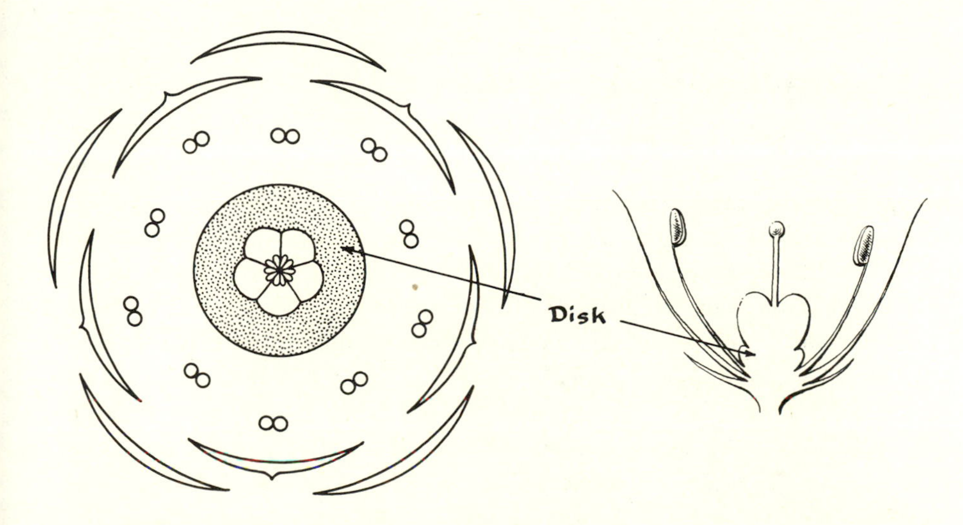

243 *Ruta*, floral diagram of terminal flower, which is 5-merous (the lateral flowers are 4-merous), and generalized longitudinal section of flower.

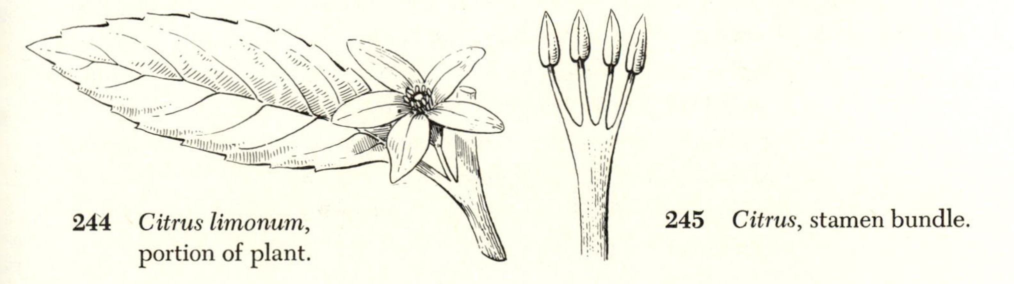

244 *Citrus limonum*, portion of plant.

245 *Citrus*, stamen bundle.

OXALIDACEAE: Wood Sorrel Family

Annual or perennial, scapose or leafy-stemmed herbs from rhizomes or scaly bulbs, the juice sour because of the presence of oxalic acid. Leaves (in ours) 3-foliolate, often long-petioled, the leaflets obcordate and characteristically folding at night. Flowers sometimes dimorphic or trimorphic because of differences in relative lengths of stamens and styles, hypogynous, perfect, often on long peduncles, solitary or in few-flowered umbel-like cymes, the corolla yellow, white, or purple. Stamens united by their filaments at the base (monadelphous), usually 10 and in two sets, one set longer and one set shorter, the outer set opposite the petals, or occasionally with an additional outer set of sterile stamens. Pistil 1, of 5 united carpels, the styles usually distinct. Fruit a loculicidal capsule or (in the Asiatic genus *Averrhoa*) a berry. Seeds 2 or more in each carpel, often discharged by the elastic separation of an outer layer (aril) of the seed coat from an inner harder layer.

The family includes about 7 genera and 900 species, mainly in tropical and subtropical regions, a few extending into temperate regions.

EXAMPLES: *Oxalis:* Wood Sorrel, with about 800 species, including some weeds in North America, and a few ornamentals.

Averrhoa: a tree, cultivated in the tropics for its edible berry, which tastes like a gooseberry.

Biophytum: with 30 species in the tropics, having leaves that are sensitive to touch.

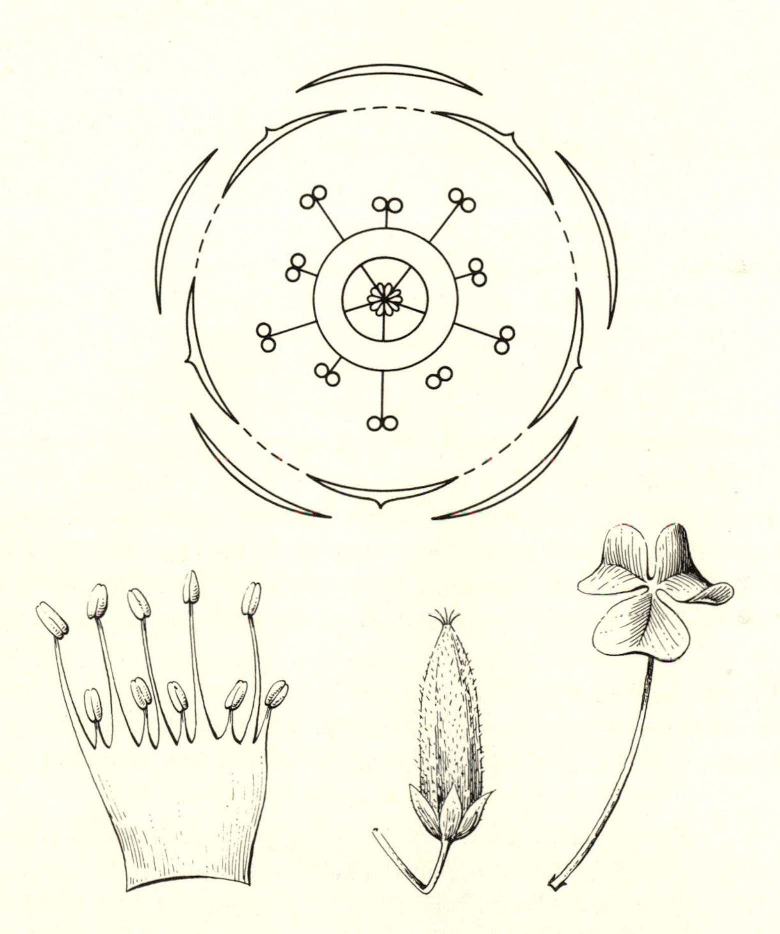

246 *Oxalis*, floral diagram, stamens with the tube opened up, capsule, and leaf.

GERANIACEAE: Geranium Family

Herbs or soft-woody semishrubs, with mostly alternate and often palmately veined leaves. Pubescence often glandular. Flowers regular to somewhat irregular, hypogynous, perfect, 5-merous, with clawed petals. Stamens mostly 5 or 10, rarely 15, some of them often sterile. Pistil of 5 united carpels, the styles long and united, the stigmas distinct. Fruit a long-beaked elastic capsule in which the styles split away from the central axis from the base upward, each carpel with 1 or 2 seeds.

The family includes 11 genera and about 650 species, in temperate regions.

EXAMPLES: *Geranium:* Wild Geranium, or Crane's-bill, with numerous species in Europe and North America.

Pelargonium: cultivated Geraniums, often grown in window boxes, as house plants, and for borders, with numerous species in South Africa.

Erodium: Stork's-bill, several species of common weeds.

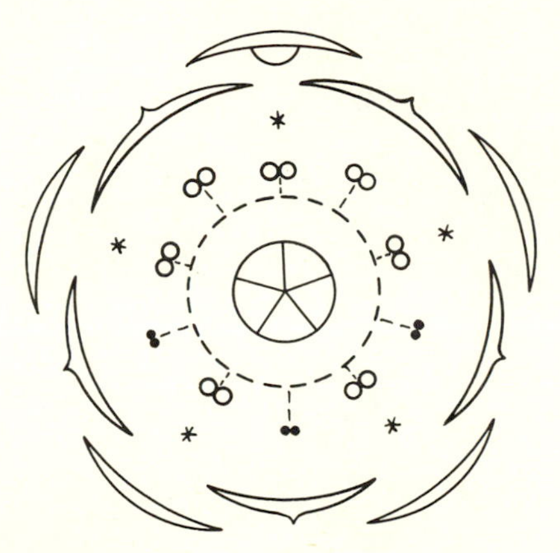

247 *Pelargonium*, floral diagram, the upper sepal spurred, and the corolla slightly irregular.

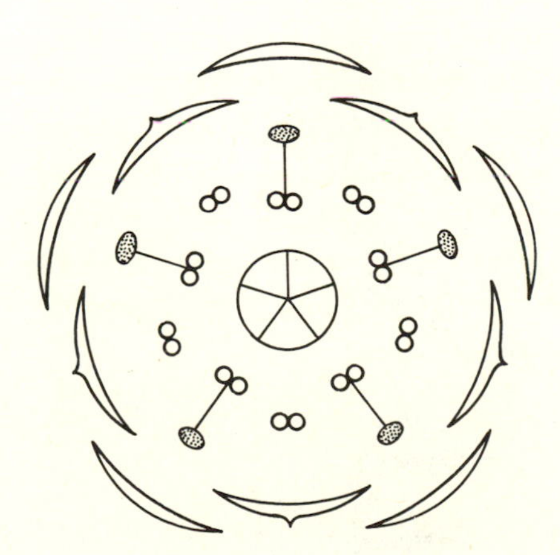

248 *Geranium*, floral diagram, the outer set of stamens reduced to nectar glands, which lie at the base of the filaments of the inner stamens.

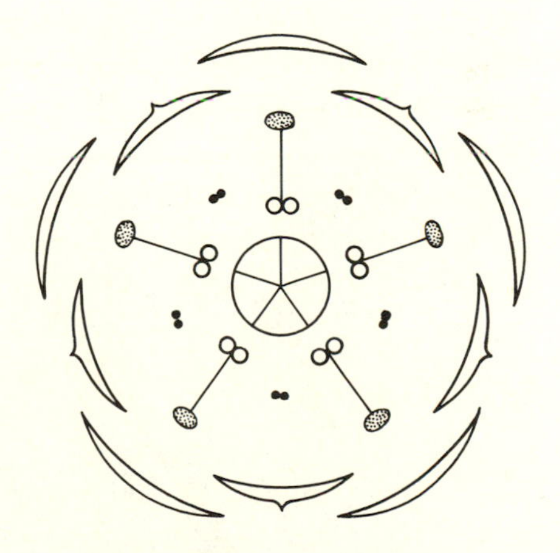

249 *Erodium*, floral diagram, showing five sterile stamens and five nectar glands, which probably represent a reduced outer set of stamens, as in *Geranium*.

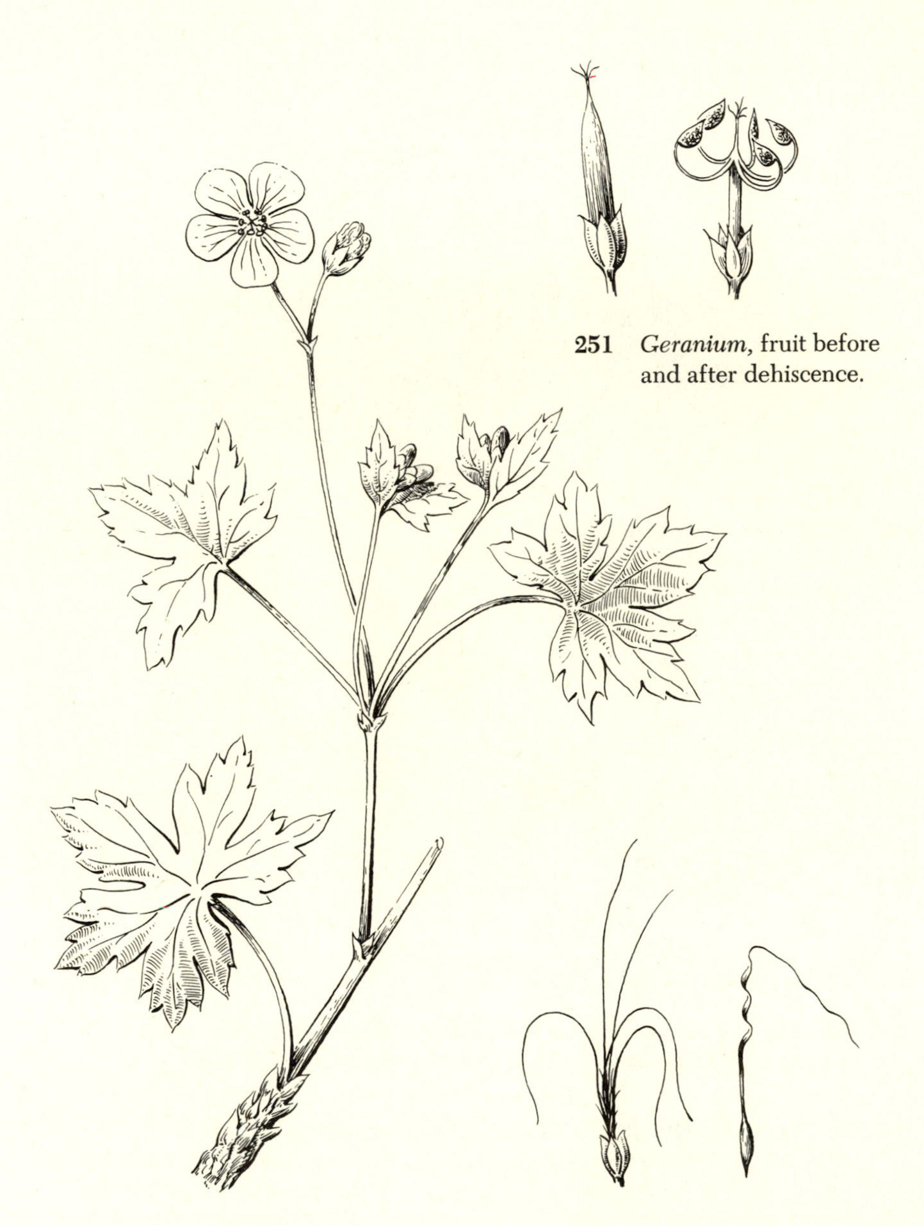

251 *Geranium*, fruit before and after dehiscence.

250 *Geranium richardsonii*, portion of plant.

252 *Erodium cicutarium*, fruit.

253 *Erodium cicutarium* (L.) L'Her., Stork's-bill, or Filaria, a common weed. Long-beaked capsules are shown on the left and right.

ZYGOPHYLLACEAE: Caltrop Family

Herbs or shrubs, with mostly opposite, compound, stipulate leaves and entire leaflets. Flowers hypogynous, perfect, and regular, 5-merous, with usually 10 distinct stamens in 2 whorls, and a single pistil of 2–6 united carpels, the styles united into a short column. Fruit a capsule, berry, or drupe, or sometimes (in *Tribulus*) forming a bur-like schizocarp.

The family includes 25 genera and about 160 species, in warm and dry regions.

EXAMPLES: *Guaiacum officinale:* a small tree of tropical America, which furnishes a gum used in medicine.

Tribulus terrestris: Puncture Vine, a weed in dry areas of the Southwest, extending as far north as Wyoming.

Larrea divaricata: the common Creosote Bush of the deserts of the Southwest.

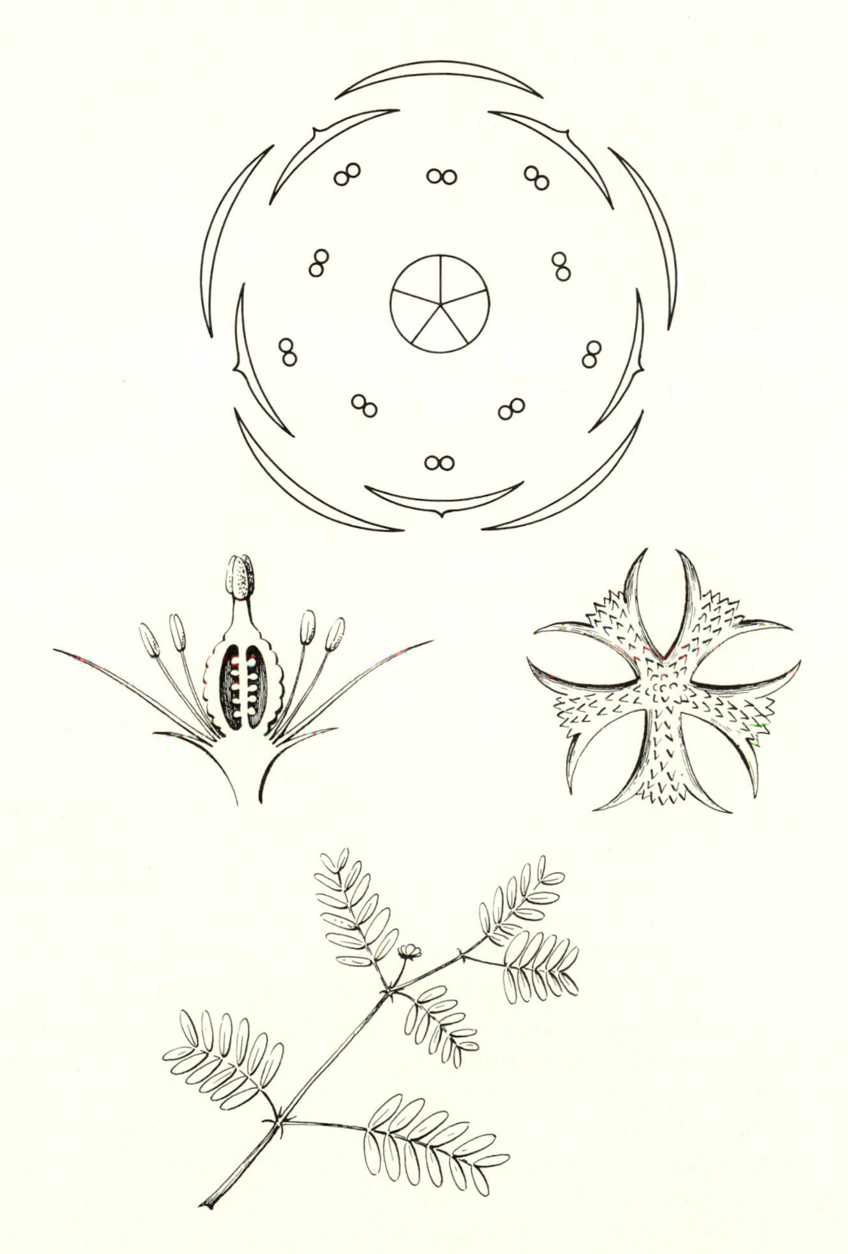

254 *Tribulus*, floral diagram, flower section, fruit, and portion of plant.

LINACEAE: Flax Family

Mostly herbs, with alternate entire leaves and perfect, regular, hypogynous, 5-merous flowers. Petals fugacious, clawed, in ours usually blue or yellow, rarely white. Stamens 5, in ours, alternate with the petals, united by their filaments below into a short tube (monadelphous), these fertile stamens often alternating with very short staminodia. Pistil 1, of 5 united carpels, the styles distinct or united below, the ovary 10-celled because of a false septum in each carpel. Ovules and seeds normally 2 in each carpel. Fruit a septicidal capsule.

The family includes 9 genera and about 150 species, in temperate and tropical regions.

EXAMPLES: *Linum usitatissimum:* Flax, cultivated for the stem fibers, which make linen, and for the seeds, which yield linseed oil.
Linum grandiflorum: a red-flowered cultivated ornamental.
Hugonia: a genus of tropical trees and shrubs with flowers having 10–25 stamens.

POLYGALACEAE: Milkwort Family

Herbs, shrubs, woody vines, or trees, with simple, alternate, opposite, or whorled leaves. Flowers appearing as if papilionaceous, perfect, hypogynous, irregular. Calyx of 5 unequal sepals, the 2 inner ones (wings) often petaloid and larger than the 3 smaller outer ones. Corolla of 5 or often 3 petals, which are more or less connate, the lowest one often concave, appearing like a keel, and with or without a fringed crest. Stamens 3–8 (often 8), monadelphous, the tube often split along the upper side, the anthers usually 1-celled and opening by a terminal or subterminal pore. Pistil 1, usually of 2 united carpels, with axillary placentation. Fruit usually a loculicidal capsule with a single seed in each carpel.

The family includes about 12 genera and 1,000 species, cosmopolitan in distribution but not extending into arctic regions.

EXAMPLES: *Polygala senega:* Seneca Snakeroot.
Polygala paucifolia: Flowering Wintergreen.

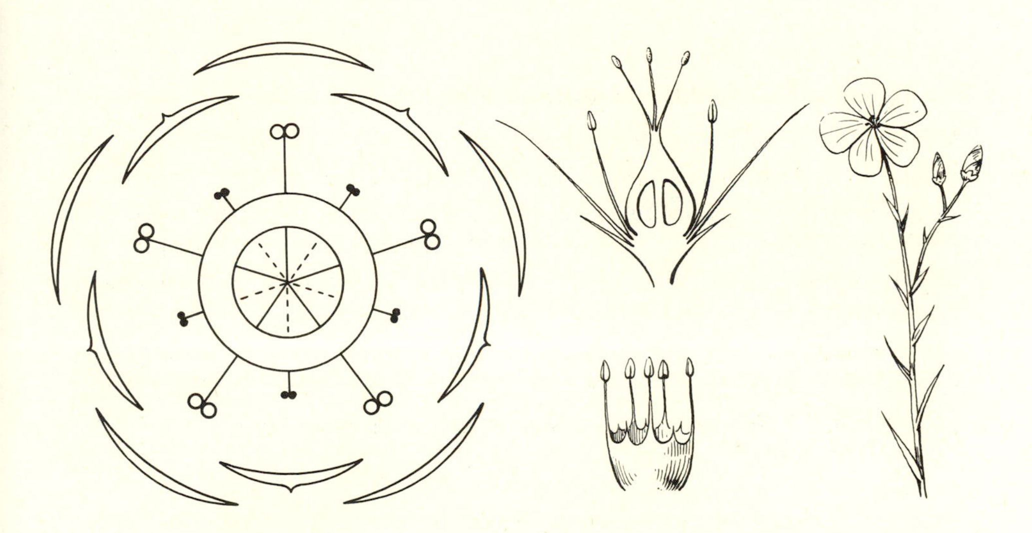

255 *Linum*, floral diagram, longitudinal section of flower, and stamen structure.

256 *Linum lewisii*, portion of plant.

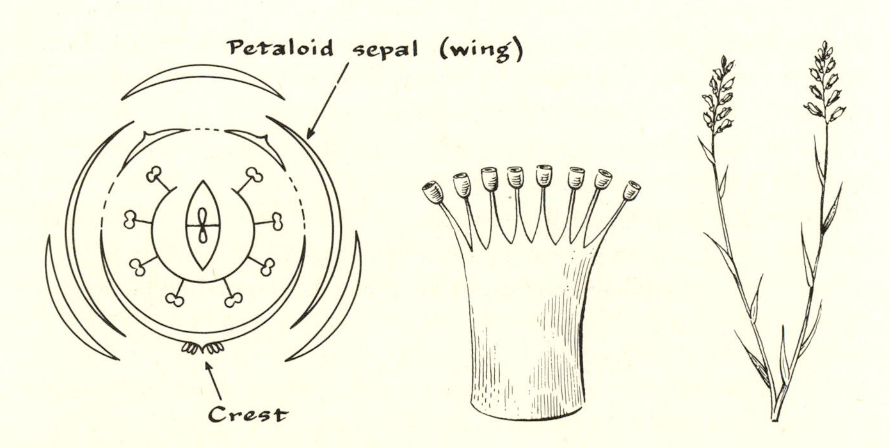

257 *Polygala*, floral diagram, group of eight stamens and stamen tube, and portion of plant.

TROPAEOLACEAE: Nasturtium Family

Annual or perennial, succulent, often twining herbs with pungent juice and orbicular, exstipulate, peltate leaves. Flowers showy, solitary, axillary on long peduncles, irregular, 5-merous, the calyx petaloid, the upper sepal with a long straight spur, the 3 lower petals bearded. Stamens 8. Pistil 1, of 3 weakly united carpels, the ovary 3-lobed, the style single, and stigmas 3. Fruit nut-like, of 1-seeded carpels, one or two often aborting.

The family includes the single genus *Tropaeolum,* ranging from Mexico southwards, in the Andes to Chile, and with 3 species in southern Brazil.

EXAMPLE: *Tropaeolum majus:* Nasturtium, a native of Peru.

BALSAMINACEAE: Jewelweed Family

Herbs with succulent stems, thin, alternate, exstipulate, pinnately veined leaves, and showy, bilabiate, axillary flowers. Dorsal sepal large, petaloid, and with a curved spur, two of the lateral sepals small and greenish, and the two lower lateral sepals much reduced or lacking. Petals 5, the lateral ones united. Stamens 5, often united by their filaments (monadelphous) or by their anthers (syngenesious). Pistil 1, of 5 united carpels, the style short or lacking, the stigma 5-toothed or 5-lobed. Fruit usually an oblong, succulent, elastically dehiscent capsule with several seeds.

The family includes 2 genera and about 400 species, mainly in tropical Asia and Africa, with a few in North America.

EXAMPLES: *Impatiens:* Jewelweed, or Touch-me-not, with about 400 species, chiefly in India, a few species cultivated.

Hydrocera triflora: of southern Asia, a monotypic genus.

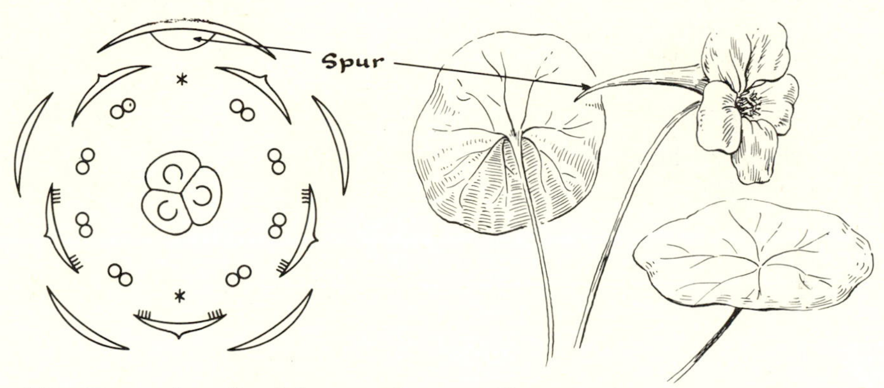

258 *Tropaeolum*, floral diagram, flower and leaves.

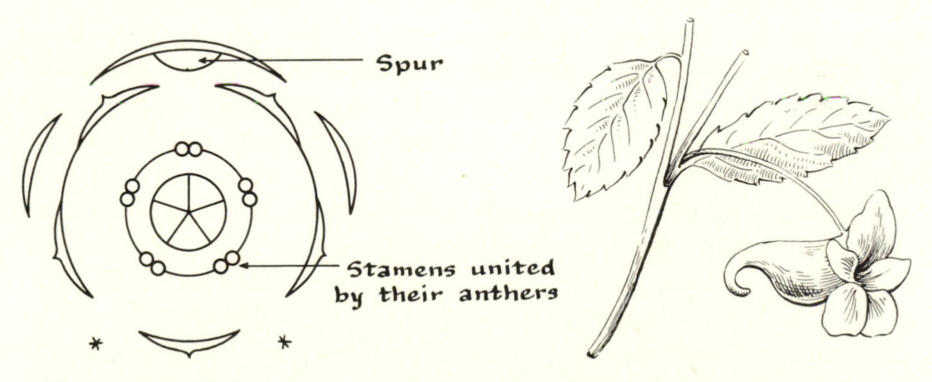

259 *Impatiens balsamina*, floral diagram.

260 *Impatiens biflora*, portion of plant.

ORDER MALVALES

Herbs, shrubs, and trees, with mostly perfect, hypogynous, polypetalous flowers. Stamens numerous, monadelphous or polyadelphous. Carpels numerous or as few as three, often weakly united and forming a ring, the placentation axillary.

The order probably represents an early modification from the *Ranales,* with tendencies toward united stamens and a characteristic ring of carpels. The following families are representative.

TILIACEAE: Basswood Family

Mostly tropical trees, shrubs, or herbs, often with strong phloem fibers. Leaves usually alternate and simple, with stipules, the blade often oblique at the base. Flowers regular, usually perfect, usually 5-merous, in cymes or panicles. Stamens 10 to many, often united by the basal part of their filaments into 5 or 10 bunches (polyadelphous), some of the stamens sometimes modified into petaloid staminodia. Pistil 1, of 2–10 carpels, the ovary 2–10-celled with 1–several ovules in each cell, the placentation usually axillary, the style single. Fruit variable, fleshy or dry, and dehiscent or indehiscent.

The family includes about 40 genera and 400 species, mostly tropical.

EXAMPLES: *Tilia:* Basswood, or Linden, a common tree of Europe and North America, often planted as a street tree, and the wood extensively used commercially. Nectar from the flowers yields a characteristically flavored honey.

Corchorus: herbaceous, often cultivated for its fiber, which is known commercially as jute.

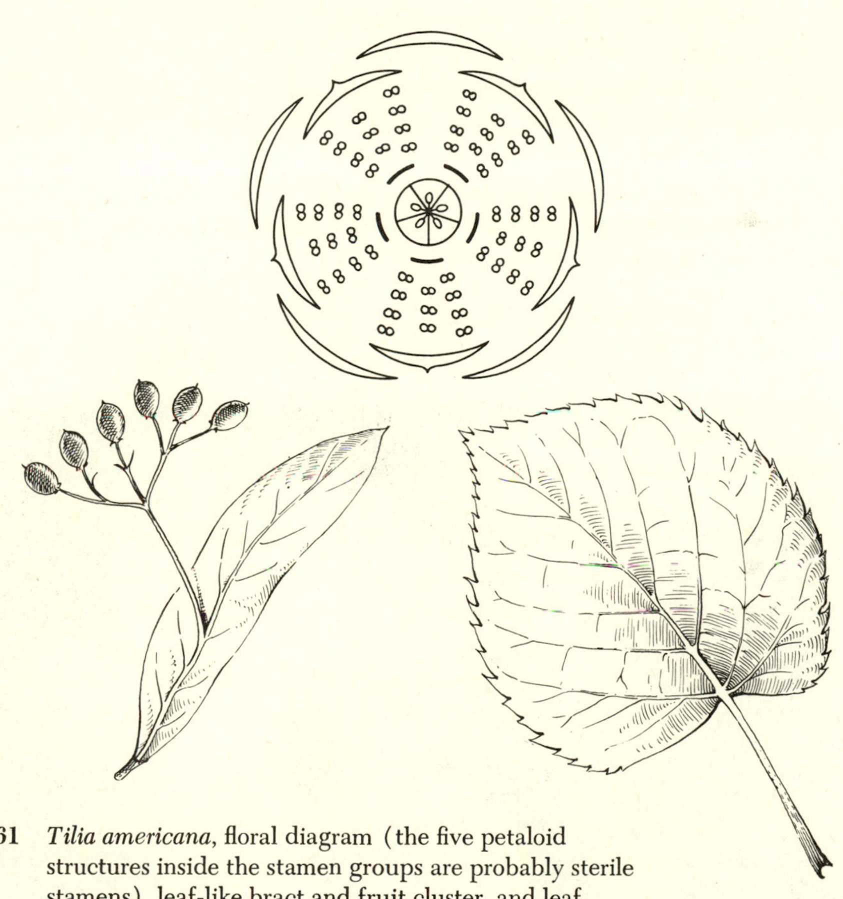

261 *Tilia americana,* floral diagram (the five petaloid structures inside the stamen groups are probably sterile stamens), leaf-like bract and fruit cluster, and leaf.

MALVACEAE: Mallow Family

Herbs (some shrubs and trees in tropical regions) with alternate, simple, and usually palmately veined leaves. Flowers often showy, 5-merous, hypogynous, the petals often attached to the tube of the numerous monadelphous stamens. Anthers 1-celled. Pistil of 3 to many carpels, these often forming a ring and each containing 1–several seeds. Fruit a capsule or schizocarp.

The family includes about 50 genera and 1,200 species, widely distributed in temperate and tropical regions.

EXAMPLES: *Gossypium:* Cotton, with several important species.
Hibiscus esculentus: Okra (other species ornamental or weedy).
Althaea rosea: Hollyhock, a common ornamental.
Malva: Mallows, or Cheeses, common weeds.

262 *Hibiscus*, portion of plant.

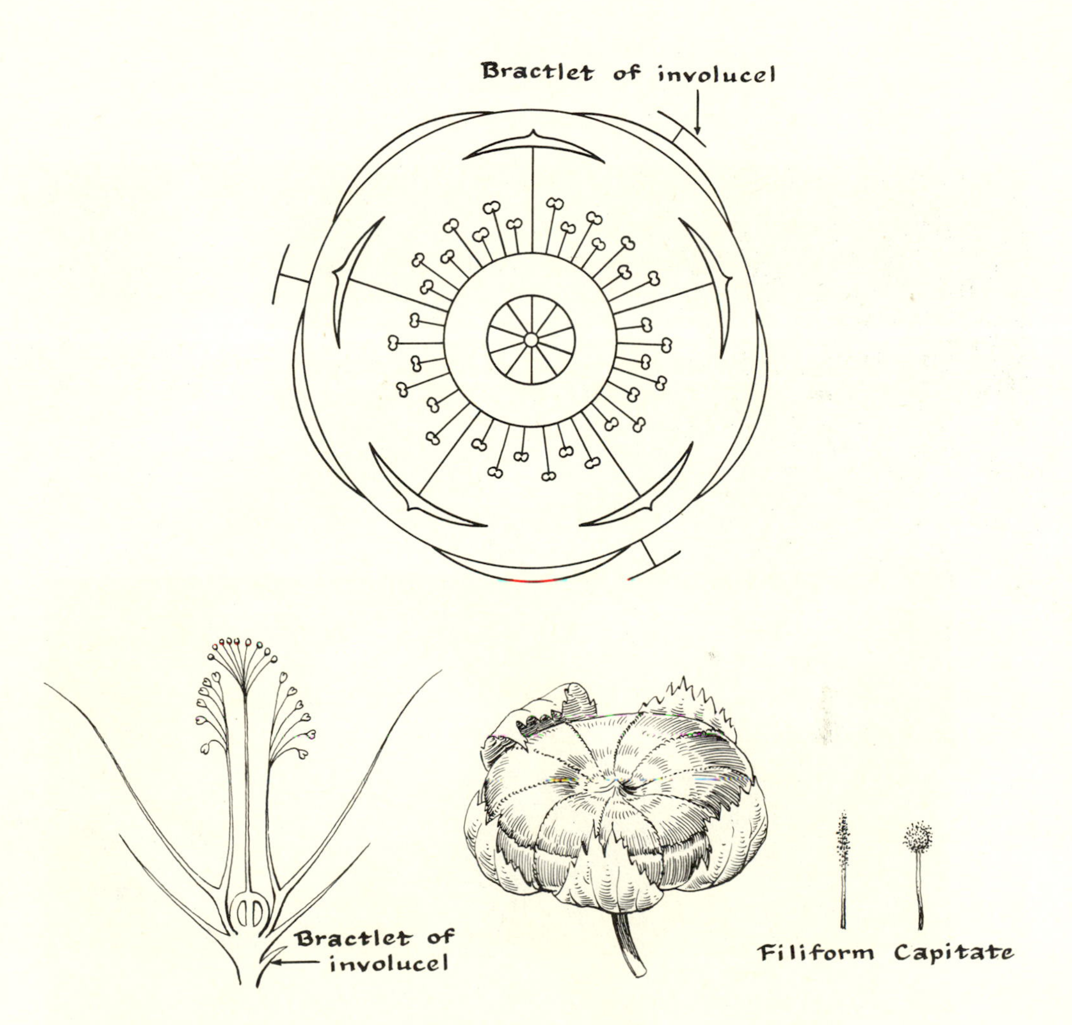

263 *Malva,* floral diagram, longitudinal section of flower, and fruit and persistent calyx. Two stigma types found in the family—filiform and capitate—are shown at the right.

264 *Iliamna rivularis* (Dougl.) Greene, a common western Mallow. The plants are about one meter high, and have white to pink or lavender flowers.

265 *Iliamna rivularis* (Dougl.) Greene, detail of a single flower showing the characteristic numerous monadelphous stamens, and the style branches protruding beyond them.

ORDER GUTTIFERALES

Herbs, shrubs, or trees, with opposite or whorled, simple leaves, which are often punctate or black-dotted. Flowers hypogynous, polypetalous, regular, 4–5-merous, usually with numerous stamens, which are united in bunches by their filaments (polyadelphous), the single pistil of 3–5 united carpels, and the ovary usually with parietal placentation.

The order, as defined by Hutchinson, includes 4 families and is mainly tropical. By others it is combined with a larger number of families in the order *Parietales,* which is generally regarded as a heterogeneous assemblage and not a natural group. The following family is representative.

HYPERICACEAE: St.-John's-wort Family

Herbs or shrubs with simple, opposite or whorled, entire leaves, which are punctate or black-dotted, without stipules. Flowers usually in cymes (dichasia), usually yellow or orange, 4–5-merous, regular. Stamens often numerous but sometimes definite in *Hypericum,* usually united below by their filaments into 3–5 (6–8) bunches. Pistil 1, of 3–5 carpels, the styles distinct or partly united, the ovary 1-celled and with 3–5 parietal placentae, or 3–5-celled by the intrusion of the placentae, and the placentation then apparently axillary. Fruit usually a septicidal, many-seeded capsule, sometimes a berry.

The family, as defined here, includes 3 genera and about 311 species of temperate and tropical regions.

EXAMPLES: *Hypericum:* St.-John's-wort, a genus of about 300 species, about 40 of which occur in the United States. Some are weeds, some are poisonous to livestock, and some are ornamentals.

Ascyrum hypericoides: St.-Andrew's-cross, an evergreen shrub, sometimes cultivated.

Triadenum: Marsh St.-John's-wort.

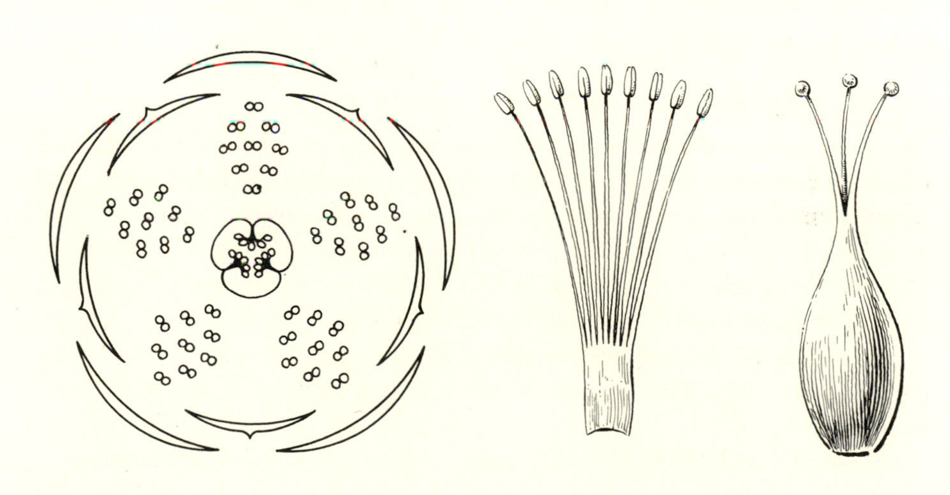

266 *Hypericum*, floral diagram, stamen bundle, and pistil.

ORDER SAPINDALES

Trees or shrubs with simple or compound leaves. Flowers numerous and often small, regular to irregular, hypogynous or perigynous, polypetalous, often with a disk below the ovary. Gynoecium 2–3-carpeled. Fruit often 1-seeded.

The order has been variously regarded by different workers as consisting of from 9 families (Hutchinson) to 23 families (Engler and Diels). Its relationship to other orders is not clear, but the perigynous tendency suggests an ancient connection with the *Rosales,* from which it may have been derived by reduction. The group is mainly tropical. The following families are representative, and may be distinguished by the following key.

Key to the Families of Sapindales

Leaves alternate, usually compound

 Plants resinous . . . . ANACARDIACEAE

 Plants not resinous . . . . SAPINDACEAE

Leaves opposite, usually simple . . . . ACERACEAE

ANACARDIACEAE: Cashew or Sumac Family

Resinous shrubs or small trees, sometimes poisonous, the leaves simple or often pinnately compound. Flowers small and numerous, perfect or unisexual, mostly regular, somewhat perigynous, usually 5-merous, but the stamens often of twice the number of the sepals (in *Mangifera* reduced to a single fertile stamen, the others sterile), and the carpels often 3, but only 1 of these functional in most genera. Fruit usually a drupe.

The family includes about 60 genera and 600 species, mostly in tropical regions but with some representatives in temperate Europe, Asia, and America.

EXAMPLES: *Anacardium occidentale:* Cashew nut of tropical America.

Mangifera indica: Mango, of Asia but cultivated in warm regions elsewhere for the large, delicious fruits.

Pistacia vera: Pistachio nuts of the Mediterranean region.

Rhus toxicodendron and *R. radicans:* Poison Ivy and Poison Oak.

Rhus vernix: Poison Sumac.

Cotinus obovatus: Smoke tree of southern and southeastern United States.

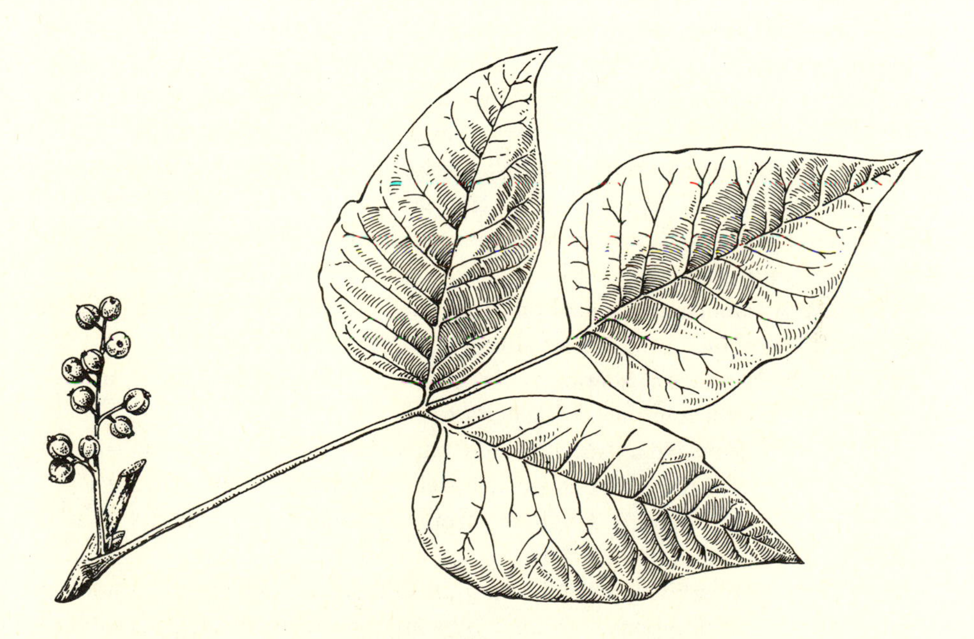

267 *Rhus radicans*, Poison Ivy. The fruits are pale greenish-white.

SAPINDACEAE: Soapberry Family

Trees, shrubs, or sometimes lianas, with mostly alternate, compound, and usually exstipulate leaves. Flowers commonly numerous and small, often polygamo-dioecious and 5-merous, regular or irregular, the petals 3–5, separate, or often lacking. Stamens usually in 2 series, but some often reduced, giving the flower 8 or fewer stamens, inserted inside a disk. Pistil 1, usually of 3 carpels, with a single style and 3-celled ovary having a single ovule and seed in each cell. Fruit various: a large capsule or nut, often winged, or a fleshy berry or drupe. Seeds often with a prominent aril and without endosperm, the embryo curved, folded, or twisted.

A large and important tropical or subtropical family of about 100 genera and 1,600 species, including some valuable timber trees.

EXAMPLES: *Sapindus:* Soapberry, a genus of about 11 species, in the warmer parts of Asia and America.

Litchi chinensis: Litchi, a native of China, widely cultivated for the sweet, fleshy aril that completely surrounds the seed.

Serjania and *Paullinia:* tropical lianas with coiled tendrils, about 300 species.

Aesculus: Horse Chestnut and Buckeye, with about 13 species in temperate regions, by some placed in the *Hippocastanaceae.*

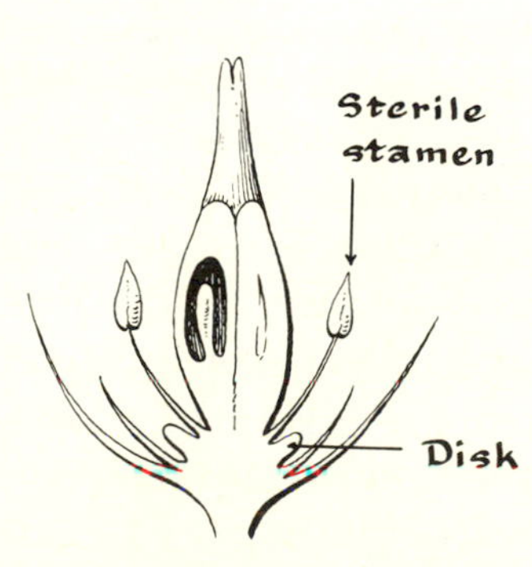

268 *Sapindus*, female flower in longitudinal section.

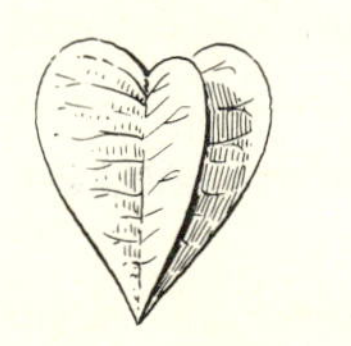

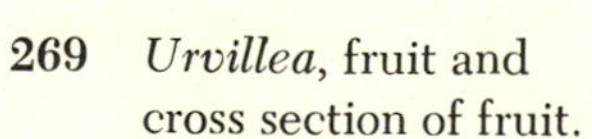

269 *Urvillea*, fruit and cross section of fruit.

270 *Sapindus*, leaf.

ACERACEAE: Maple Family

Trees or shrubs with sweetish, watery sap and opposite, exstipulate leaves with long petioles, the blades varying from simple and entire to palmately lobed, trifoliolate, or pinnately compound, usually palmately veined. Flowers regular, dioecious, polygamous, or sometimes perfect, usually on a short, leafy shoot, appearing before the leaves, somewhat perigynous, often with a disk. Sepals 4–5. Petals 4–5 or none. Stamens 4–12 (usually 7–8), attached to the edge of the disk, which may represent a reduced hypanthium. Pistil 1, of 2 united carpels (occasionally 3), each flattened and becoming winged, the placentation axillary, and seeds 1–2 in each carpel. Fruit a winged schizocarp, which splits into 2 single-winged mericarps (the fruit often described as a paired samara), the wing subterminal in *Acer*, surrounding the carpel in *Dipteronia*.

The family includes 2 genera and about 125 species, mostly in north temperate or mountainous regions. It includes valuable timber trees, which may also produce maple sugar, and several species are valuable shade trees.

EXAMPLES: *Acer saccharum:* Sugar Maple, or Hard Maple.

Acer saccharinum: Silver Maple, with several ornamental cut-leaved varieties.

Acer negundo: Box Elder, with pinnately compound leaves.

Acer macrophyllum: Big-leaf Maple of the Pacific Northwest.

Dipteronia: with two species, in central China.

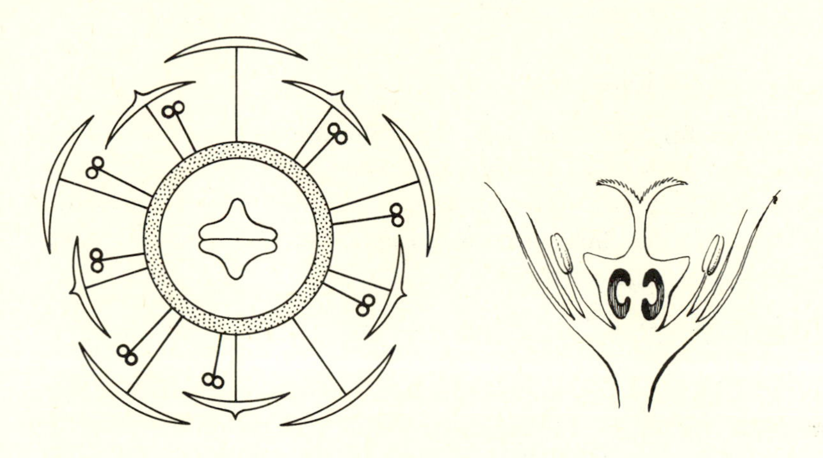

271 *Acer glabrum*, floral diagram and longitudinal section of perfect flower, the latter much enlarged.

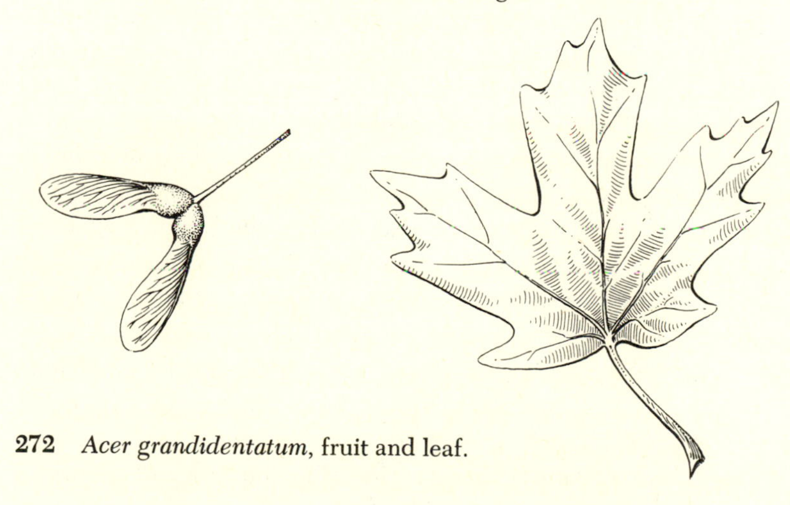

272 *Acer grandidentatum*, fruit and leaf.

273 *Dipteronia*, fruit. [After Pax, in Engler and Prantl, *Nat. Pflanz.* 3^5:269 (1896).]

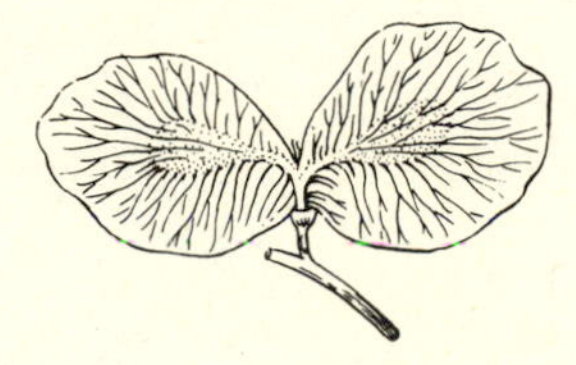

ORDER EUPHORBIALES

An anomalous group, including only the following family. Hutchinson suggests a possible origin from *Malvales* and *Sapindales*. The plants seem to be highly specialized, as evidenced by very reduced flowers in many, the peculiar and flower-like inflorescence in *Euphorbia,* and the milky juice and attendant complex secretory tissue.

EUPHORBIACEAE: Spurge Family

Herbs, shrubs, or trees, often with milky juice. Leaves simple or compound, usually alternate, sometimes reduced to spines. Flowers monoecious or sometimes dioecious, commonly in cymes, with or without a perianth, sometimes with a corolla. Staminate flowers variable, often reduced to a single stamen. Pistillate flowers rather uniform, consisting of a single pistil of mostly 3 carpels, 3-lobed, and forming a capsule or schizocarp which splits into three 1-seeded nutlets in fruit. In specialized forms, such as *Euphorbia,* the inflorescence may simulate a flower, being reduced to one or more cymules or cyathia, these sometimes with ornamental bracts.

The family includes about 283 genera and 7,300 species, widely distributed in temperate and tropical regions, the Indo-Malayan region and Brazil being the chief centers of distribution. Some African species of *Euphorbia* resemble cactus plants. Economically important products include food, drugs, and rubber. Many species are poisonous, the genus *Toxicodendron* of South Africa including some of the most poisonous plants known.

EXAMPLES: *Euphorbia:* the largest genus, with 1,600 species estimated, including *E. pulcherrima,* the Poinsettia.

Hevea braziliensis: an important rubber tree.

Manihot utilissima: Cassava, furnishing tapioca and arrowroot starch from the tuberous roots.

Ricinus communis: Castor Bean, the source of castor oil.

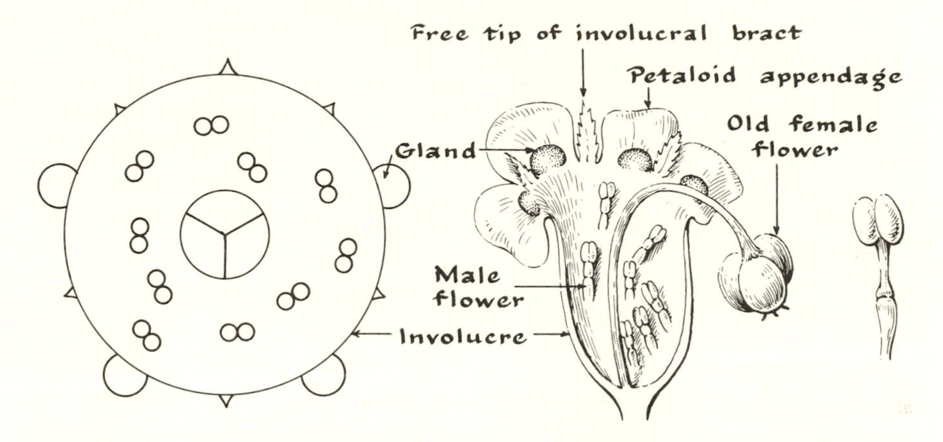

274 *Euphorbia*, diagram of cyathium (cymule), cyathium cut open lengthwise to show structure, enlarged, and single staminate flower on a pedicel.

275 *Euphorbia pulcherrima* Willd., Poinsettia, or Christmas Plant. The leaf-like bracts subtending the inflorescence are red.

ORDER RHAMNALES

Trees, shrubs, or lianas, with simple or palmately compound, mostly stipulate, alternate or opposite leaves. Flowers regular, perfect or unisexual, perigynous, the stamens of the same number as the petals and opposite them or in that position if the petals are lacking. Carpels 1–5, rarely more, united.

The order probably represents an early modification from the *Rosales*, being characterized by having perigynous flowers, but the stamens are definite and in an unusual position. The following families are representative.

RHAMNACEAE: Buckthorn Family

Mostly shrubs or trees, sometimes spiny, with simple, alternate or sometimes opposite 3–5-nerved or pinnately veined leaves, which usually have small and deciduous stipules. Flowers small, in cymes, the corolla inconspicuous or 4–5-merous, perfect or polygamous, perigynous, the hypanthium sometimes becoming adnate to the base of the ovary. Stamens of the same number as the sepals and alternate with them (opposite the petals when these are present). Pistil of 2–4 carpels (commonly 3), forming a capsule or a few-seeded berry in fruit.

The family includes about 50 genera and 600 species, widely distributed, and commoner in warm regions.

EXAMPLES: *Rhamnus purshiana:* Cascara, found in the Northwest.
Rhamnus cathartica: often planted for ornament.
Ceanothus: ornamentals and chaparral shrubs in the Southwest.

276 *Rhamnus*, floral diagram and longitudinal section of flower, enlarged.

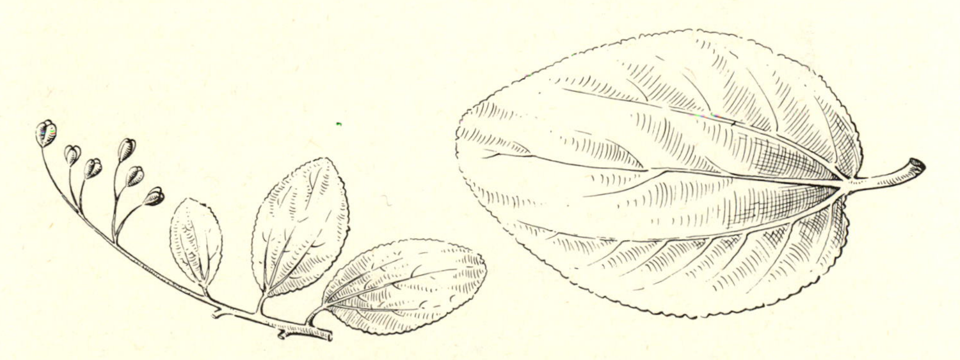

277 *Ceanothus velutinus*, fruiting twig and lower leaf.

ELAEAGNACEAE: Russian Olive Family

Trees or shrubs with alternate or opposite, entire, exstipulate leaves, these as well as the young stems and often the fruit covered with stellate hairs or scales, which often give the plants a silvery appearance. Flowers unisexual or perfect, 4-merous, apetalous, regular, and perigynous, the hypanthium enclosing the ovary, which thus appears inferior, and in fruit the hypanthium adnate to the ovary. Stamens mostly 4 or 8. Pistil 1, with a single slender style and capitate stigma, the ovary 1-celled and 1-seeded, becoming drupaceous in fruit because of the adherent fleshy or leathery hypanthium.

A small family of 3 genera and about 20 species, found chiefly in north temperate and subtropical regions, a few species occurring in the Indo-Malayan region.

EXAMPLES: *Elaeagnus angustifolia:* Russian Olive, widely cultivated.
Elaeagnus commutata: Silver-berry, common along western stream courses at lower elevations.
Hippophae rhamnoides: Sea Buckthorn.
Shepherdia canadensis: Canadian Buffalo-berry, a common shrub in wooded areas of North America.

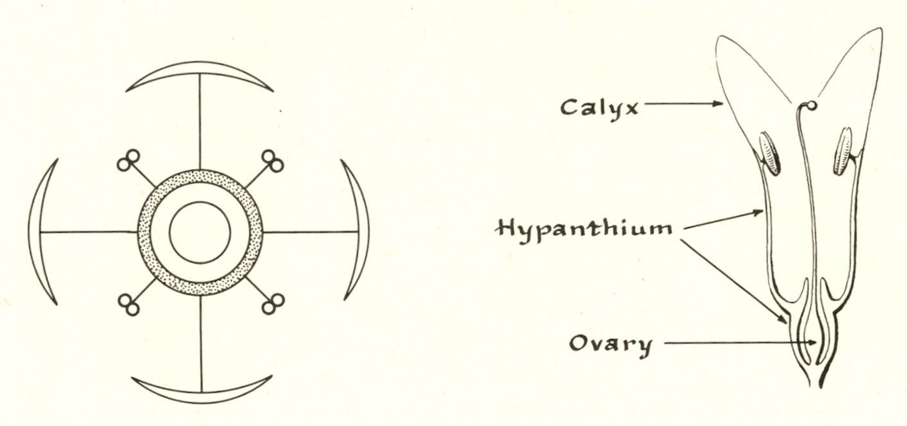

278 *Elaeagnus*, floral diagram and longitudinal section of flower.

279 *Elaeagnus*, scale from leaf, greatly enlarged.

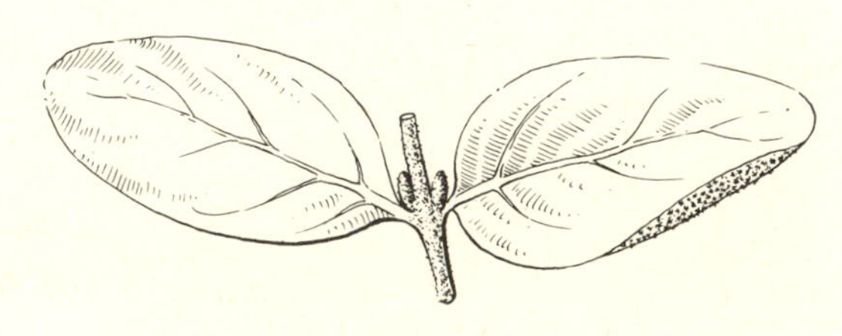

280 *Shepherdia canadensis*, portion of branch.

281 *Elaeagnus commutata* Bernh., Silver-berry, showing flowers and a single fruit from the preceding year.

ORDER VIOLALES

Herbaceous or woody plants with mostly alternate, stipulate leaves, and usually irregular, hypogynous or perigynous, polypetalous flowers. Stamens usually definite (as many as 40 in *Resedaceae*) and free. Carpels 2–6 (often 3), usually united, the ovary with parietal placentation.

The order probably represents an early and terminal development from the *Rosales*. The following family is representative.

VIOLACEAE: Violet Family

Mainly annual or perennial herbs in temperate latitudes, but in tropical regions often shrubs or trees. Leaves usually alternate or basal, with prominent stipules. Flowers solitary or variously clustered, regular or irregular, perigynous (the hypanthium short), 5-merous, the lower petal often spurred. Stamens 5, connivent around the pistil, the lower pair often spurred, the filaments usually dilated and sometimes connate in a short tube. Pistil 1, tricarpellate, the ovary 1-celled with 3 parietal placentae, the style single and terminal, the stigma variously shaped. Fruit an elastic, loculicidal, 3-valved capsule with several to many seeds. Cleistogamous flowers are sometimes found in addition to the normal flowers, or the showy flowers may be sterile.

The family includes about 18 genera and 450 species, widely distributed in temperate and tropical regions.

EXAMPLES: *Viola:* Violet and Pansy, including some 600 species and numerous hybrids. Flowers irregular.

Rinorea: 60 species of tropical trees and shrubs with regular flowers.

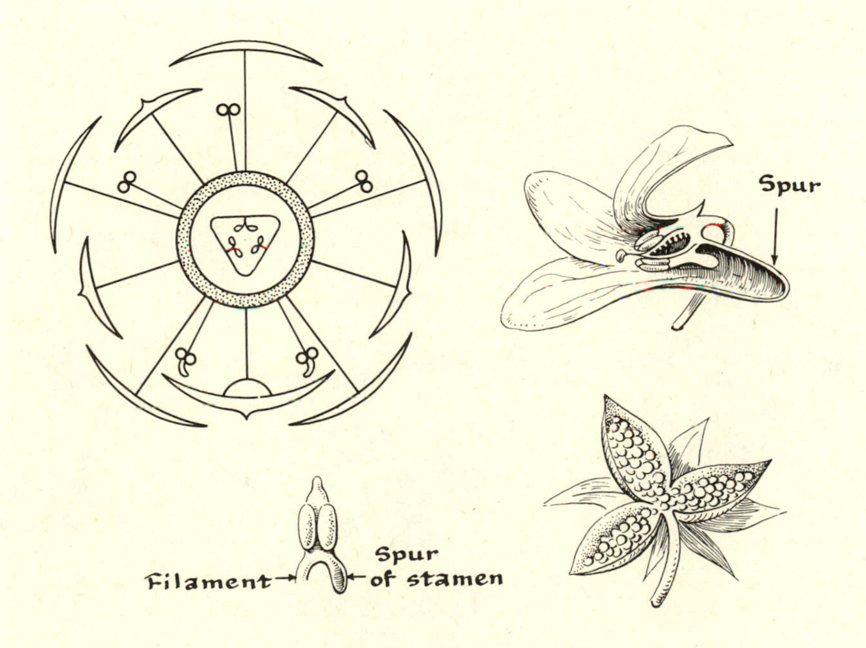

282 *Viola*, floral diagram, longitudinal section of flower enlarged, single stamen of lower pair, enlarged, and open capsule, enlarged.

283 *Viola canadensis* L., a wide-ranging species with leafy stems and white or pale lavender flowers.

284 *Viola nuttallii* Pursh, Yellow Prairie Violet, a common American species from the Great Plains westward.

ORDER LOASALES

Mostly herbs with well-developed exstipulate leaves and regular, mostly 5-merous, perigynous or epigynous, polypetalous flowers. Stamens numerous to definite, often in bunches. Pistil 1, usually of 3 carpels, the ovary with parietal or, rarely, axillary placentation.

The order includes the *Turneriaceae* (perigynous) and the *Loasaceae* (epigynous), both characteristically American families. The relationships are rather obscure, but Hutchinson suggests affinities with the *Papaverales*. There may also be relationships with the *Myrtales*.

LOASACEAE: Loasa Family

Mostly herbaceous plants having the herbage covered with rough, bristly, or sometimes stinging hairs. Flowers epigynous, with little or no hypanthium, regular, usually 5-merous. Stamens numerous and in bunches opposite the petals, or sometimes definite, the outer filaments sometimes dilated and petaloid. Ovary usually 1-celled and with parietal placentation. Fruit a many-seeded capsule.

The family includes about 15 genera and 250 species, chiefly in the drier parts of the southwestern United States, and in Mexico and South America, with 1 monotypic genus in South Africa and Arabia.

EXAMPLES: *Mentzelia:* Blazing Star, or Stickleaf, with numerous species, some of them with large, handsome flowers.
Loasa vulcania: an ornamental annual of South America.

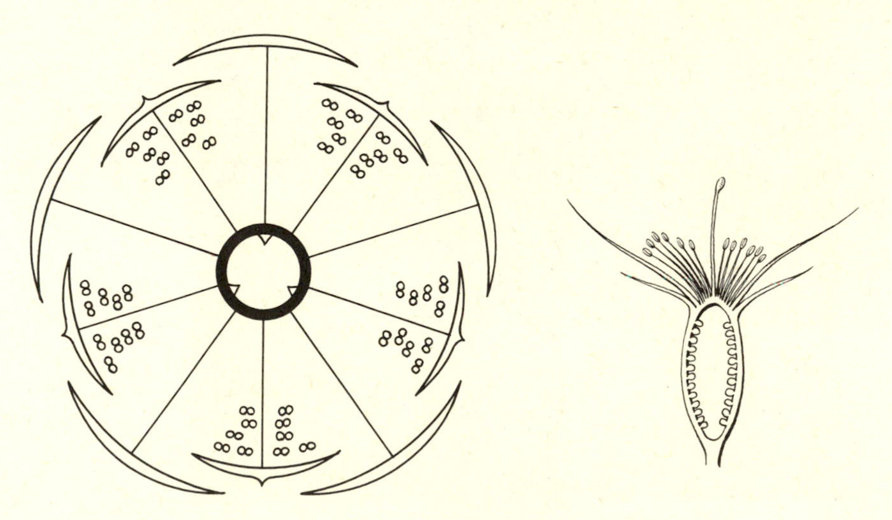

285 *Mentzelia,* floral diagram and longitudinal section of flower.

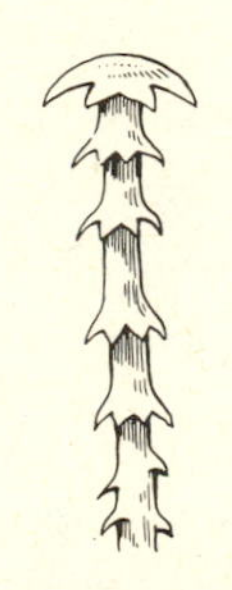

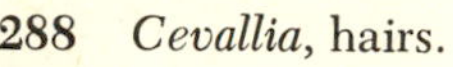

286 *Mentzelia*, hair. **287** *Eucnide*, hair. **288** *Cevallia*, hairs.

289 *Mentzelia nuda* (Pursh) T. and G., var. *stricta* (Osterh.) Harrington, a Rocky Mountain plant known as Blazing Star or Stickleaf. The petals are creamy to white.

ORDER CACTALES

The order includes only the following family and therefore has its characteristics. Divergent views have been expressed about the relationship of the *Cactales* to other orders: Bessey and Hutchinson considered the order to be related to the *Cucurbitales* (Pumpkin and Squash Order); Wettstein, Hallier, and others have suggested, because of anatomical and embryological evidence, that it is related to the *Caryophyllales* (Chickweed or Pink Order). On the basis of floral morphology, epidermal structures, and serological evidence gathered by Mez, it might also be related to the *Loasales* and the complex *Parietales*. Proliferation, which results in the development of new flowers on top of old fruits, occurs in some members and suggests the possibility that the ovary is embedded in stem tissue.

CACTACEAE: Cactus Family

Herbs, shrubs, and trees, mostly fleshy and succulent, generally without true leaves on mature stems, the stems often globose, cylindrical, or flattened, and provided with sharp bristles or spines, which emerge from areolae. Flowers solitary, perfect, regular to slightly irregular, epigynous, with or without a hypanthium. Sepals and petals not clearly differentiated, often in several series, and often merging into bracts. Stamens numerous and sometimes thigmotropic. Pistil 1, of 3 or more carpels, the ovary 1-celled and with as many parietal placentae as there are carpels and stigmas. Ovules numerous. Fruit a somewhat dry or fleshy berry.

The family includes about 120 genera and perhaps 1,500 species (there is disagreement on generic limits), and it is chiefly American, being found from Argentina and Chile northward to British Columbia. The genus *Rhipsalis* of tropical America has found its way to Africa and Ceylon; and several species of *Opuntia* occur on the shores of the Mediterranean Sea and in South Africa and Australia.

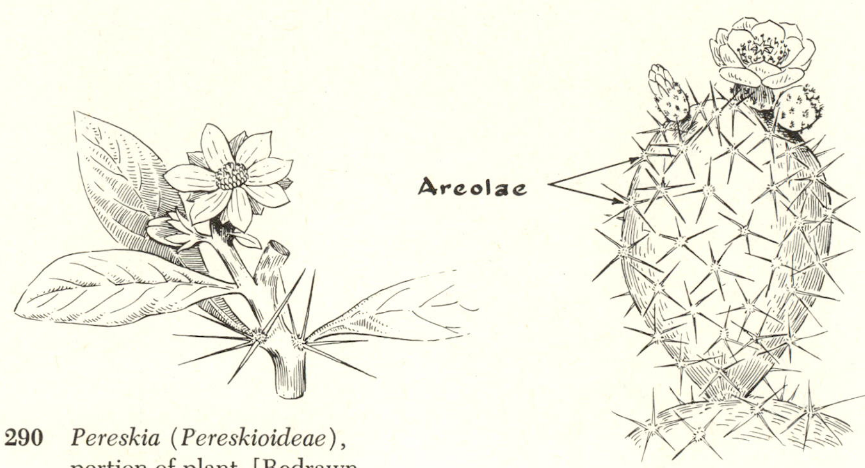

290 *Pereskia* (*Pereskioideae*), portion of plant. [Redrawn from Schumann in Engler & Prantl, *Nat. Pflanz.* 3^6:204 (1894).]

291 *Opuntia* (*Opuntioideae*), portion of plant.

292 *Opuntia polyacantha* Haw., a species of semi-desert regions (subfamily *Opuntioideae*).

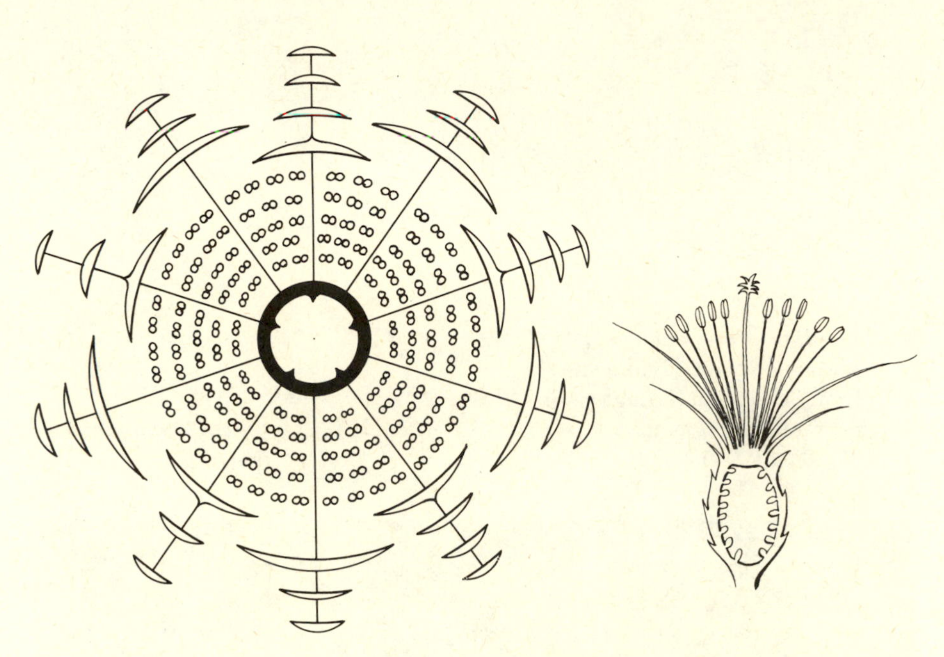

293 *Pereskia* (*Pereskioideae*), floral diagram and longitudinal section of flower. This is also the general situation in the subfamily *Opuntioideae*.

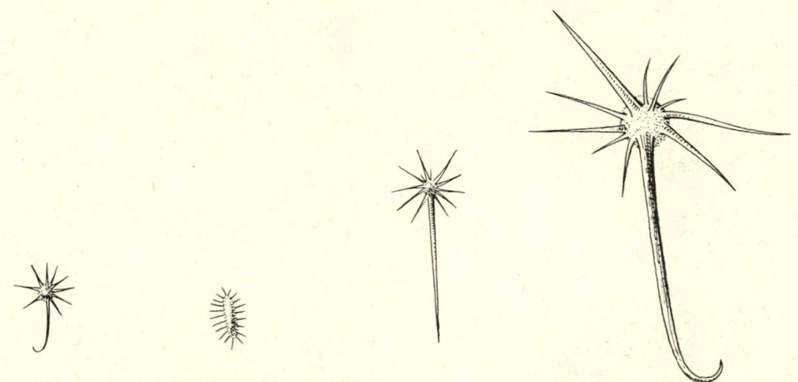

294 Various types of spines of cacti.

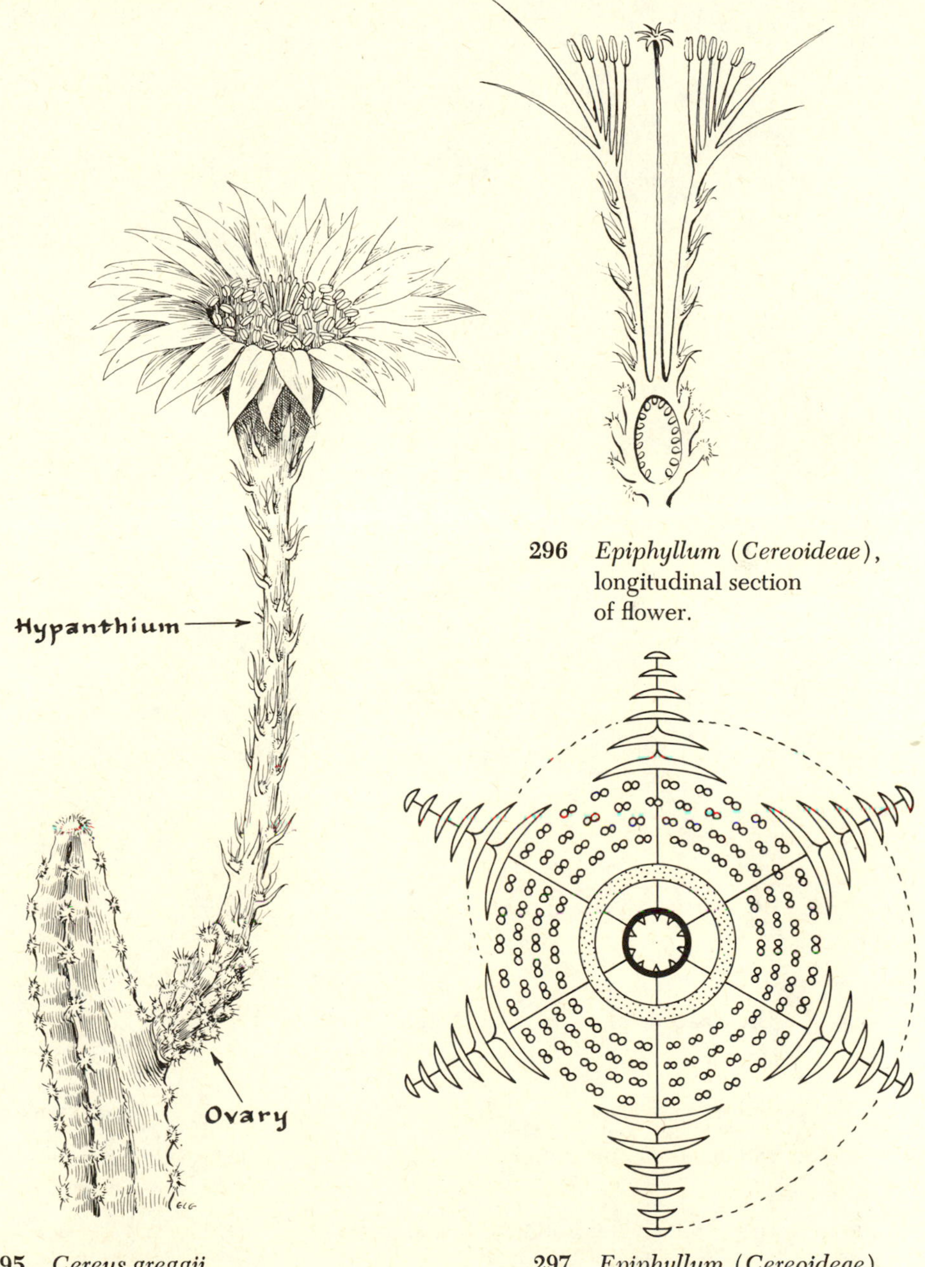

295 *Cereus greggii* (*Cereoideae*), portion of plant. [Modified from Engelmann, *Cactaceae of the Mexican Boundary*, pl. 63 (1859).]

296 *Epiphyllum* (*Cereoideae*), longitudinal section of flower.

297 *Epiphyllum* (*Cereoideae*), floral diagram.

Key to the Subfamilies of Cactaceae

Flowers without a hypanthium; plants leafy or leafless

Plants with normal leaves .... PERESKIOIDEAE

Plants without leaves on mature stems, but new shoots and seedlings may be leafy .... OPUNTIOIDEAE

Flowers with a tubular hypanthium; plants leafless .... CEREOIDEAE

298 *Coryphantha vivipara* (Nutt.) Britt. and Rose, a Nipple Cactus with bright purplish flowers (subfamily *Cereoideae*).

ORDER LYTHRALES

Herbaceous or woody plants, sometimes aquatic, with reduced flowers, with simple and mostly exstipulate leaves. Flowers often 4-merous, regular or nearly so, perigynous or epigynous, the latter with or without a hypanthium. Stamens as many or twice as many as the petals, free. Gynoecium syncarpous, with axillary placentation.

The order, which probably evolved from the *Rosales*, includes 8 families as defined by Hutchinson, the following being representative.

ONAGRACEAE: Evening Primrose Family

Mostly herbs with simple, alternate or opposite, exstipulate leaves. Flowers axillary or interminal racemes, regular, perfect, 2-merous or usually 4-merous, epigynous, with or without a tubular epigynous hypanthium. Stamens as many or twice as many as the petals. Ovary inferior, mostly 2- or 4-celled, the style single and the stigma 4-lobed, capitate, or discoid. Fruit usually a capsule, rarely a berry (*Fuchsia*) or indehiscent and nut-like (*Circaea* and *Gaura*). Seeds sometimes with a tuft of silky hairs (comatose).

The family includes about 38 genera and 500 species, mainly in temperate and subtropical regions, the chief center of distribution being western North America.

EXAMPLES: *Oenothera:* Evening Primrose, with about 100 species in North and South America.

Epilobium: Willow-herb or Fire-weed, with about 160 species, widely distributed except in the tropics.

Clarkia and *Godetia:* commonly cultivated ornamentals.

Fuchsia: about 60 species of woody plants in Central and South America and New Zealand.

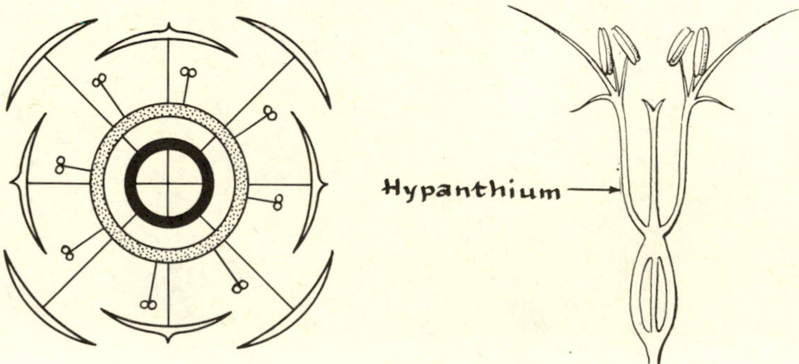

299 *Oenothera*, floral diagram and flower section.

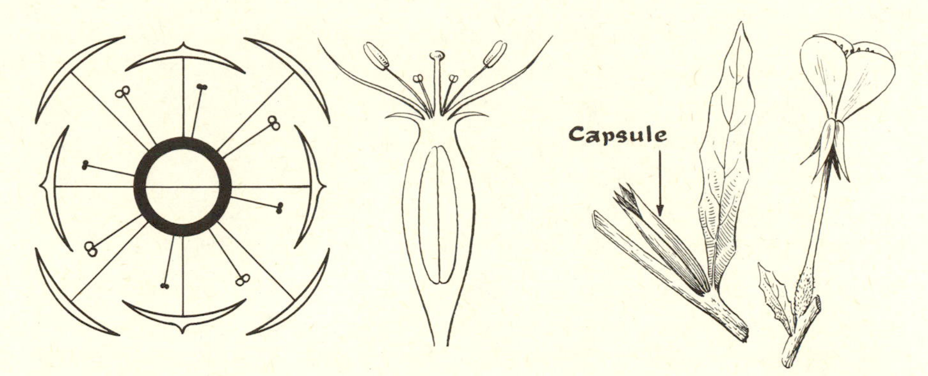

300 *Gayophytum,* floral diagram and flower section.

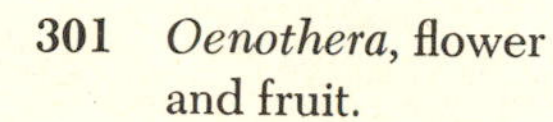

301 *Oenothera,* flower and fruit.

302 *Oenothera caespitosa* Nutt., an Evening Primrose common in the Rocky Mountains. The flowers are white and fade pink.

ORDER APIALES (UMBELLALES)

Woody or herbaceous plants with simple or usually compound or dissected leaves, with or without stipules. Flowers small, epigynous, without a hypanthium, regular (or sometimes the marginal flowers of the inflorescence irregular), in capitate clusters or often in simple or compound umbels. Sepals much reduced. Stamens of the same number as the petals and alternate with them. Pistil of 2–5 carpels, each usually 1-seeded.

This, the highest order of the polypetalae, probably evolved from an ancestral stock near the *Rosales.* The following families are the chief ones in the North American flora.

Key to the Families of Apiales

Leaves simple and usually opposite; flowers mostly 4-merous; style usually single . . . . CORNACEAE

Leaves compound (rarely simple) and alternate or basal; flowers 5-merous; styles usually 2 or 5

- Styles 5; fruit a berry or drupe . . . . ARALIACEAE
- Styles 2; fruit a schizocarp (2 mericarps) . . . . APIACEAE (UMBELLIFERAE)

CORNACEAE: Dogwood Family

Mostly trees and shrubs, occasionally herbaceous, with opposite or rarely alternate, simple, and usually entire leaves without stipules. Flowers small, in umbels, cymes, or heads, often with a showy involucre of pink or white bracts, epigynous, usually 4-merous, regular, with stamens of the same number as the petals and alternate with them. Pistil 1, mostly of 2 carpels, with a single style and generally a lobed stigma. Fruit a drupe or a few-seeded berry.

The family includes about 11 genera and 90 species, chiefly in the northern hemisphere.

EXAMPLES: *Cornus florida:* Flowering Dogwood, a small tree.

Cornus canadensis: Bunchberry, exceptional in being herbaceous.

Cornus stolonifera: Red Osier Dogwood, a common shrub.

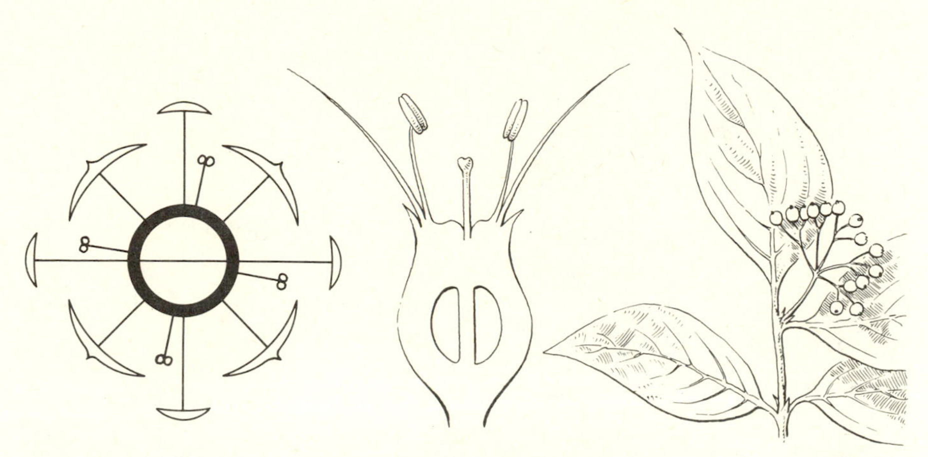

303 *Cornus*, floral diagram, flower section, and fruiting branch.

304 *Cornus stolonifera* Michx., Red Osier Dogwood, a common shrub. A portion of the plant is shown in fruiting condition. The fruits are white.

ARALIACEAE: Ginseng Family

Trees, shrubs, or herbs, with mostly alternate and divided or compound leaves, with petioles often sheathing at the base. Flowers small, in heads or umbels, regular, epigynous, with 5 stamens. Pistil 1, of 2–5 (usually 5) carpels, the styles usually as many as the carpels. Fruit a drupe or berry.

The family includes about 65 genera and 750 species, mainly tropical, with distributional centers in tropical America and the Indo-Malayan region.

EXAMPLES: *Hedera helix:* Ivy, often grown on walls.
Panax quinquefolium: Ginseng.
Aralia spinosoa: Hercules'-club.
Oplopanax horridus: Devil's-club.

APIACEAE (UMBELLIFERAE): Celery or Parsnip Family

Aromatic herbs with alternate or basal and usually compound leaves having petioles dilated and sheathing the stem at the nodes. Flowers small, mostly in compound umbels, regular (or the marginal flowers of an umbellet sometimes irregular), epigynous, without a hypanthium, 5-merous, the sepals very small and sometimes obsolete. Stamens 5, alternate with the petals. Pistil 1, of 2 united carpels, with 2 styles, and an inferior 2-celled ovary, which ripens into 2 small 1-seeded fruits, each known as a mericarp and collectively known as a schizocarp, each carpel indehiscent but separating from the other carpel.

The family includes about 200 genera and 3,000 species, found mainly in the temperate regions of the world or in the mountains of the tropics.

EXAMPLES: *Apium graveolens:* Celery.
Petroselinum hortense: Parsley.
Daucus carota: Carrot.
Cicuta: Water Hemlock, a stock-poisoning marsh herb.
Conium maculatum: Poison Hemlock, or Snakeweed, very poisonous.

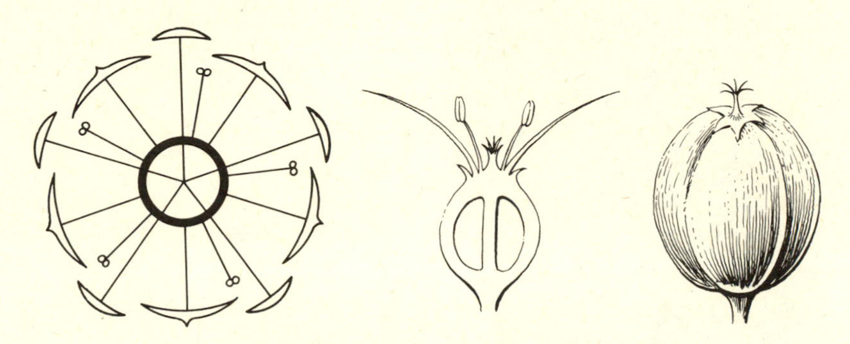

305 *Aralia*, floral diagram, flower section, and fruit, enlarged.

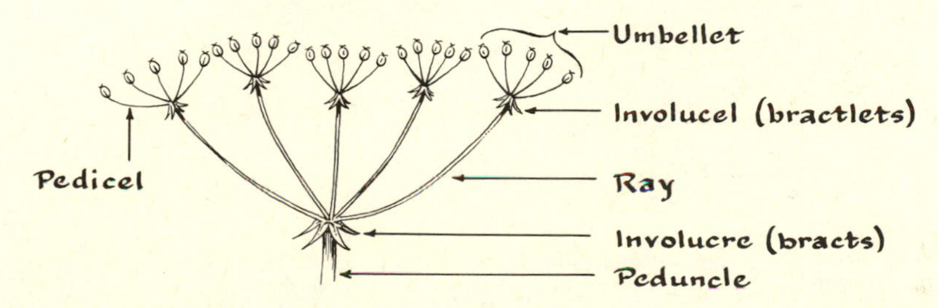

306 Parts of a typical compound umbel.

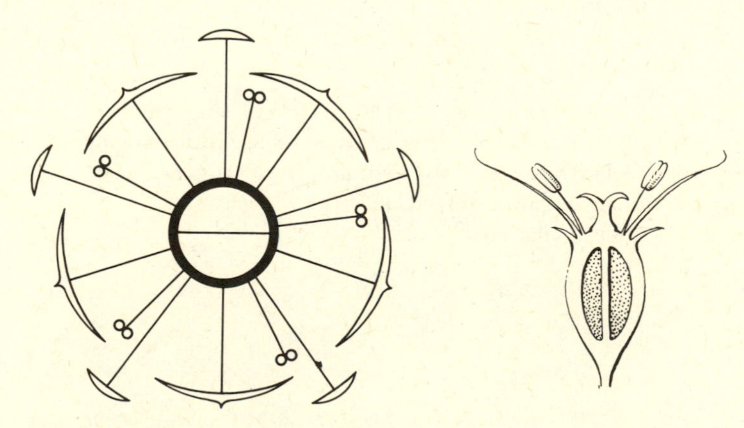

307 Generalized floral diagram and flower section for the *Apiaceae*.

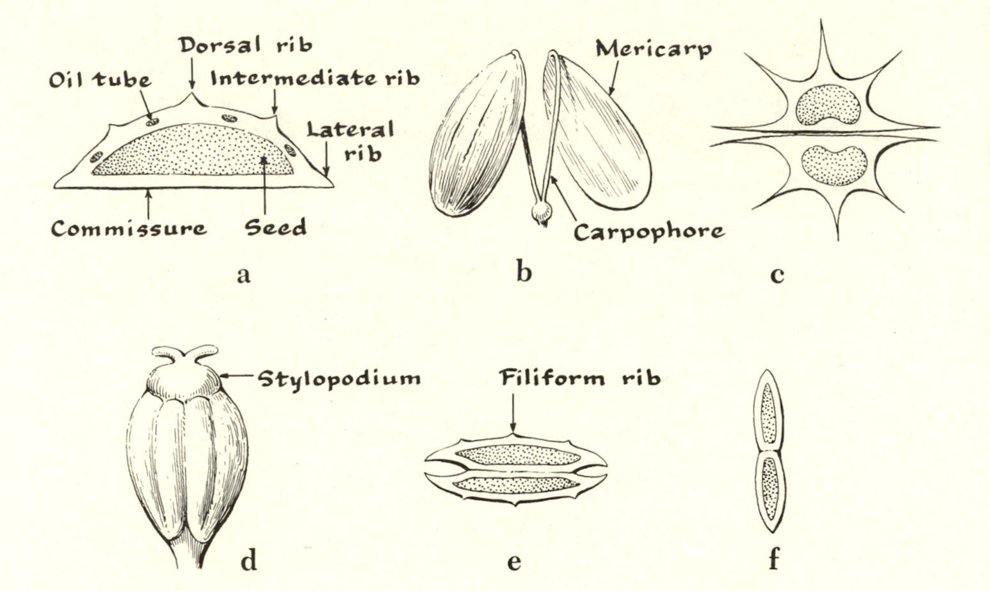

308 (**a**) Cross section of a mericarp. (**b**) Fruits from a single flower (two mericarps). (**c**) *Cymopterus*, cross section of fruit showing winged ribs. (**d**) Fruit with a stylopodium. (**e**) *Pastinaca*, cross section of a dorsally flattened fruit. (**f**) *Hydrocotyle*, cross section of a laterally flattened fruit.

309 *Sium suave* Walt., Water Parsnip, a swamp plant with white flowers. The species is regarded as poisonous to livestock.

THE SYMPETALAE

This group includes all families of dicotyledonous flowering plants whose flowers show a tendency toward united petals. It includes two major lines of development, which are somewhat parallel: one a hypogynous (rarely epigynous) line, probably derived from the *Ranales* and showing relationships to the *Caryophyllales;* the other a strictly epigynous line, probably evolved from the *Rosales* and showing relationships to the *Apiales* and *Lythrales.* The sympetalae are thus probably biphyletic in origin.

The perigynous condition of the flower, which was strongly developed in the polypetalous *Rosales* and a few related orders, does not appear in the sympetalae.

Figure 310 illustrates most of the common forms of flowers having united petals and the chief parts of sympetalous corollas.

The Hypogynous Sympetalae

These orders and families, perhaps, have evolved from a hypogynous ancestry, and may have been derived from the *Ranales* through the *Caryophyllales.* The epigynous condition occurs only rarely, as in the subfamily *Vaccinioideae* (Huckleberries) of the family *Ericaceae.*

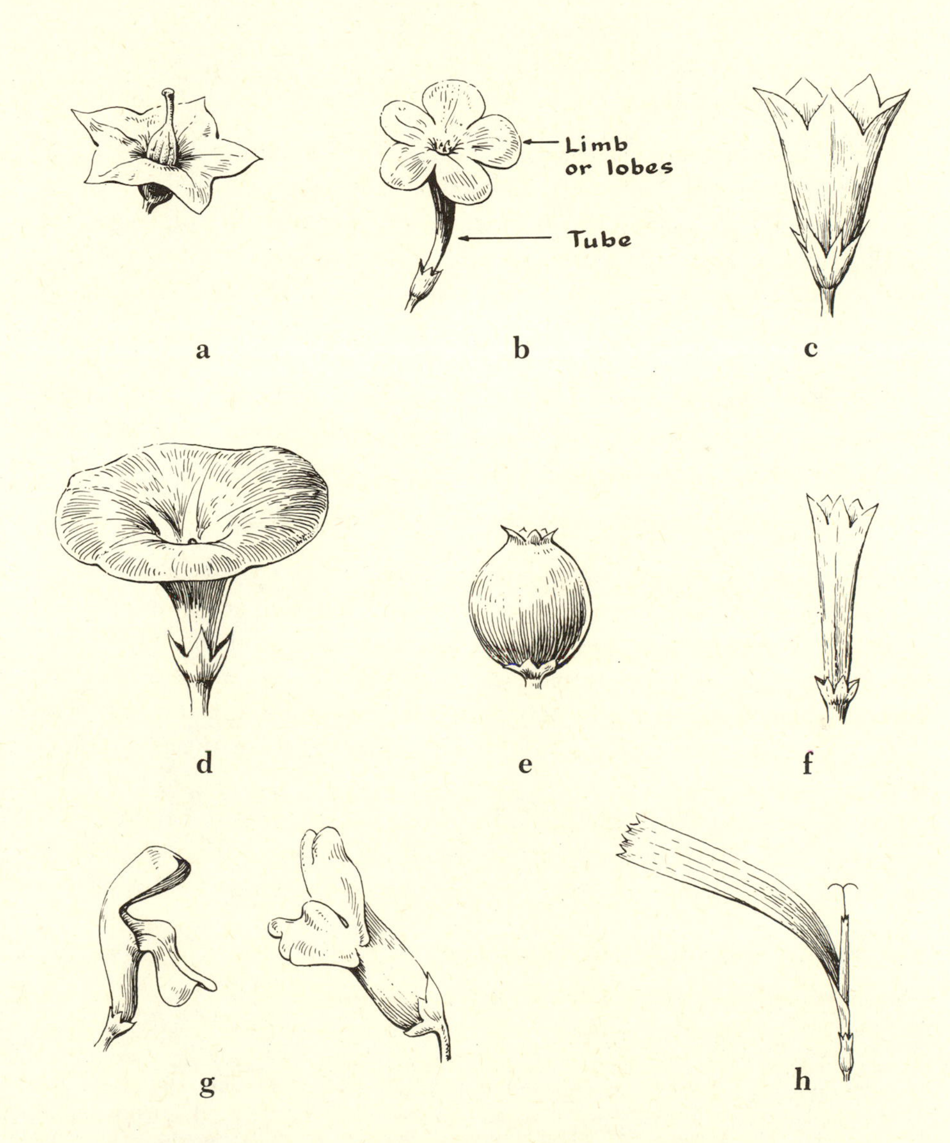

310 (**a**) Rotate corolla of *Solanum*. (**b**) Salverform corolla of *Phlox*. (**c**) Campanulate corolla of *Campanula*. (**d**) Funnelform corolla of *Ipomoea*. (**e**) Urceolate, or urn-shaped, corolla of *Gaultheria*. (**f**) Tubular corolla of *Nicotiana*. (**g**) Bilabiate corollas of *Lamium* (left) and *Antirrhinum* (right). (**h**) Ligulate corolla of *Cichorium*.

ORDER ERICALES

Woody or sometimes herbaceous plants, occasionally saprophytic or parasitic, with simple, alternate, and usually leathery exstipulate leaves. Flowers usually perfect, regular or nearly so, hypogynous or epigynous, the corolla (with a few exceptions) of united petals. Stamens free from the corolla, usually twice as many as its lobes, the anthers often opening by terminal pores and often with awn-like appendages. Pistil 1, of few to several united carpels, the ovary with axillary placentation, the style single.

The order includes the families *Clethraceae, Ericaceae* (with 4 subfamilies as treated here), *Epacridaceae,* and *Diapensiaceae.* The following family is representative and well represented in North America.

ERICACEAE: Heath Family

Herbs, shrubs, or trees, sometimes saprophytic, with alternate or sometimes opposite, simple, and often leathery leaves. Flowers perfect, regular or nearly so, hypogynous or epigynous, 4–5-merous, the petals usually united. Stamens as many or twice as many as the lobes of the corolla (petals), the anthers commonly opening by terminal pores and often appendaged. Pistil 1, of 4–5 carpels, with a single style and capitate stigma. Fruit a capsule or berry.

As treated here, the family includes about 90 genera and 1,700 or more species, found chiefly in temperate and cold regions or in the mountains of the tropics. The plants are often bog xerophytes, preferring acid soil.

Key to the Subfamilies of Ericaceae

Flowers hypogynous; fruit a capsule or berry

  Plants green, with normal leaves

    Herbs with a somewhat scapose habit; petals distinct or only slightly united . . . . 1. PYROLOIDEAE

    Shrubs or trees; petals usually united . . . . 2. ERICOIDEAE

  Plants saprophytic, without green leaves . . . . 3. MONOTROPOIDEAE

Flowers epigynous; plants shrubby, sometimes slender and trailing; fruit a berry . . . . 4. VACCINIOIDEAE

SUBFAMILY 1. PYROLOIDEAE: *Pyrola Subfamily*

Green leafy herbs with a somewhat scapose habit. Inflorescence usually a raceme or the flowers solitary. Petals 5, distinct or slightly united. Anthers opening by terminal pores, not appendaged, the pollen in tetrads. Ovary superior, incompletely septate. Style commonly declined. Fruit a loculicidal capsule.

The subfamily includes 3 genera and about 30 species, mainly in forested regions of the north temperate and arctic zones.

EXAMPLES: *Pyrola:* Pyrola, or Wintergreen.
Chimaphila: Pipsissewa, or Prince's-pine.
Moneses uniflora: Moneses.

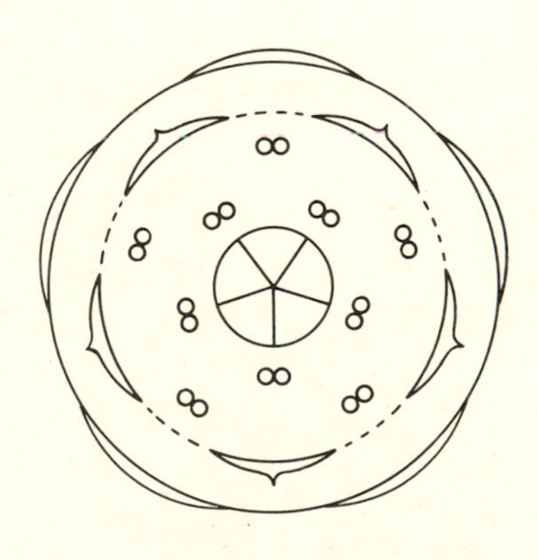

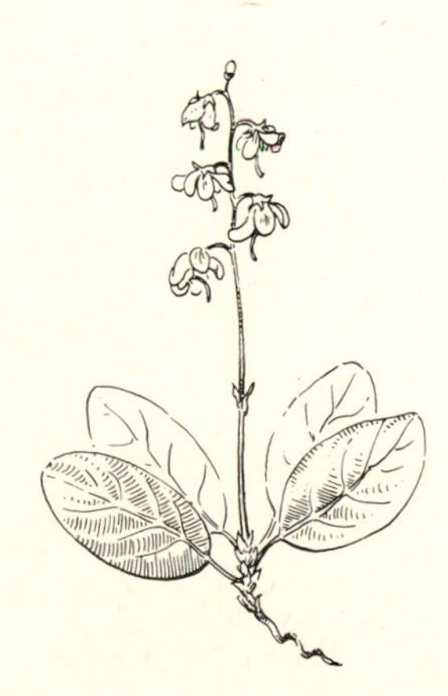

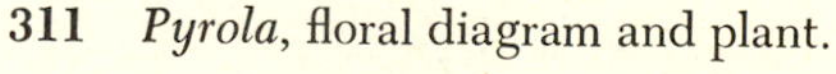

311 *Pyrola*, floral diagram and plant.

SUBFAMILY 2. ERICOIDEAE: *Heath Subfamily*

Shrubs or trees with mostly leathery leaves. Petals united or occasionally distinct. Anthers opening by terminal pores or longitudinal slits, often awned. Ovary superior, completely septate. Fruit a septicidal capsule or sometimes a berry.

The subfamily includes about 54 genera and 1,700 species, in temperate and cold regions and in the mountains of the tropics.

EXAMPLES: *Rhododendron* (including *Azalea*): shrubs and trees, often ornamentals, with about 800 species, many in southern Asia.

Erica: Heath, with about 500 species, mainly in South Africa and the Mediterranean region.

Arbutus menziesii: Madroño tree of the Pacific Coast.

Arctostaphylos: Bearberry and Manzanita, about 20 species, mostly in North and Central America.

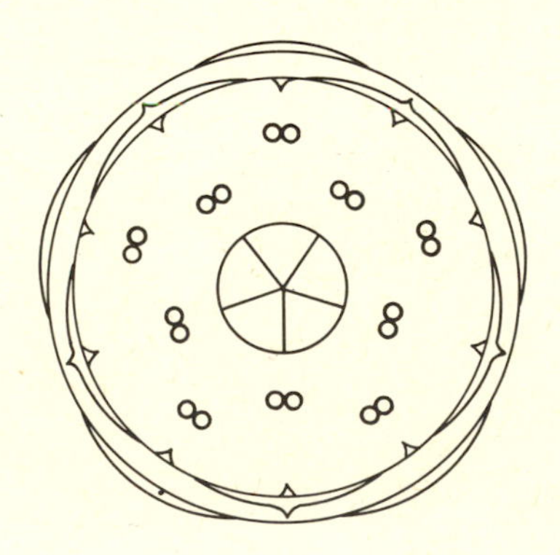

312 *Kalmia*, floral diagram. The corolla has ten pleats lengthwise.

313 *Chamaedaphne*, flowering branch.

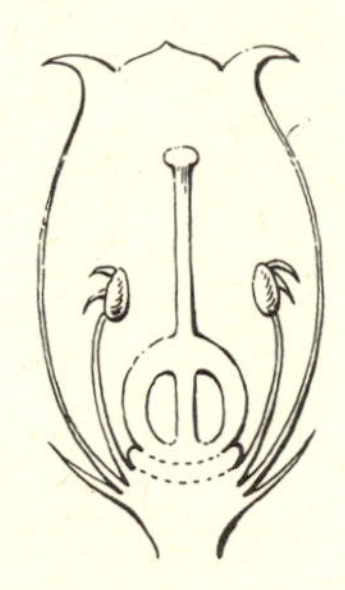

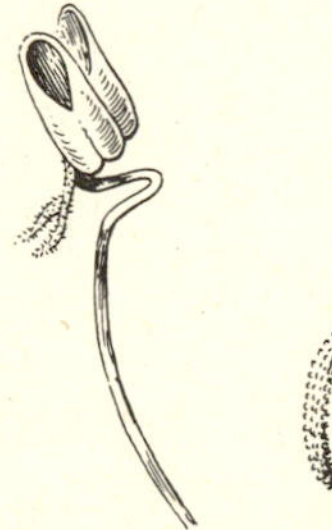

314 *Arctostaphylos*, longitudinal section of flower, enlarged.

315 *Erica*, flowers, stamen, and pistil. [Redrawn from Drude in Engler & Prantl, *Nat. Pflanz.* 4:25 (1897).]

316 *Arctostaphylos uva-ursi* (L.) Spreng., Bearberry, a common sprawling shrub (subfamily *Ericoideae*).

SUBFAMILY 3. MONOTROPOIDEAE: *Indian Pipe Subfamily*

Saprophytic or parasitic fleshy herbs without green color, the leaves reduced to scales. Petals distinct or united. Anthers opening by longitudinal slits, the pollen simple. Ovary superior, incompletely septate, the upper portion commonly 1-celled. Fruit a loculicidal capsule.

The subfamily includes 5 genera and 7 species, found in moist woods of the northern hemisphere.

EXAMPLES: *Monotropa uniflora:* Indian Pipe.
Pterospora andromedea: Pinedrops.
Sarcodes sanguinea: Snow Plant of Pacific North America.

SUBFAMILY 4. VACCINIOIDEAE: *Huckleberry Subfamily*

Shrubs, sometimes slender and trailing, with alternate leaves and epigynous flowers. Corolla of united petals, cylindric, urn-shaped, or nearly globose. Anthers opening by terminal pores in their tubular tips. Ovary completely septate and forming a berry in fruit.

The subfamily includes 23 genera and about 330 species, in the north temperate zone and in the mountains of the tropics. Several species are circumpolar, and some are epiphytes in tropical mountains.

EXAMPLES: *Vaccinium oxycoccos:* Cranberry.
Gaylussacia: Huckleberries.

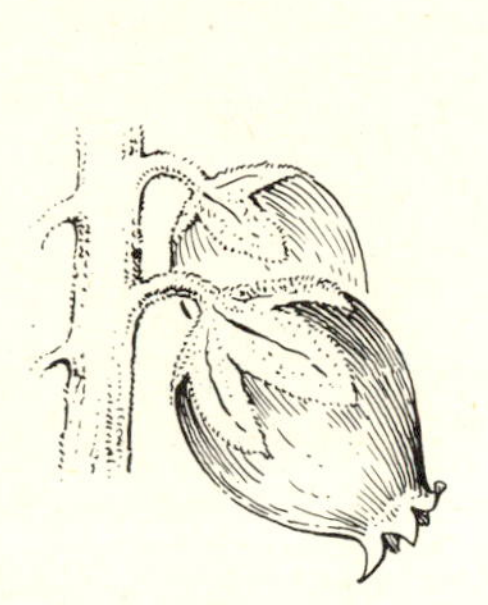

317 *Pterospora,* portion of raceme, enlarged.

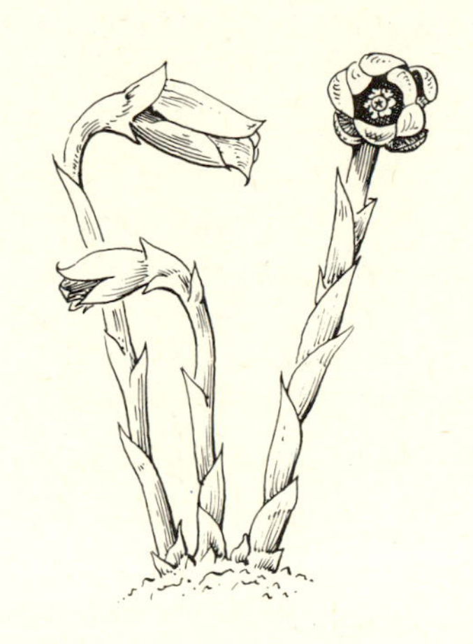

318 *Monotropa uniflora,* plants.

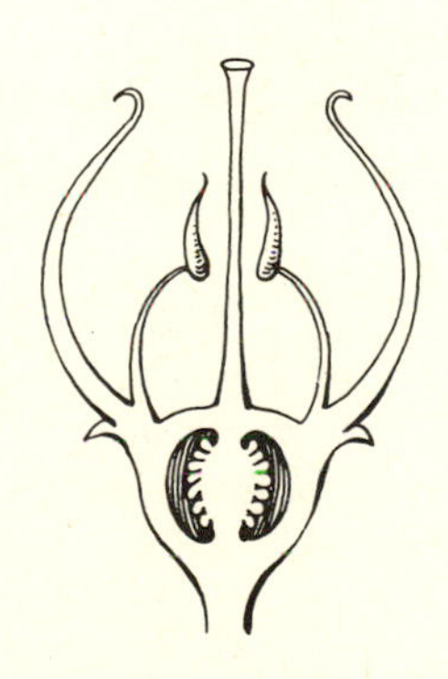

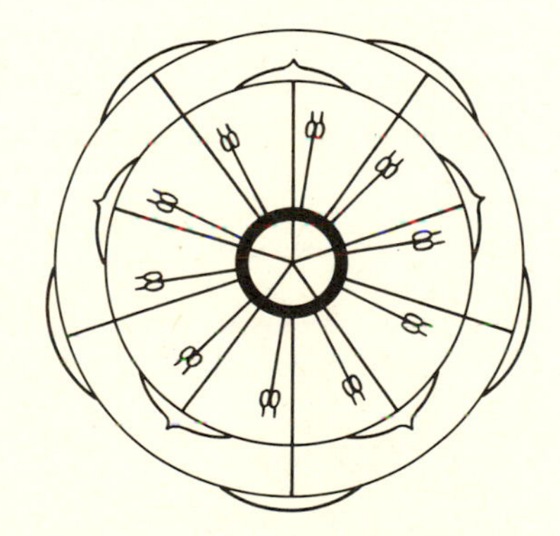

319 *Vaccinium,* longitudinal section of flower, floral diagram, and stamen, enlarged.

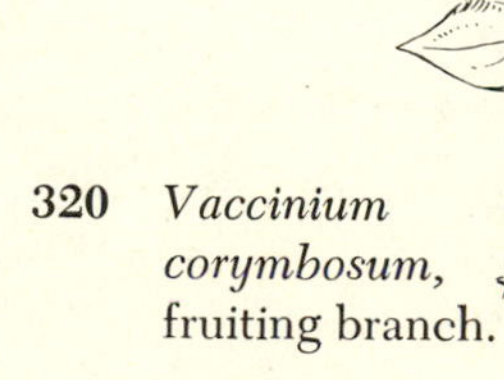

320 *Vaccinium corymbosum,* fruiting branch.

ORDER PRIMULALES

Mostly scapose herbs, sometimes leafy-stemmed, with umbellate or unilateral inflorescences, and regular, hypogynous, sympetalous, 5-merous flowers. Stamens of the same number as the lobes of the corolla and opposite them, adnate to the corolla tube. Pistil 1, of united carpels, the styles united or distinct. Ovules 1–many, from a free-central placenta.

The order includes the families *Primulaceae* and *Plumbaginaceae*. It is generally regarded as a sympetalous development from the *Caryophyllaceae*, the free-central placentation identifying this order with the *Centrospermae*. The following family is representative.

PRIMULACEAE: Primrose Family

Mostly scapose but sometimes caulescent herbs with simple leaves. Flowers hypogynous, 5-merous, the corolla regular and of united petals. Stamens 5, attached to the tube of the corolla opposite its lobes. Pistil 1, probably of 5 united carpels, but the style and stigma single and the ovary with free-central placentation. Fruit a capsule, which is usually dehiscent by 5 teeth or valves, less commonly circumscissile.

The family includes about 22 genera and 600 species, cosmopolitan in distribution, but commonest in the temperate and cooler parts of the northern hemisphere, many being found in arctic and alpine regions.

EXAMPLES: *Primula:* Primrose, several species being cultivated house plants.
Cyclamen: with about 20 species of Mediterranean origin.
Dodecatheon: Shooting-star, common in moist meadows.

ORDER GENTIANALES

Herbs, shrubs, or trees, with mostly opposite or whorled, exstipulate, simple or pinnately compound leaves. Flowers hypogynous, sympetalous or occasionally apetalous, 4–5-merous, the corolla limb often convolute. Stamens as many as the lobes of the corolla, inserted on the corolla tube alternate with its lobes, or sometimes the stamens half as many. Gynoecium of 2 carpels, which are weakly or completely united. Internal phloem occurs except in *Oleaceae*.

The order may have evolved from a stock ancestral to the *Primulales* and *Polemoniales* and probably represents an offshoot that has terminated a line of evolution, as evidenced by the very specialized floral structures of the *Asclepiadaceae* (Milkweed Family). By some the group is split into three orders.

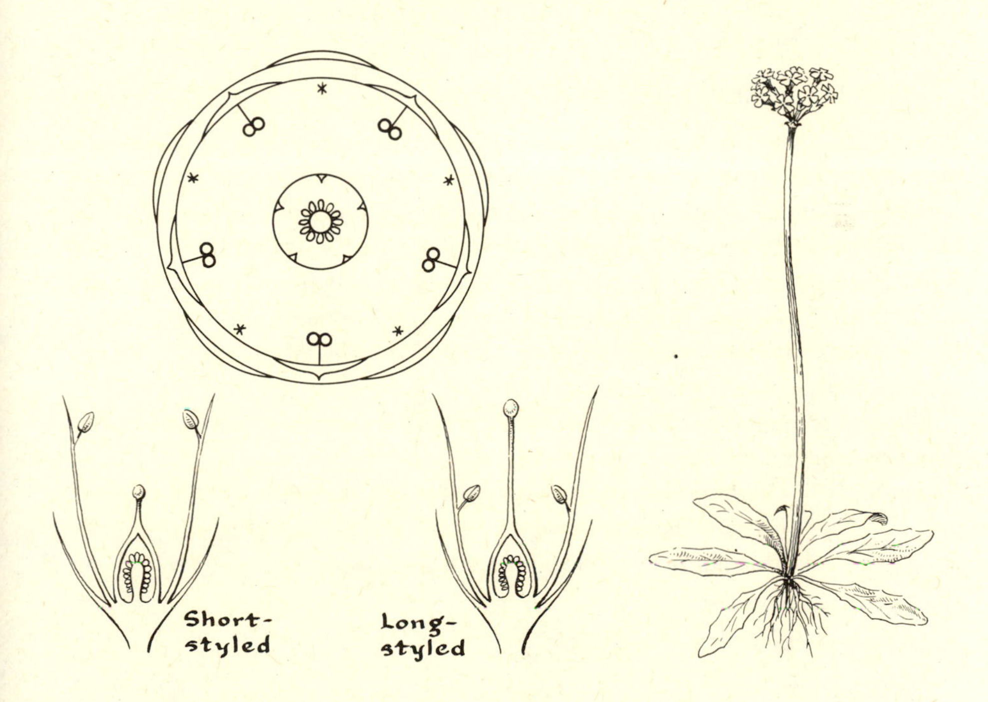

321 *Primula.* (Above) Floral diagram. (Below) Dimorphic flowers in longitudinal section, one short-styled and the other long-styled. This favors cross pollination.

322 *Primula farinosa*, plant.

Key to the Families of Gentianales

Juice not milky; flowers 4–5-merous; fruit not a follicle; seeds not comose

Mostly herbs; stamens 4–5 .... GENTIANACEAE

Mostly trees and shrubs; stamens 2 .... OLEACEAE

Juice milky; flowers 5-merous; fruit a follicle; seeds comose

Stamens free from each other, seldom coherent with the stigma; corona none or inconspicuous; follicles linear .... APOCYNACEAE

Stamens united by their anthers and to the stigma; corona conspicuous; follicles lanceolate .... ASCLEPIADACEAE

GENTIANACEAE: Gentian Family

Mostly glabrous annual or perennial herbs, with opposite or whorled, simple, sessile, entire, and exstipulate leaves. Flowers often showy, perfect, regular, 4–5-merous, the corolla of united petals, and the stamens as many as the lobes of the corolla, alternate with them, and inserted on the corolla tube. Pistil 1, of 2 carpels, the ovary usually 1-celled, with 2 large and often 2-lobed parietal placentae, the style single or none, and the stigma single or commonly 2-lobed. Fruit a 2-valved septicidal capsule with numerous seeds.

The family includes about 70 genera and 800 species, world-wide in distribution, but mainly in temperate regions.

EXAMPLES: *Gentiana:* Gentian, with about 400 species.
Frasera: Elkweed.
Menyanthes: Buckbean, with compound leaves.

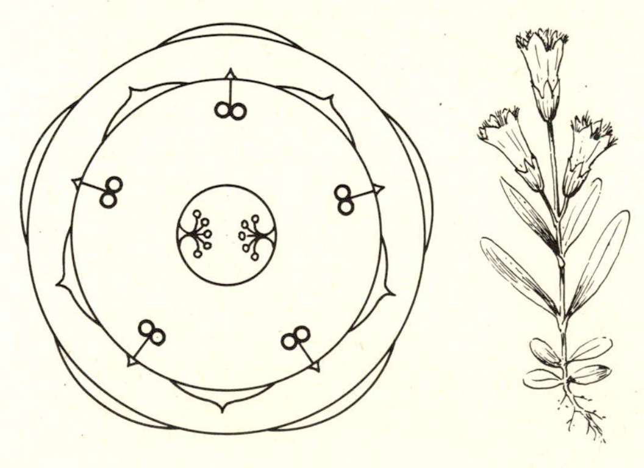

323 *Gentiana,* floral diagram and plant. The flowers may be either 4-merous or 5-merous.

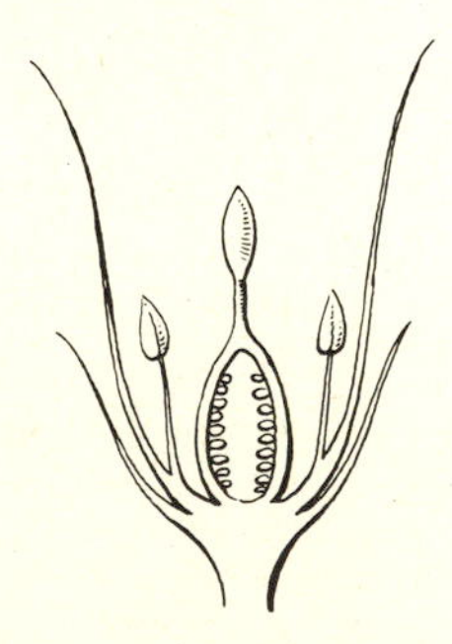

324 *Eustoma,* longitudinal section of flower.

325 *Gentiana detonsa* Rottb., var. *unicaulis* (A. Nels.) C. L. Hitchc., a common Gentian of the Rocky Mountains. The flowers are deep blue.

OLEACEAE: Olive Family

Trees or shrubs with mostly opposite, simple or odd-pinnate leaves without stipules. Flowers perfect, polygamous, or dioecious, hypogynous, regular, usually 4-merous, the corolla of united petals or lacking. Stamens usually 2. Pistil 1, of 2 carpels, the ovary 2-celled, the style single, and the stigma 2-lobed. Ovules few (generally 2 in each cell, and sometimes only 1 maturing). Fruit a capsule, drupe, berry, or samara.

The family includes 22 genera and about 400 species, in temperate and tropical regions, chiefly in eastern and southern Asia.

EXAMPLES: *Syringa:* Lilac, with 10 species.

Jasminum: Jasmin, with 200 species, some cultivated.

Olea europea: Olive of southeastern Europe and Asia Minor.

Fraxinus: Ash, with about 39 species, several in North America.

Ligustrum: Privet, with about 35 species, some used for hedges.

Forsythia: Golden-bell, an ornamental, early-blooming shrub.

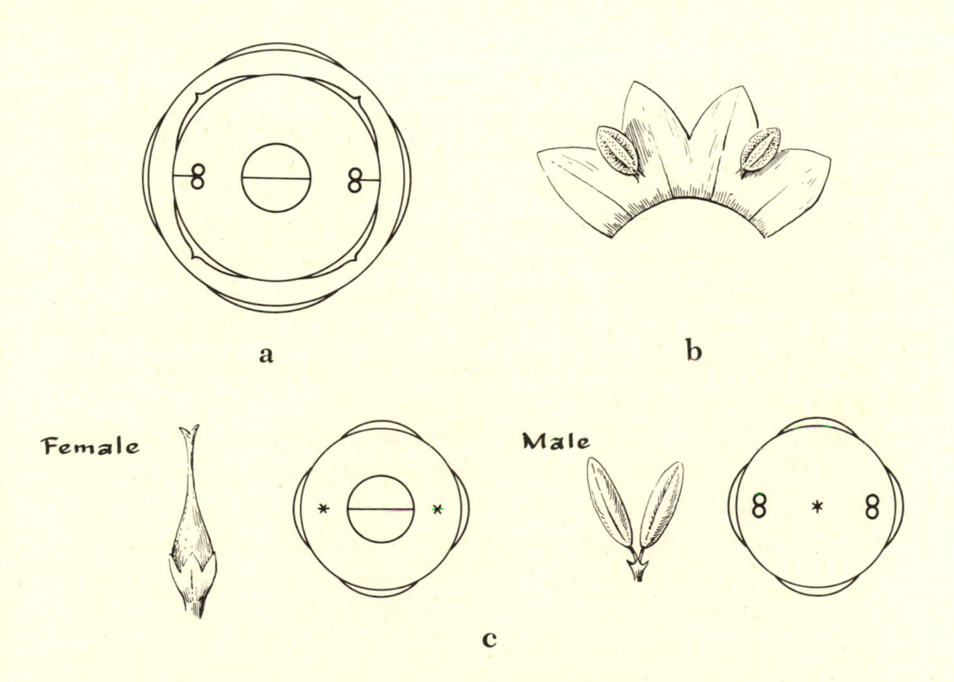

326 (**a**) *Forsythia*, floral diagram. (**b**) *Olea*, corolla and stamens, enlarged. (**c**) *Fraxinus americana*, female and male flowers and floral diagrams, the flowers enlarged.

APOCYNACEAE: Dogbane Family

Herbs, shrubs, or trees, with milky juice and simple, entire, and often opposite leaves, usually without stipules. Flowers hypogynous, perfect, regular, usually 5-merous, the corolla of united petals and salverform, tubular, or funnelform. Stamens of the same number as the corolla lobes, alternate with them, and attached to the corolla tube, the anthers sometimes adhering to the stigma. Carpels 2, the ovaries separate or connate, their styles and stigmas united. Fruit commonly 2 follicles, the seeds often hairy.

The family includes about 155 genera and 1,000 species, mainly in warm regions. The plants are often poisonous to livestock.

EXAMPLES: *Nerium oleander:* Oleander, a large, poisonous shrub of the Mediterranean region, often grown for ornament.

Vinca: Periwinkle, an ornamental trailing plant.

Apocynum: Dogbane, or Indian Hemp, with several species in North America.

Plumeria: Frangipani, a fragrant-flowered tropical genus, the corollas white, yellowish, or pink, and often used in making leis in Hawaii.

Rauwolfia serpentina: a source of tranquilizing drugs, native of India.

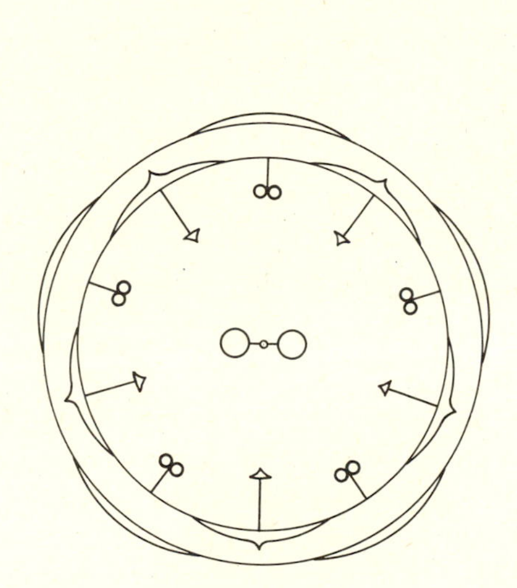

327 *Apocynum,* floral diagram.

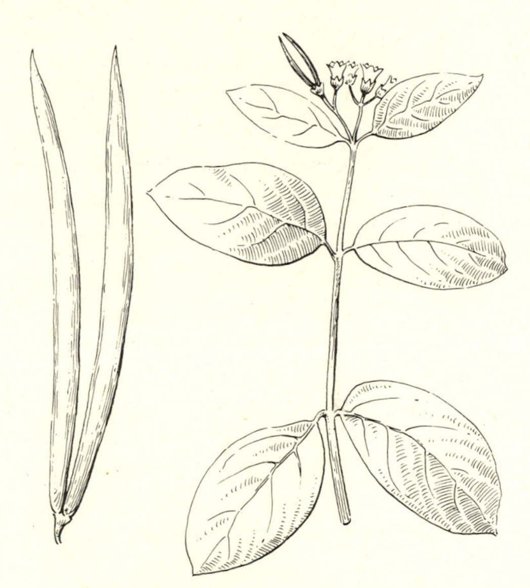

328 *Apocynum androsaemifolium,* pair of follicles and flowering shoot.

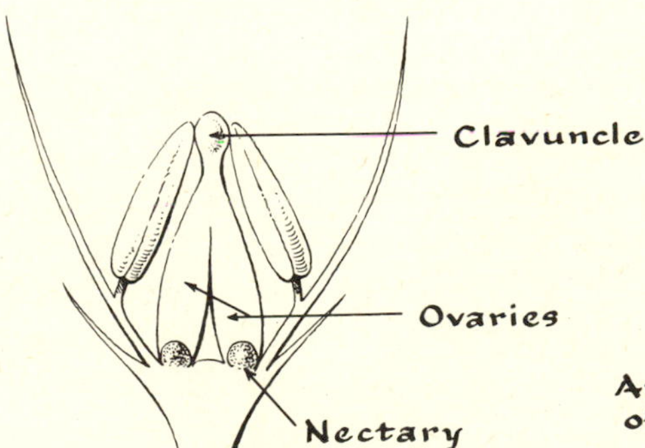

329 *Apocynum,* diagrammatic longitudinal section of flower, enlarged. A *clavuncle* is a thickened fleshy style that is common to the two carpels, whose ovaries are separate at the base.

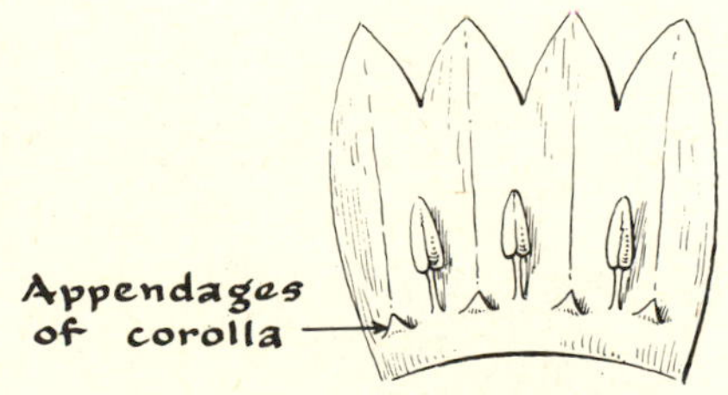

330 *Apocynum,* part of corolla and stamens, enlarged. A *corona* is a set of appendages on the corolla and may be small and inconspicuous, as in this family, or very conspicuous, as in the following family (*Asclepiadaceae*).

ASCLEPIADACEAE: Milkweed Family

Perennial herbs or shrubby climbers, shrubs, or trees, sometimes succulent, with milky juice and simple, usually entire, mostly opposite or whorled, exstipulate leaves. Flowers usually in umbels or cymes, hypogynous, 5-merous, regular, and perfect. Corolla rotate or salverform, often with a set of appendages arising from the back of the stamens or from the corolla, which form a petaloid corona. Stamens 5, the anthers united to form a cone, which is commonly attached to the stigma, the pollen often united into waxy pollinia. Carpels 2, the ovaries separate but the styles and stigmas united, each carpel forming a follicle in fruit (or sometimes 1 carpel aborting). Seeds numerous and often with a tuft of silky hairs at one end.

The family includes about 280 genera and 1,800 species, mostly in the tropics, but with several representatives in North America.

EXAMPLES: *Asclepias:* Milkweed or Butterfly-weed. Some species are poisonous to livestock.

Hoya carnosa: Wax Plant, a vine of tropical regions often grown as a house plant.

Stapelia: Starfish-flower, succulents often grown as house plants.

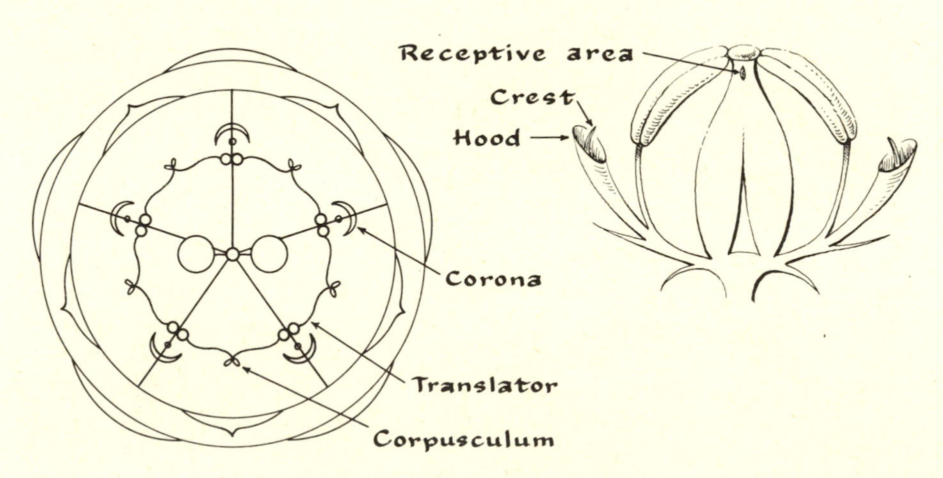

331 *Asclepias*, floral diagram and diagrammatic longitudinal section of flower.

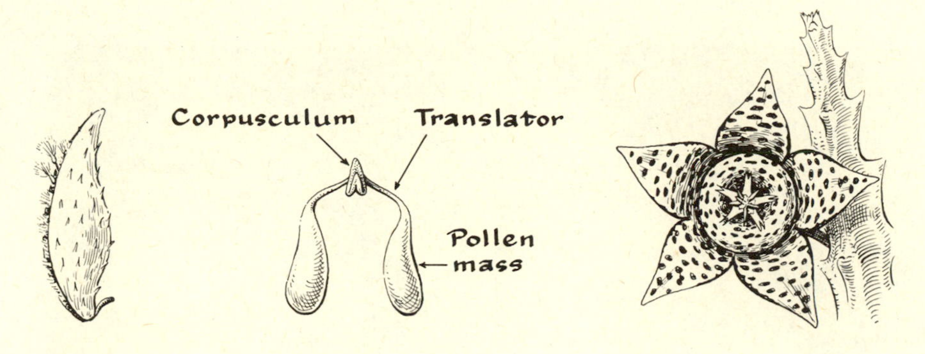

332 *Ascepias speciosa,* follicle and pollinium, enlarged.

333 *Stapelia,* flower and branch.

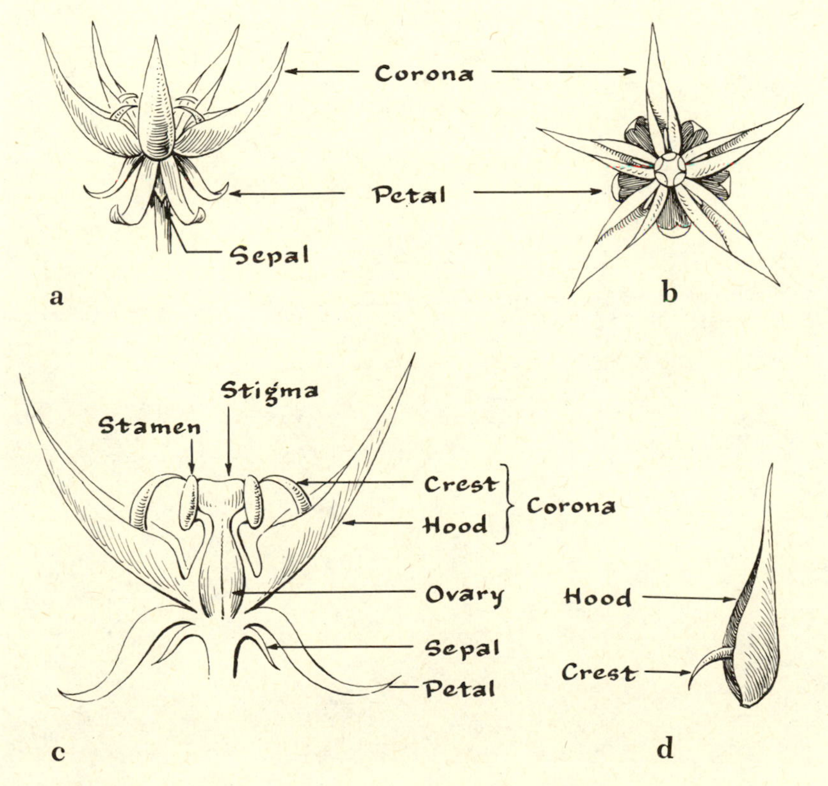

334 *Asclepias speciosa,* (**a**) side view of flower, (**b**) face view of flower, (**c**) longitudinal section of flower, (**d**) side view of a segment of the corona.

335 *Asclepias speciosa* Torr., a Milkweed, showing the conspicuous corona of the flowers.

336 *Asclepias speciosa* Torr., two opened follicles with comose seeds.

The "Tubiflorae"

This series of familes and orders is sometimes treated as one large order having the following general tendencies: the development of a deeply 4-lobed ovary with a consequent fruit of 4 nutlets (*Boraginales* and *Lamiales*), and the development of irregular corollas and a simultaneous modification of the stamens by the reduction of 1 or 3 of them (*Lamiales*), and the retention of the capsular type of fruit, with many seeds, and the development of irregular corollas and modified stamens (*Scrophulariales*). The *Plantaginales* may represent an offshoot in which the scapose habit, parallel-veined leaves, and small, spicate flowers with a scarious corolla are the chief distinguishing features.

Key to the Families of the "Tubiflorae"

Corolla regular; stamens of the same number as the corolla lobes

- Flowers 5-merous; corolla not scarious
  - Corolla scarcely lobed or the plants parasitic . . . . CONVOLVULACEAE
  - Corolla definitely lobed; plants green
    - Carpels 3, stigmas 3 . . . . POLEMONIACEAE
    - Carpels 2, stigmas 1 or 2
      - Ovary deeply 4-lobed . . . . BORAGINACEAE
      - Ovary not deeply 4-lobed
        - Styles 2 or the style 2-cleft . . . . HYDROPHYLLACEAE
        - Style 1 or the stigma sessile . . . . SOLANACEAE
- Flowers 4-merous; corolla scarious . . . . PLANTAGINACEAE

Corolla irregular, sometimes only slightly so; stamens often fewer than the corolla lobes or some of them sterile

- Aromatic herbs with square stems and opposite leaves; ovary usually deeply 4-lobed and fruit 4 nutlets; style arising from between the lobes of the ovary . . . . LAMIACEAE

Plants seldom aromatic; sometimes with square stems and opposite leaves; ovary not deeply 4-lobed; fruit usually a many-seeded capsule; style arising from the apex of the ovary.

Plants parasitic, without green color . . . . OROBANCHACEAE

Plants not parasitic, green leaves present

Stems woody; leaves often compound . . . . BIGNONIACEAE

Stems herbaceous, rarely woody; leaves never compound but sometimes deeply cleft

Aquatic herbs, often insectivorous, the submerged leaves usually finely divided; flowers with 2 stamens and the ovary with free-central placentation . . . . LENTIBULARIACEAE

Mostly terrestrial plants, sometimes in wet places, the leaves not finely divided; flowers with mostly 4 stamens, the ovary never with free-central placentation

Flowers in spikes or heads; fruit 2–4 nutlets . . . . VERBENACEAE

Flowers not in spikes or heads but sometimes congested; fruit a many-seeded capsule . . . . SCROPHULARIACEAE

ORDER POLEMONIALES

Herbs or low shrubs with alternate or opposite leaves. Flowers sympetalous, hypogynous, regular, 5-merous (rarely 4-merous). Stamens of the same number as the lobes of the corolla, alternate with them, and inserted on the corolla tube. Pistil 1, of 2–3 united carpels, the placentation axillary. Seeds numerous or few. Fruit a capsule.

The order may have evolved from the same ancestral stock as the *Gentianales,* and this same line of development may have given rise to the more advanced *Boraginales, Lamiales, Scrophulariales,* and *Plantaginales.* As defined here, the order includes the following two families.

POLEMONIACEAE: Phlox Family

Mostly annual or perennial herbs, occasionally shrubby, with alternate or opposite leaves without stipules. Flowers often showy, commonly in cymes or solitary, regular, hypogynous, 5-merous, the corolla of united petals and usually salverform or tubular, the lobes contorted in bud. Stamens 5, inserted on the corolla tube alternate with its lobes. Pistil 1, of 3 united carpels, the ovary 3-celled, the style slender and single, the stigmas or style-branches 3. Fruit a loculicidal capsule with several to many seeds.

The family includes about 15 genera and 275 species and is especially well represented in western North America. There are numerous alpine species.

EXAMPLES: *Phlox:* Phlox, with about 50 species, some of them cultivated.
Polemonium: Jacob's-ladder, with about 20 species.
Gilia: Gilia, with about 100 species, but some of these often placed in segregate genera.

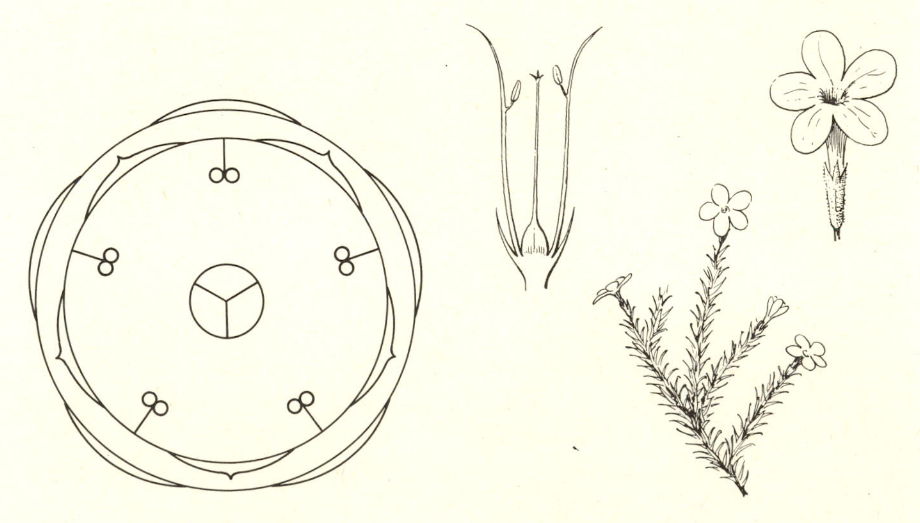

337 *Phlox*, floral diagram.

338 *Phlox hoodii,* flower section, portion of plant, and flower.

339 *Phlox multiflora* A. Nels., a mat-forming species with white to pinkish or lavender flowers, common in the Rocky Mountains.

CONVOLVULACEAE: Morning-glory Family

Mostly annual or perennial herbs, often twining (occasionally shrubby or arborescent in the tropics), with simple or rarely pinnate, alternate, exstipulate leaves; or the plants leafless, twining, parasitic herbs. Flowers often showy, solitary and axillary or in axillary clusters, usually perfect, regular, hypogynous, 5-merous, with a funnelform or tubular corolla, which is usually twisted in the bud. Stamens 5, inserted on the tube of the corolla alternate with its lobes. Pistil 1, of 1–3 (usually 2) carpels, the ovary usually 2-celled with 1–2 ovules in each cell, the style usually single and terminal, and the stigmas 2. Fruit a loculicidal capsule.

The family includes about 47 genera and 1,100 species, mostly in warm regions. A few are troublesome weeds, and a few are cultivated for their flowers or tuberous roots. The parasitic members (the single genus *Cuscuta*, or Dodder) are destructive of certain crops and may spread disease. By some these are treated as a separate family, the *Cuscutaceae*.

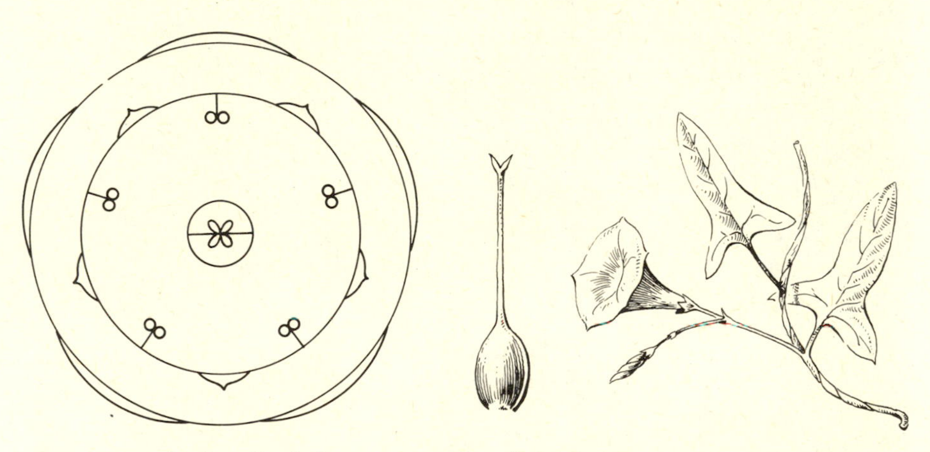

340 *Convolvulus*, floral diagram; pistil, enlarged; and portion of C. *arvensis*.

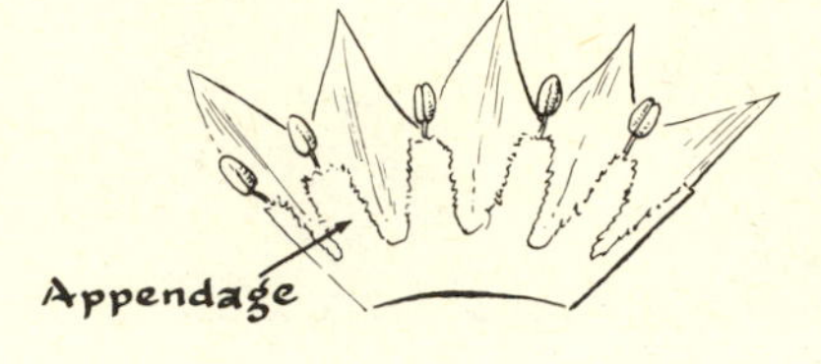

341 *Cuscuta arvensis*, corolla opened up to show appendages, enlarged.

342 *Convolvulus arvensis* L., a common weedy species with white to pinkish flowers.

Key to the Subfamilies of Convulvulaceae

Plants green, with normal leaves . . . . 1. CONVOLVULOIDEAE
Plants parasitic and leafless, not green . . . . 2. CUSCUTOIDEAE

SUBFAMILY 1. CONVOLVULOIDEAE: *Morning-glory Subfamily*

EXAMPLES: *Convolvulus:* with about 200 species, includes wild Morning glory and Bindweed.

Ipomoea: with about 400 species, includes *I. purpurea,* the commonly cultivated Morning glory, and *I. batatas,* the Sweet Potato.

SUBFAMILY 2. CUSCUTOIDEAE: *Dodder Subfamily*

Includes only the genus *Cuscuta,* or Dodder, with about 100 species. In this subfamily the corolla is often very small and appendaged within as shown in Figure 341.

ORDER BORAGINALES

Herbs, or sometimes shrubs or trees, with mostly alternate leaves and more or less coiled inflorescences. Flowers sympetalous, hypogynous, regular, 5-merous. Stamens of the same number as the lobes of the corolla and alternate with them, inserted on the corolla tube. Pistil 1, of 2 carpels, the ovary shallowly lobed and with numerous ovules or deeply lobed and the ovules 4 or less. Fruit a capsule or 1–4 nutlets.

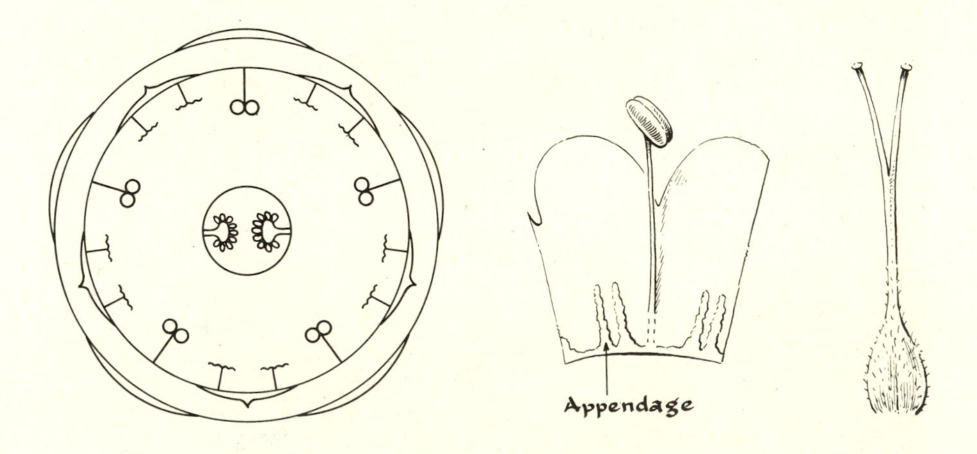

343 *Phacelia,* floral diagram; portion of corolla and stamens, enlarged; and pistil, enlarged.

The order shows a progressive development from the *Polemoniales* in the tendency toward a reduction in the number of carpels and ovules and in the specialized inflorescence, while maintaining a regular corolla and isomerous stamens. It includes the following two families.

HYDROPHYLLACEAE: Water-leaf Family

Herbs, or occasionally undershrubs, with various exstipulate leaves, which often form a basal rosette. Flowers small or sometimes showy, often in coiled cymes, perfect, regular, hypogynous, 5-merous, the corolla of united petals and often with scale-like appendages alternate or opposite the lobes. Stamens 5, on the lower part of the corolla tube, alternate with the lobes, usually well exserted. Pistil 1, of 2 united carpels, the ovary not deeply lobed, usually 1-celled with 2 large parietal placentae, or sometimes 2-celled with axillary placentation, the styles 2 and distinct or partly united, usually exserted, the stigmas usually capitate. Fruit commonly a loculicidal capsule or sometimes berry-like.

The family includes about 18 genera and 200 species, mostly in North America.

EXAMPLES: *Hydrophyllum:* Water-leaf.
Phacelia: Phacelia, with more than 100 species.

344 *Hydrophyllum fendleri* (A. Gray) Heller, a species of Water-leaf.

BORAGINACEAE: Borage Family

Mostly hispid, bristly or hairy herbs, rarely shrubs or trees, with simple, generally alternate leaves without stipules. Flowers often in coiled cymes, perfect, hypogynous, 5-merous, regular or nearly so, the corolla of united petals, tubular, salverform, or funnelform, the limb commonly spreading, and the throat often closed by petaloid appendages, which form a small corona. Stamens 5, inserted on the tube of the corolla alternate with its lobes, and included. Pistil 1, of 2 united carpels, each usually deeply 2-lobed, the ovary thus 2-celled and usually deeply 4-lobed, with 1 ovule in each lobe. Style single or occasionally the styles 2, usually originating from the base of the ovary between the lobes (gynobasic), the stigmas 1 or 2, or sometimes 4. Fruit commonly consisting of four 1-seeded nutlets, sometimes fewer by abortion, and sometimes a drupe.

The family includes about 88 genera and 1,600 species, mostly in warm and temperate regions, especially in the Mediterranean basin, tropical America, and the western United States.

EXAMPLES: *Borago:* Borage.
Cynoglossum: Hound's-tongue.
Lappula: Stickseed.
Myosotis: Forget-me-not.

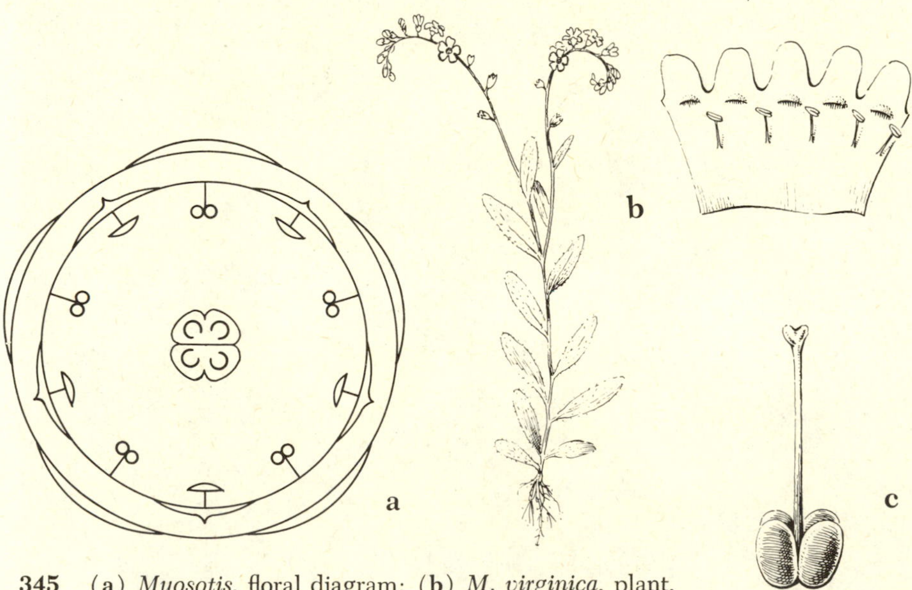

345 (**a**) *Myosotis,* floral diagram; (**b**) *M. virginica,* plant, and corolla and stamens, the corolla opened; (**c**) *Myosotis,* pistil, showing gynobasic style.

346 *Cryptantha bradburyana* Payson. Notice the coiling of the inflorescence and the bristly hairs of the plant, both characteristic of the family.

ORDER LAMIALES

Mostly herbs, often aromatic, with opposite or whorled leaves and irregular, often 2-lipped flowers. Stamens reduced to 4 or 2, rarely 5; sterile filaments sometimes present. Carpels usually 2, each with 2 ovules. Fruit a drupe, a berry, or, more commonly, four 1-seeded nutlets.

VERBENACEAE: Vervain Family

Herbs, or sometimes shrubs or trees, with opposite or whorled leaves without stipules, and mostly small, perfect, 4–5-merous, hypogynous flowers produced in spikes or heads. Corolla of united petals, somewhat irregular. Stamens 4, in two pairs (didynamous), inserted on the corolla tube alternate with its lobes. Pistil 1, of 2 united carpels, with a terminal style and 1 or 2 stigmas, the ovary not deeply lobed, and forming 2 or 4 nutlets in fruit or sometimes a drupe or berry.

The family includes about 90 genera and 3,000 species, widely distributed but most abundant in tropical and subtropical regions.

EXAMPLES: *Verbena:* Vervain, with about 270 species, some cultivated.
Lantana: ornamental shrubs and climbers.
Tectona grandis: Teak, of southeastern Asia, a valuable timber tree with very heavy, durable wood.

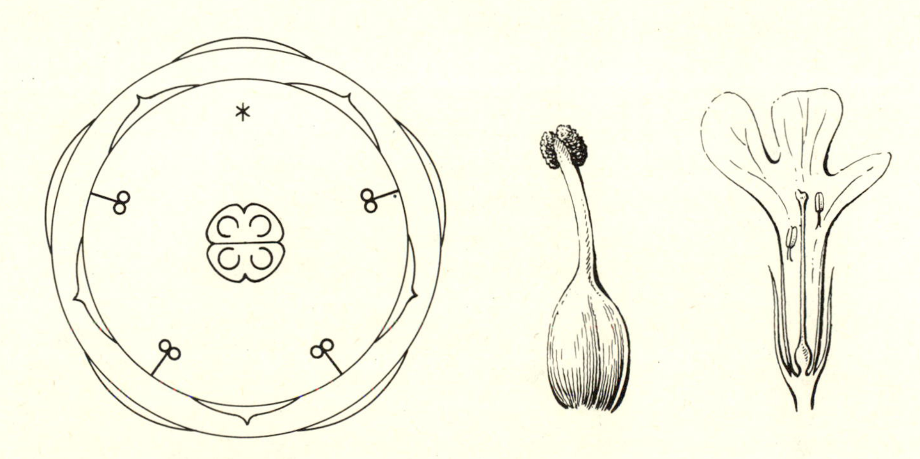

347 *Verbena*, floral diagram, pistil, and flower section.

348 *Verbena hastata*, portion of plant.

349 *Phyla cuneifolia*, portion of plant.

LAMIACEAE (LABIATAE): Mint Family

Mostly aromatic herbs, occasionally shrubs or trees, with square stems and opposite, simple, exstipulate leaves. Flowers perfect, hypogynous, 5-merous, the calyx persistent and from nearly regular to bilabiate, the corolla usually strikingly irregular and bilabiate. Stamens 4 and didynamous (two long and two short) or reduced to 2, inserted on the corolla tube alternate with its lobes, the fifth (upper) stamen usually lacking or sometimes present as a staminodium. Pistil 1, of 2 united carpels, each usually deeply 2-lobed, the ovary thus deeply 4-lobed and 2-celled, with 1 ovule in each lobe. Style single, terminal or usually gynobasic, bifid at the summit into unequal branches. Fruit four 1-seeded nutlets enclosed by the persistent calyx.

The family includes about 160 genera and 3,500 species, in temperate and warm regions, the chief center of distribution being in the Mediterranean basin.

EXAMPLES: *Mentha piperita:* Peppermint.
Mentha spicata: Spearmint.
Salvia officinalis: Sage, often used for seasoning.
Thymus vulgaris: Thyme.
Lavandula spica: Lavender, source of a common perfume.
Lamium: Dead Nettle.

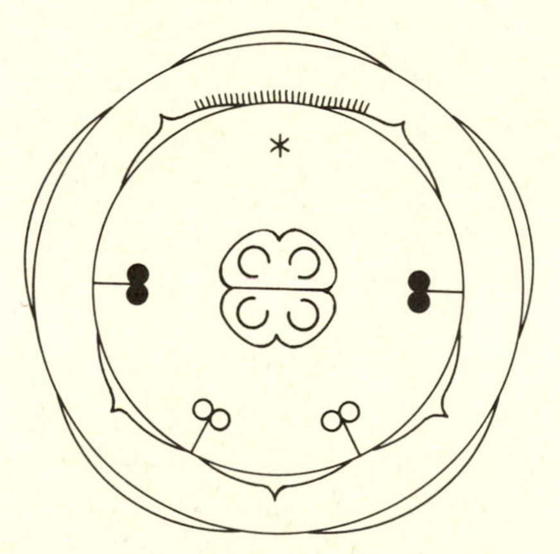

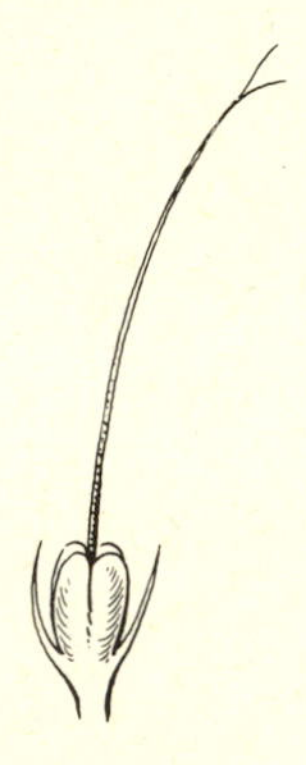

350 *Monarda*, floral diagram and pistil.

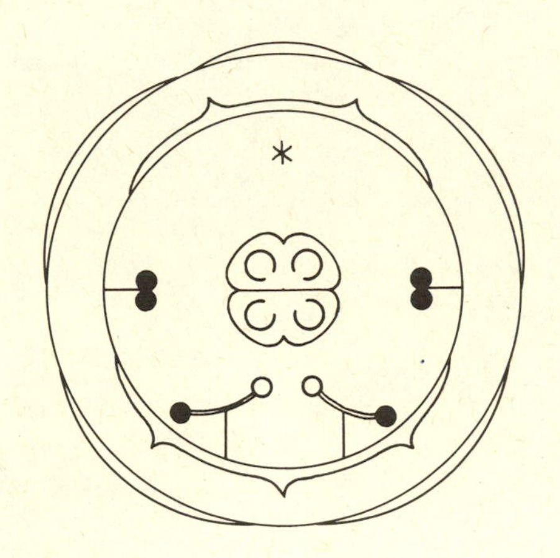

351 *Salvia,* floral diagram.

352 *Scutellaria brittonii,* portion of plant.

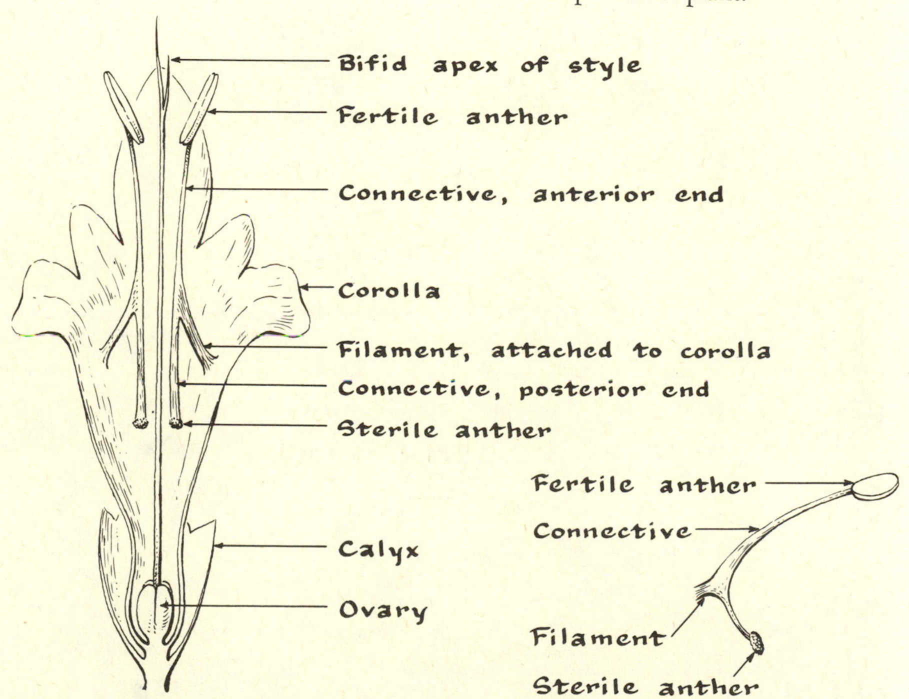

353 *Salvia,* flower opened up to show structure, enlarged. This is a highly specialized flower type, modified for insect pollination.

354 *Salvia,* single stamen, enlarged. Insect visitors receive pollen on their backs by pushing against or stepping on the pedal-like sterile anther.

355 (Left) *Stachys palustris* L., Hedge Nettle, a wide-ranging species of moist situations, and having a 2-lipped corolla. (Right) *Mentha arvensis* L., a common species of Mint across North America in moist places, having a nearly regular corolla.

356 *Stachys palustris* L., detail of part of the inflorescence.

357 *Mentha arvensis* L., detail of part of the inflorescence.

ORDER SCROPHULARIALES

Plants of various habit but predominantly herbaceous. Leaves various but mostly alternate. Flowers hypogynous, sympetalous, mostly 5-merous, the corolla from nearly or quite regular to mostly strongly irregular or bilabiate. Stamens 5 and all fertile, or more commonly reduced to 4 and didynamous, with or without a fifth sterile stamen, or sometimes with only 2 fertile stamens. Pistil 1, of 2 united carpels, the ovary not deeply lobed, with numerous ovules, 2-celled, or rarely 1-celled with free-central placentation. Fruit commonly a capsule, sometimes a berry.

This, the largest order of the *Tubiflorae,* represents another divergent line of evolution from the *Polemoniales,* which culminates in the *Scrophulariaceae.* The unspecialized gynoecium and fruit are retained, but there is a strong tendency toward irregularity in the corolla and with it a reduction in the number of fertile stamens.

SOLANACEAE: Potato Family

Plants chiefly herbaceous, sometimes climbing, and occasionally woody, the stems with bicollateral bundles. Leaves mostly alternate, without stipules. Flowers hypogynous, 5-merous, regular to somewhat irregular, perfect, the corolla of united petals, usually plicate in bud. Stamens 5, inserted on the tube of the corolla alternate with the lobes, rarely 4 or 2. Pistil 1, of 2 united carpels, the ovary not deeply lobed, usually 2-celled with an oblique placenta, but sometimes nearly 4-celled by the development of additional placental lobes, the style single or none, and the stigma single or slightly 2-lobed. Fruit a capsule or berry.

The family includes about 85 genera and 2,200 species, about half of which are in the large genus *Solanum*. The chief center of distribution is in South America, but the general distribution is world-wide in temperate and warm regions. Many of the plants contain poisonous alkaloids, some are important drug plants, some ornamentals, and some valuable food plants.

EXAMPLES: *Solanum tuberosum:* Potato.

Solanum dulcamara: Nightshade, a plant with poisonous berries.

Lycopersicon esculentum: Tomato.

Nicotiana tabacum: Tobacco.

Capsicum frutescens: Peppers.

Atropa belladonna: source of the drug atropine.

Petunia: species of garden ornamentals.

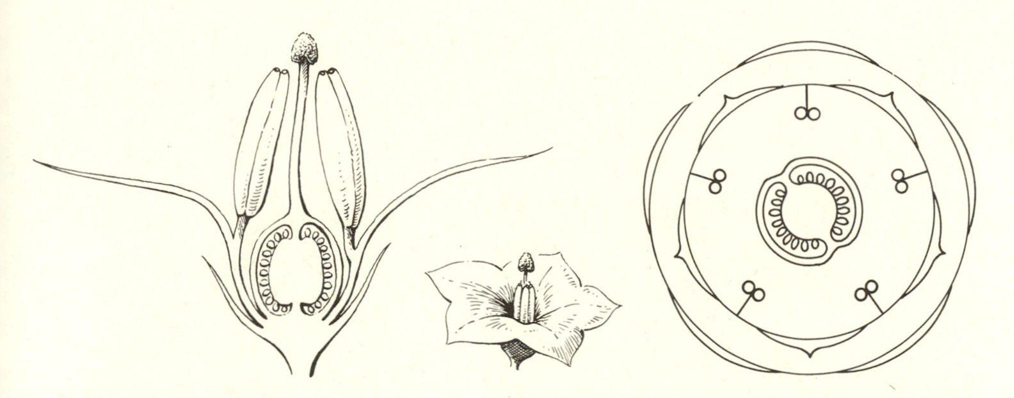

358 *Solanum*, diagrammatic flower section, flower, and floral diagram.

359 *Hyoscyamus niger* L., Henbane, an introduced weed with some medicinal properties. The greenish-yellow flowers are slightly irregular.

360 *Datura meteloides* DC., native of the southwestern United States and Mexico. It is notable for its large flowers, which are white and up to 2 dm long. Like many members of the *Solanaceae*, it is poisonous.

BIGNONIACEAE: Bignonia Family

Mostly woody plants, often lianas in the tropics, with simple or compound and often opposite leaves. Flowers showy, with the general structure of those in the *Scrophulariaceae,* but the ovary either 2-celled with two placentae in each cell or 1-celled and with parietal placentae. Stigma 2-lipped. Fruit a capsule, often elongated, or sometimes fleshy and indehiscent. Seeds often winged.

The family includes about 100 genera and 800 species, widely distributed in tropical or subtropical regions, a few extending into temperate regions. The chief center of distribution is Brazil.

EXAMPLES: *Catalpa:* Catalpa, trees often cultivated for fence posts and as shade trees.

Campsis radicans: Trumpet Creeper, a common ornamental vine.

Chilopsis linearis: Desert Willow, a common shrub of the deserts of the southwestern United States.

OROBANCHACEAE: Broomrape Family

Root parasites with reduced leaves and little or no green color. Flower structure essentially like that in the *Scrophulariaceae,* but the ovary 1-celled and with 2–6 parietal placentae.

The family includes about 14 genera and 170 species, mostly in the northern hemisphere, especially in the Old World.

EXAMPLE: *Orobanche:* Broomrape, with about 100 species, is fairly common in dry sandy areas, attached to roots of various host plants.

LENTIBULARIACEAE: Bladderwort Family

Mostly aquatic, insectivorous herbs, the submerged leaves usually finely divided into capillary divisions and commonly bearing minute bladder-like traps. Flowers essentially like those in the *Scrophulariaceae,* the corolla spurred, the stamens 2, the stigma sessile, and the ovary with free-central placentation.

The family includes 5 genera and about 300 species, widely distributed in temperate and tropical regions.

EXAMPLES: *Utricularia:* Bladderwort, with about 300 species, many tropical.

Pinguicula: Butterwort, with about 45 species, mainly in the north temperate zone and in the mountains of tropical America.

SCROPHULARIACEAE: Figwort Family

Mostly herbs, occasionally shrubby or arborescent, and sometimes partly parasitic, but the leaves green, alternate, opposite, or whorled, without stipules. Flowers in cymes or racemes, or sometimes spicate, hypogynous, perfect, 5-merous, the corolla of united petals (rarely lacking) and from nearly regular to strongly bilabiate. Stamens usually 4 and didynamous, with a sterile upper fifth stamen, but occasionally 5, and sometimes reduced to 2, inserted on the corolla tube alternate with its lobes. Pistil 1, of 2 united carpels, the ovary not deeply lobed, 2-celled, with axillary placentation and numerous ovules. Style 1- or 2-lobed. Fruit usually a capsule, sometimes a berry.

The family includes about 205 genera and 2,600 species, cosmopolitan in distribution, with many species in the western United States.

EXAMPLES: *Verbascum:* Mullein.

Scrophularia: Figwort.

Digitalis: Foxglove, a common drug plant.

Penstemon: Beardtongue, with about 225 species in North America.

Castilleja: Indian Paintbrush.

Antirrhinum majus: Snapdragon, a common ornamental.

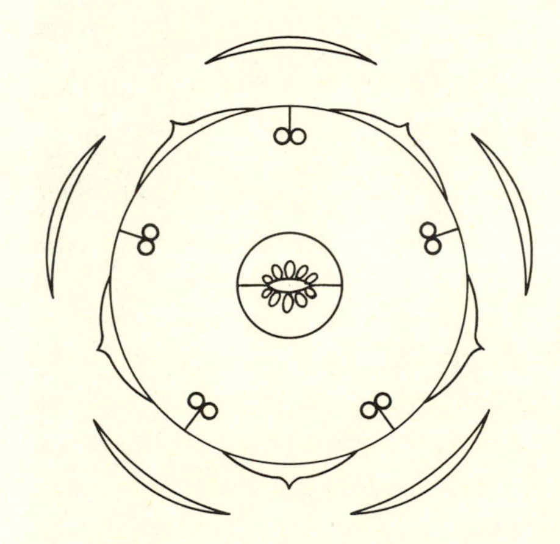

361 *Verbascum*, floral diagram, the corolla nearly regular.

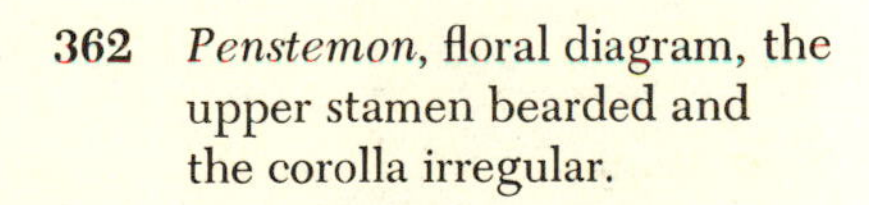

362 *Penstemon*, floral diagram, the upper stamen bearded and the corolla irregular.

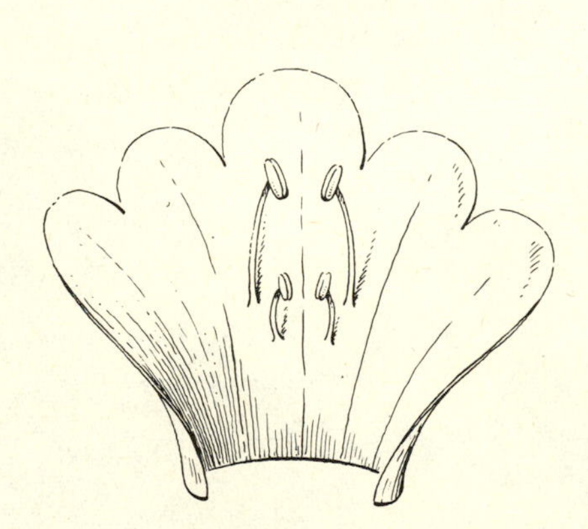

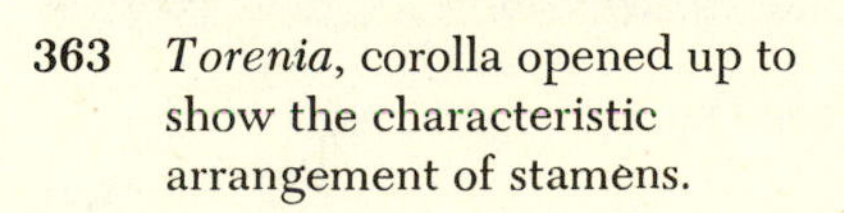

363 *Torenia*, corolla opened up to show the characteristic arrangement of stamens.

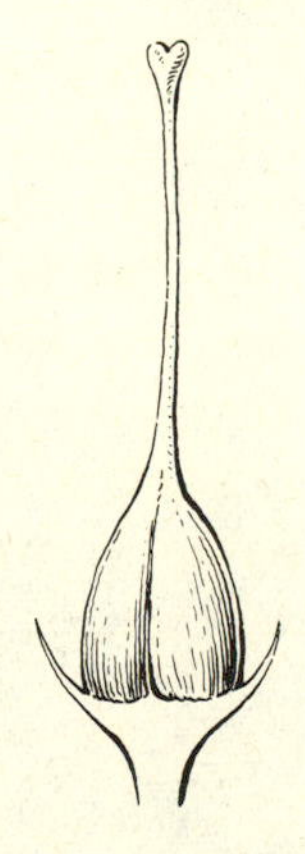

364 Pistil type found in the *Scrophulariaceae*.

365 *Penstemon eriantherus* Pursh, flower opened up to show the bearded, sterile stamen.

366 *Penstemon secundiflorus* Benth., a Beardtongue having lavender to pinkish-violet flowers.

367 *Pedicularis groenlandica* Retz., a Lousewort that ranges from Greenland to Alaska, and southward in the western mountains. The flowers are purplish.

368 *Castilleja chromosa* A. Nels., one of many species of Indian Paintbrush. The bracts are brilliant red.

ORDER PLANTAGINALES

PLANTAGINACEAE: Plantain Family

Mostly scapose herbs with prominently parallel-veined basal leaves and spicate, bracteate inflorescences. Flowers hypogynous, sympetalous, 4-merous, usually perfect, regular, and small. Calyx herbaceous and persistent. Corolla scarious or membranaceous, usually salverform. Stamens mostly 4 or 2, rarely 1, inserted on the corolla tube alternate with its lobes, or sometimes free from the corolla, often long-exserted. Pistil 1, the ovary 1–4-celled, the style single and filiform, longitudinally stigmatic. Fruit a pyxis, or indehiscent and nut-like.

The family includes 3 genera and about 200 species, mainly in temperate regions. It probably represents an offshoot from the *Polemoniales* or perhaps from the *Primulales*.

EXAMPLES: *Plantago:* Plantain, with about 200 species, some of them noxious weeds.

Littorella: Shore-grass, with 2 species having unisexual flowers, one in Europe and North America, the other in South America.

Bougueria: a monotypic genus of the high Andes.

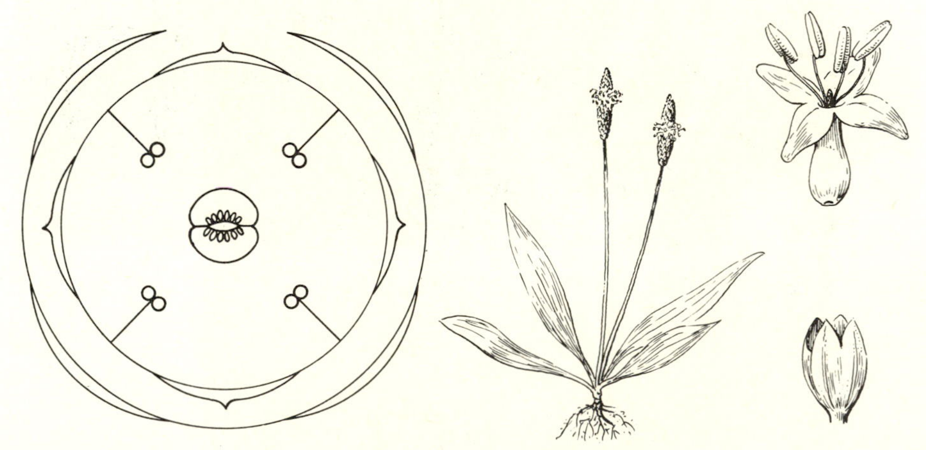

369 *Plantago lanceolata*, floral diagram, plant, and calyx and flower with calyx removed. [Calyx and flower after Harms, in Engler and Prantl, *Nat. Pflanz.* 4^{3b}:366 (1895).]

370 *Plantago patagonica* Jacq., Pursh's Plantain.

The Epigynous Sympetalae

These orders and families, perhaps, have evolved along lines that parallel those of the hypogynous sympetalae. They may have come from a perigynous or epigynous polypetalous ancestry such as that which produced the *Lythrales* and *Apiales,* and not from a hypogynous ancestry, although serodiagnostic evidence indicates a relationship between the *Rubiales* and the *Polemoniales.* Along with the development of a corolla of united petals have gone tendencies toward a reduction in, and often a late development of, the calyx, which eventually becomes an organ of dissemination of the fruit. Irregularity in the corolla occurs in some members, and there is a general tendency toward a reduction in the number of carpels and ovules. In the highest groups there is also a modification of the inflorescence as well, culminating in the well-known involucrate head of flowers that is characteristic of the *Asteraceae* (*Compositae*). Some of these tendencies had already made their appearance in the *Apiaceae.*

Key to the Principal Families of the Epigynous Sympetalae

Tendril-bearing herbs; fruit a pepo (*Cucurbitales*) . . . . CUCURBITACEAE

Tendrils none; fruit not a pepo

  Stamens free from the corolla (*Campanulales*) . . . . CAMPANULACEAE

  Stamens inserted on the corolla tube

    Ovary 2–several-celled and 2–several-seeded (*Rubiales*)

      Flowers regular; leaves opposite and stipulate or appearing whorled and exstipulate . . . . RUBIACEAE [1]

      Flowers regular or irregular; leaves opposite or perfoliate and usually exstipulate . . . . CAPRIFOLIACEAE [1]

    Ovary 1-celled and 1-seeded (*Asterales*)

      Flowers not in an involucrate head; anthers distinct . . . . . VALERIANACEAE

      Flowers in an involucrate head; anthers united or distinct

        Anthers distinct . . . . DIPSACACEAE

        Anthers united (very few exceptions) . . . . ASTERACEAE

[1] Although it has long been the custom to treat the *Rubiaceae* and *Caprifoliaceae* as distinct families, they can be distinguished only arbitrarily and would be better combined.

ORDER RUBIALES

Herbaceous or woody plants with opposite or apparently whorled leaves and mostly cymose inflorescences. Flowers epigynous, sympetalous, typically perfect, regular to irregular, 4–5-merous, the stamens usually of the same number as the lobes of the corolla or sometimes fewer, inserted on the corolla tube alternate with its lobes. Carpels usually few. Seeds numerous to 1 in each carpel.

The order includes the following two families, which are usually treated as separate families but might be better combined.

RUBIACEAE: Madder Family

Mostly trees or shrubs, occasionally (especially in northern regions) herbs or climbers, with opposite or apparently whorled, simple, entire leaves, the stipules often resembling the leaves and thus giving the whorled appearance. Flowers in cymes or panicles, or sometimes capitate or solitary, epigynous, usually perfect and regular and usually 4–5-merous. Stamens of the same number as the lobes of the corolla and inserted on the corolla tube alternate with its lobes. Pistil 1, usually of 2 united carpels, the ovary usually 2-celled, the style single or divided at the summit. Fruit a capsule, berry, or drupe, or the carpels separating to form mericarps.

The family includes about 400 genera and 7,000 species, mostly in tropical regions but some members extending into temperate regions.

EXAMPLES: *Coffea arabica:* Coffee, an evergreen shrub or small tree of tropical Asia and Africa, widely cultivated.

Cinchona officinalis: Quinine, of the Andes, but grown in Asia and Australia for its valuable bark.

Gardenia florida: Gardenia, an ornamental, native of China.

Galium: Bedstraw, or Cleavers, a common North American genus.

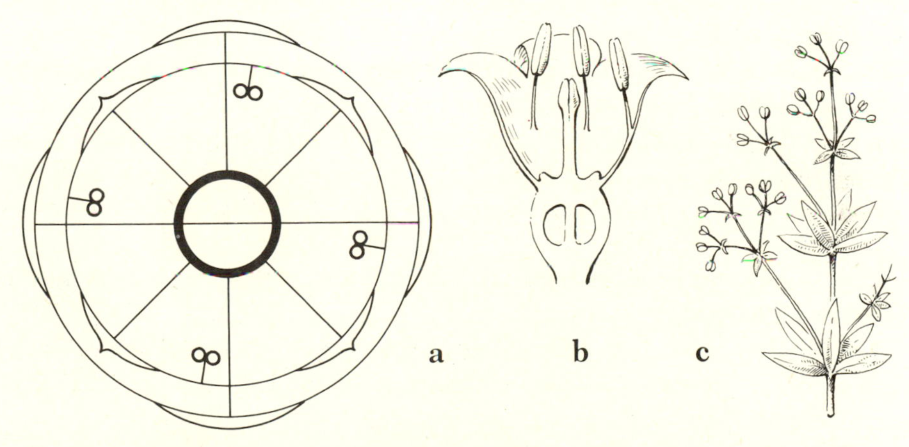

371 (**a**) *Cephalanthus,* floral diagram. (**b**) *Rubia,* longitudinal section of flower. The genus has 5-merous flowers. (**c**) *Galium triflorum*, portion of fruiting plant.

CAPRIFOLIACEAE: Honeysuckle Family

Generally shrubby plants, with opposite, simple or compound leaves, which are exstipulate or with small or reduced stipules. Flowers variously arranged but often in cymes, perfect, regular to irregular, epigynous, usually 5-merous, the corolla of united petals. Stamens usually as many as the lobes of the corolla, inserted on the corolla tube alternate with its lobes, or sometimes the upper (posterior) stamen reduced and the 4 remaining stamens didynamous. Pistil 1, of 2–5 united carpels, the ovary 1–5-celled, and the styles distinct or united. Fruit commonly a berry, but sometimes a drupe or capsule.

The family includes about 11 genera and 350 species, mainly in north temperate regions.

EXAMPLES: *Sambucus:* Elderberries, unusual in having compound leaves.
Lonicera: Honeysuckles, with about 175 species, some cultivated.
Symphoricarpos: Snowberry.
Linnaea: Twinflower, a trailing plant named for Linnaeus.
Viburnum: Arrow-wood, Sheepberry, or Nannyberry.

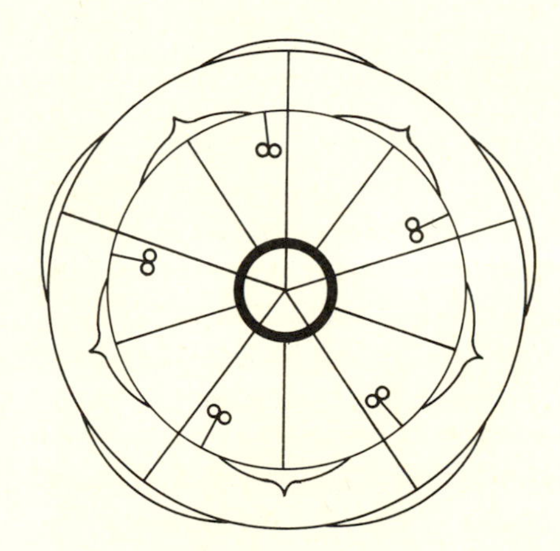

372 *Sambucus canadensis,* floral diagram. Ovary 3–5-celled, and flowers 4-merous. Stamens sometimes six or more. Flowers regular.

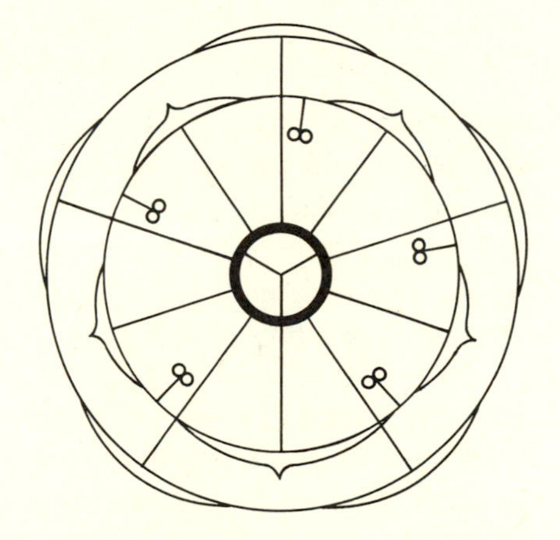

373 *Lonicera sempervirens,* floral diagram. Flowers somewhat irregular, with one larger petal.

374 *Lonicera canadensis,* flowering branch.

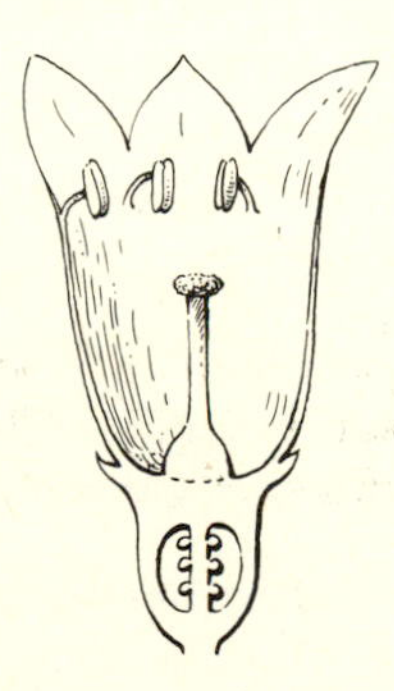

375 *Symphoicarpos racemosus,* longitudinal section of flower, enlarged.

ORDER CUCURBITALES

Mostly herbs, often climbing, with unisexual, 5-merous, epigynous flowers, the corolla of separate or united petals. Stamens numerous to few, free or variously united. Carpels 2–5, often 3, several-seeded.

The order has somewhat doubtful affinities with the *Myrtales* and *Lythrales,* and it may more properly belong with the polypetalae. It includes the families *Cucurbitaceae, Begoniaceae, Datiscaceae,* and *Caricaceae,* and is an essentially tropical or subtropical group. The following family is representative.

CUCURBITACEAE: Gourd or Pumpkin Family

Coarse, often scabrous, tendril-bearing herbs (rarely shrubs or trees), with palmately veined leaves. Flowers monoecious or dioecious, rarely perfect, epigynous, regular, the corolla of 5 free or united petals. Stamens 1–5, usually 3, often variously united, the anthers often variously folded or curved. Carpels usually 3, united, the ovary with parietal placentation or the placentae often meeting in the center, the style usually single, and stigmas 3. Fruit fleshy, often large, and several–many-seeded, with a leathery or hardened exocarp, and thus a modified berry or pepo.

The family includes about 90 genera and 700–760 species, mainly tropical.

EXAMPLES: *Citrullus lanatus:* Watermelon.
Cucurbita pepo: Pumpkin.
Cucumis melo: Muskmelon.
Cucumis sativus: Cucumber.
Lagenaria: Calabash, the fruit often used to make utensils.
Gurania: of tropical America, a genus of shrubs.
Dendrosicyos: of Africa and Socotra Island in the Indian Ocean, a genus of trees.

376 *Sicyos* (Star Cucumber), floral diagram of female flower. The number of parts is variable. The stippled area outside the ovary represents a glandular ring that may have originated from reduced stamens.

377 *Echinocystis lobata,* portion of plant showing female flower and fruit.

378 *Echinocystis lobata* (Michx.) T. & G., Wild Cucumber, a common weedy vine with monoecious flowers. The two globose and prickly fruits shown are immature.

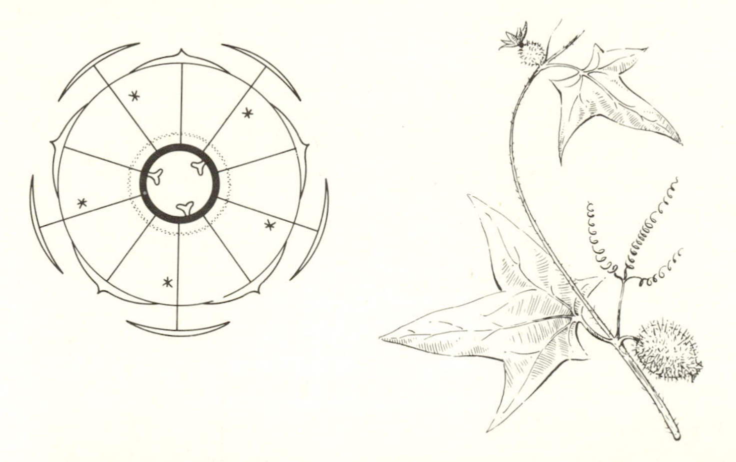

ORDER CAMPANULALES

Plants herbaceous to somewhat woody. Flowers regular to strikingly irregular, nearly always epigynous, sympetalous, and 5-merous. Stamens 5, free from the corolla or nearly so, sometimes monadelphous, and the anthers free, connivent, or united (syngenesious). Ovary 1–10-celled (usually 2–3-celled), the placentation usually axillary and the ovules numerous.

The order may be related to the *Rubiales;* but, as suggested by Hutchinson, it may have been derived from the *Gentianales* or *Polemoniales.* The following family is represented in North America; two others, the *Goodeniaceae* and *Stylidiaceae,* are mainly Australian and antarctic in distribution.

CAMPANULACEAE: Bluebell Family

Mostly herbs, rarely woody, often with milky juice, the leaves usually simple, alternate, and exstipulate. Flowers solitary or in racemes, spikes, or heads, often showy, usually perfect and 5-merous, epigynous, regular to very irregular, the calyx often persistent and the corolla of united petals, campanulate, tubular, or bilabiate. Stamens 5, free from the corolla tube or nearly so, alternate with the lobes of the corolla, the filaments commonly dilated and sometimes united, the anthers free, connate, or united. Pistil 1, of 2–5 (commonly 3) united carpels, with a single terminal style, which usually elongates in anthesis and often has a terminal, hairy, brush-like apical portion, which later splits to expose the stigmatic surfaces. Fruit usually a many-seeded capsule.

The family includes about 61 genera and 1,500 species, in temperate and subtropical regions. It is divided into the following three subfamilies, which are sometimes treated as families.

Key to the Subfamilies of Campanulaceae

Corolla regular . . . . 1. CAMPANULOIDEAE

Corolla irregular

Anthers distinct . . . . 2. CYPHIOIDEAE

Anthers united (syngenesious) . . . . 3. LOBELIOIDEAE

SUBFAMILY 1. CAMPANULOIDEAE

Flowers regular, the anthers usually free. Cosmopolitan in distribution.

EXAMPLES: *Campanula:* Bluebell.
Platycodon: Balloon-flower.

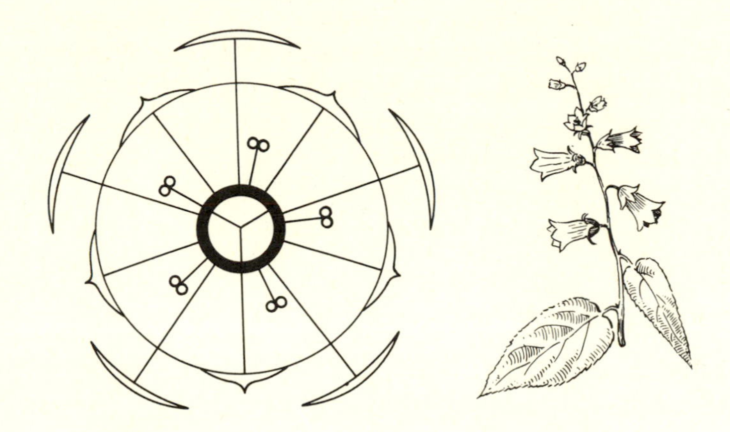

379 *Campanula rapunculoides*, floral diagram and portion of plant.

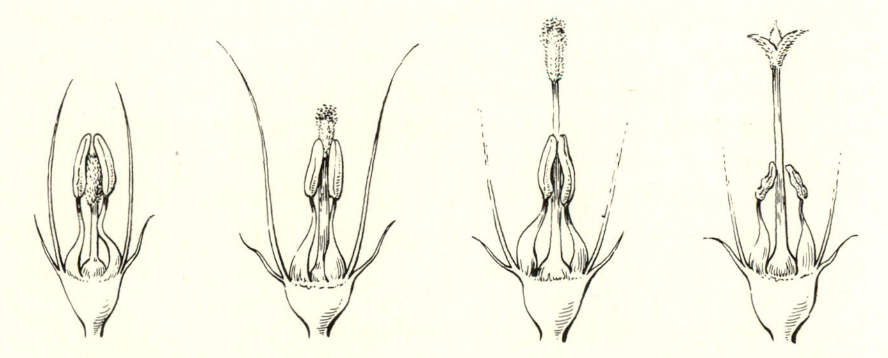

380 Successive stages in the elongation of the style of *Campanula*. Part of the flower has been cut away to show the reproductive parts. The terminal portion of the style acts like a brush to sweep out the pollen grains and carry them upward, after which this portion splits to expose the stigmatic surfaces.

SUBFAMILY 2. CYPHIOIDEAE

Flowers irregular, the stamens sometimes monadelphous but the anthers free. African and western American in distribution.

EXAMPLES: *Cyphia:* with about 20 species in Africa.
Nemacladus: with 2 species in California.
Parishella: with 1 species in California.
Cyphocarpus: with 1 species in Chile.

SUBFAMILY 3. LOBELIOIDEAE

Flowers strikingly irregular and 2-lipped, the anthers united into a tube (syngenesious). Tropical and north and south temperate regions.

EXAMPLES: *Lobelia:* with about 200 species, several in North America. *L. cardinalis* is the Cardinal Flower.
Siphocampylus: a genus with numerous species in tropical America.

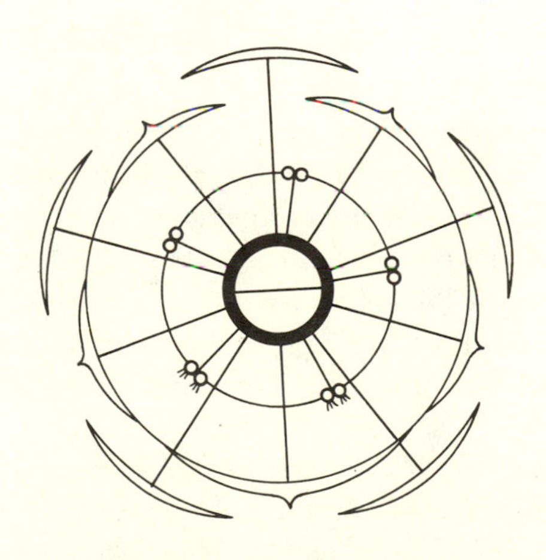

381 *Lobelia,* floral diagram.

382 *Lobelia splendens,* flower.

ORDER ASTERALES

Mainly herbaceous but sometimes woody plants, with alternate or opposite exstipulate leaves. Inflorescence commonly an involucrate head but sometimes cymose or paniculate. Flowers epigynous, sympetalous, regular or irregular, usually perfect and 5-merous, the stamens 1–5 (usually 5), inserted on the corolla tube alternate with the lobes, the anthers free or united. Ovary mostly 1-celled and containing a single ovule. Fruit an akene, which is often crowned by a late-developing calyx of scales or hairs known as the *pappus*.

The order includes the families *Adoxaceae, Valerianaceae, Dipsacaceae, Calyceraceae,* and *Asteraceae* (*Compositae*), and was probably derived from ancestral forms of the *Apiales* and *Campanulales*. The following families are representative.

VALERIANACEAE: Valerian Family

Perennial or annual herbs, rarely shrubs, with strong-smelling rhizomes and opposite or radical, often divided, exstipulate leaves. Flowers usually in cymes, perfect or unisexual by abortion, epigynous. Calyx represented in the flower by an epigynous ring, becoming enlarged and often plumose in fruit. Corolla usually salverform, 5-lobed, often saccate or spurred at the base and somewhat irregular. Stamens 1–4, usually 4, on the tube of the corolla and alternate with its lobes, the anthers free. Pistil of 3 united carpels and typically 3-celled in flower, but only 1 carpel fertile and maturing into a 1-seeded indehiscent fruit (akene) crowned by the pappus, which is the enlarged calyx.

A progressive modification of the flower is found in different genera, from *Patrinia* and *Nardostachys*, in which the corolla is nearly regular, to *Centranthus*, in which the corolla is spurred and the stamens are reduced to one. In extreme cases the corolla tube may be divided into two compartments by a septum which separates the stamens from the style and stigma.

The family includes about 10 genera and 350 species, chiefly in north temperate regions and in the Andes.

EXAMPLES: *Valeriana:* Valerian, with about 200 species.
Nardostachys jatamansi: Spikenard, the source of an unguent used especially by the Romans.

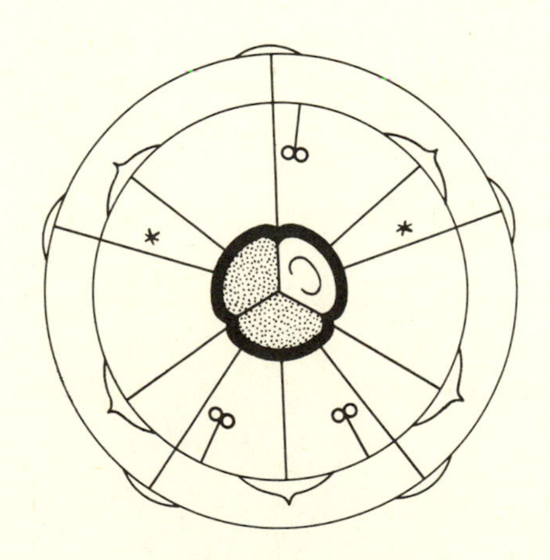

383 *Valeriana officinalis*, floral diagram. Two of the carpels abort, leaving only one in fruit.

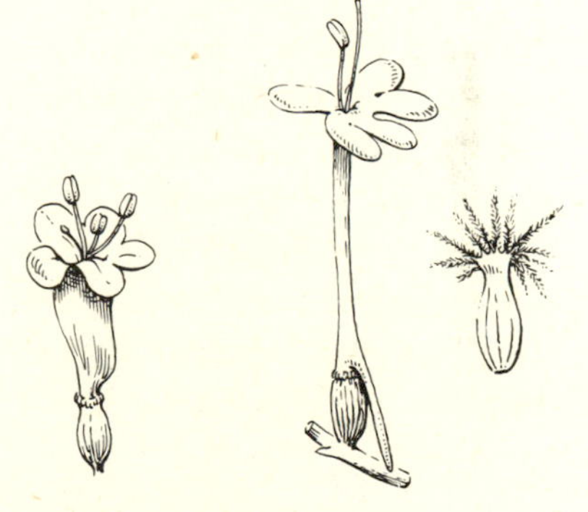

384 *Valeriana*, flower, enlarged.

385 *Centranthus*, flower and fruit, enlarged.

DIPSACACEAE: Teasel Family

Annual or perennial, often very scabrous herbs, with opposite or rarely whorled, exstipulate leaves. Flowers small, in dense bracteate heads, perfect, epigynous, sympetalous, mostly irregular, each flower surrounded by an epicalyx or involucel of small bracts. Corolla 4–5-lobed. Stamens 4, 2, or 3, inserted on the base of the corolla tube alternate with its lobes, the anthers free. Ovary 1-celled and with a single ovule, the style filiform and the stigma simple or bifid. Fruit an akene, often crowned by the persistent bristly or spiny calyx, or pappus.

In some genera, such as *Scabiosa,* the flowers show a differentiation into ray and disk flowers, both as to size and as to degree of irregularity of the corolla.

The family includes about 9 genera and 150 species and is world-wide in range, the chief center of distribution being the eastern Mediterranean region.

EXAMPLES: *Dipsacus fullonum:* Teasel, the fruiting heads of which are used commercially for raising the nap of woolen cloth.
Scabiosa atropurpurea: Scabiosa, a common ornamental from southern Europe.

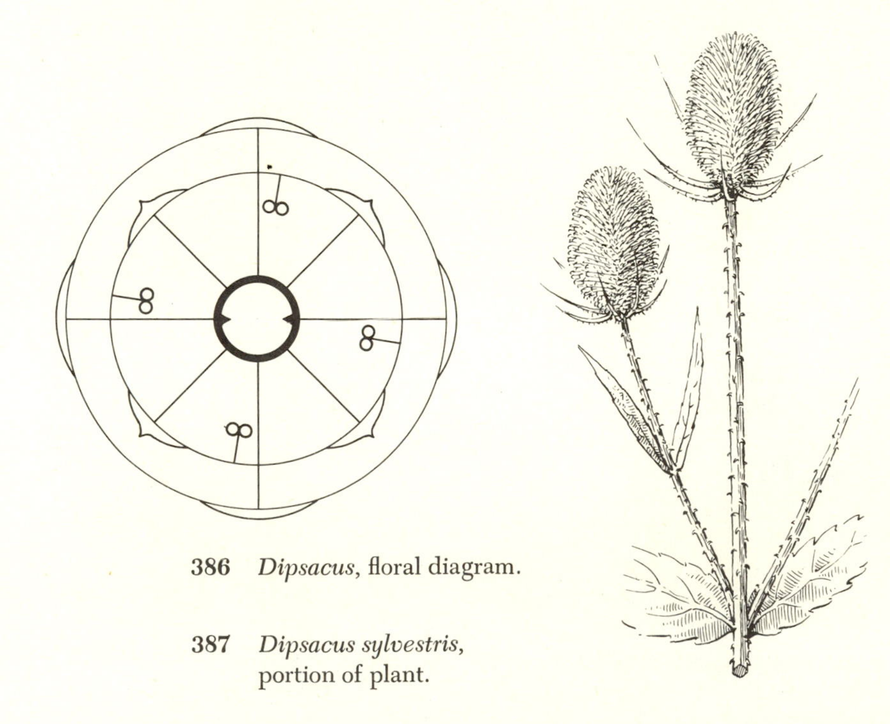

386 *Dipsacus,* floral diagram.

387 *Dipsacus sylvestris,* portion of plant.

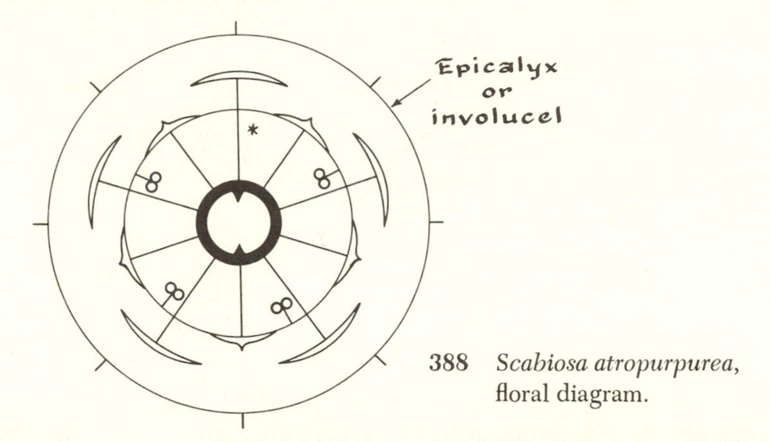

388 *Scabiosa atropurpurea*, floral diagram.

389 *Dipsacus sylvestris* Huds., the common Teasel, an introduced European plant now common in the United States. This is a flower head after the flowers have dropped.

ASTERACEAE (COMPOSITAE): Sunflower Family

Mainly herbaceous, but sometimes woody, plants, with alternate, opposite, or, rarely, whorled leaves, which may be entire to variously dissected, but never truly compound, without stipules. Flowers in dense heads, the common receptacle of each head surrounded below by an involucre of several to many bracts (the *phyllaries*), each such head appearing like a single flower but actually composed of few to many small flowers (*florets*), which are epigynous, mostly 5-merous, all tubular and regular, or the central (*disk flowers*) tubular and regular and the marginal (*ray flowers*) strap-shaped (*ligulate*) and irregular, or all the flowers in the head ligulate or all bilabiate. Stamens 5, inserted on the corolla tube, the filaments usually free but the anthers united in a ring (*syngenesious*), the arrangement alternate with the lobes of the corolla in regular flowers. Pistil 1, of 2 united carpels, the ovary 1-celled and containing a single seed at maturity. Fruit an akene, which is usually crowned by a late-developing calyx (the *pappus*), which may be composed of hairs, bristles, awns, or scales, but may be lacking.

The family includes about 950 genera and 20,000 species, ranking as one of the largest families in the plant kingdom and involving about a tenth of the species of flowering plants. It is cosmopolitan in range but especially well adapted to temperate and cooler climates.

Traditionally the family has been divided into 2 series and 13 tribes, as given in the comparison table at the end of the family treatment. There is, however, good reason for combining two of the tribes (*Heliantheae* and *Helenieae* under the former name), as pointed out by Arthur Cronquist in *Amer. Midl. Nat.* **53**:478–511, 1955.

The 2 series may be distinguished as follows:

Some of the flowers of the head with tubular or sometimes bilabiate corollas; juice not milky . . . . 1. TUBULIFLORAE

All of the flowers of the head ligulate; juice milky . . . . . . 2. LIGULIFLORAE

Series 1. *Tubuliflorae:* The larger series, including 12 tribes. Plants of various aspect, the juice never milky; flowers all tubular and regular, sometimes the flowers all bilabiate, and usually the central (disk) flowers tubular and regular and the marginal (ray) flowers ligulate and irregular.

EXAMPLES: *Aster:* the Asters.
Solidago: Goldenrod.
Helianthus: Sunflower.
Senecio: Old Man, the largest genus, with some 2,300 species of herbs, shrubs, climbers, and even a few trees.

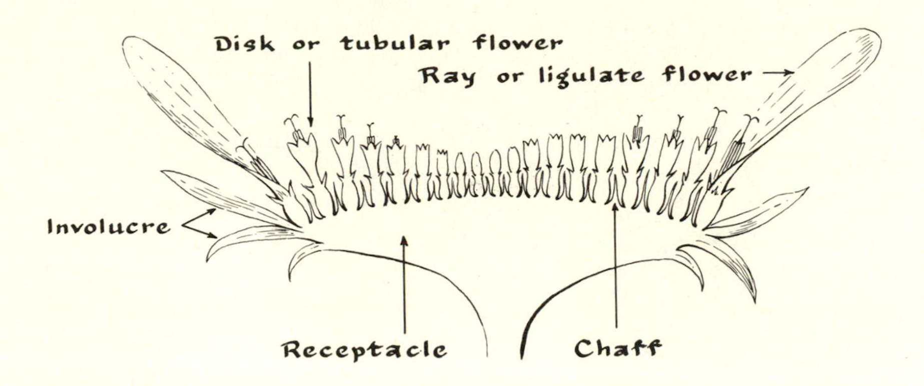

390 Typical composite head cut vertically in half.

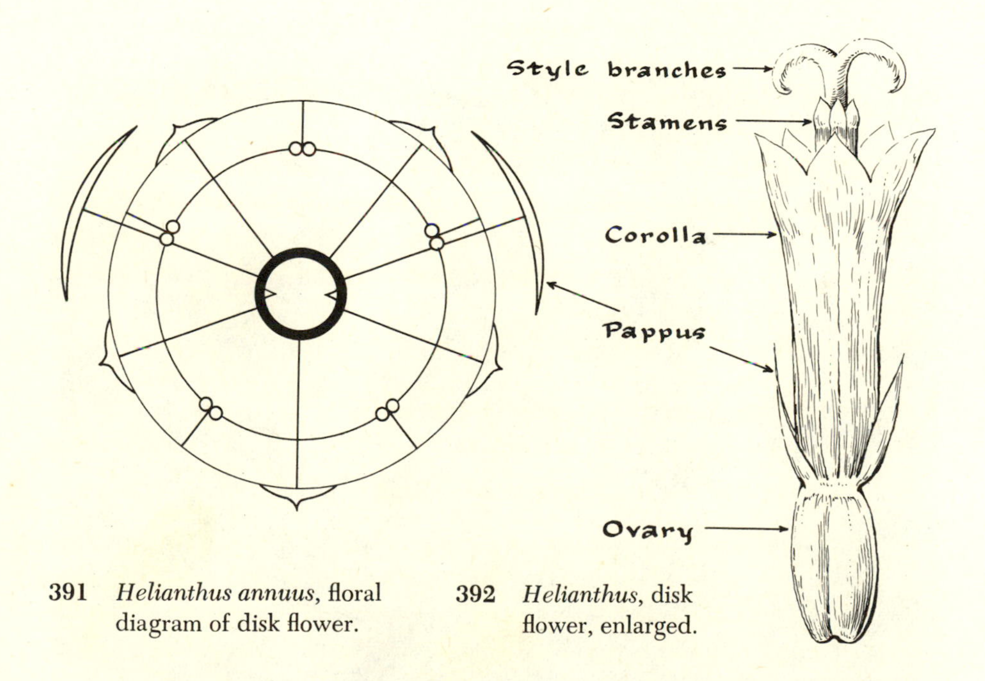

391 *Helianthus annuus,* floral diagram of disk flower.

392 *Helianthus,* disk flower, enlarged.

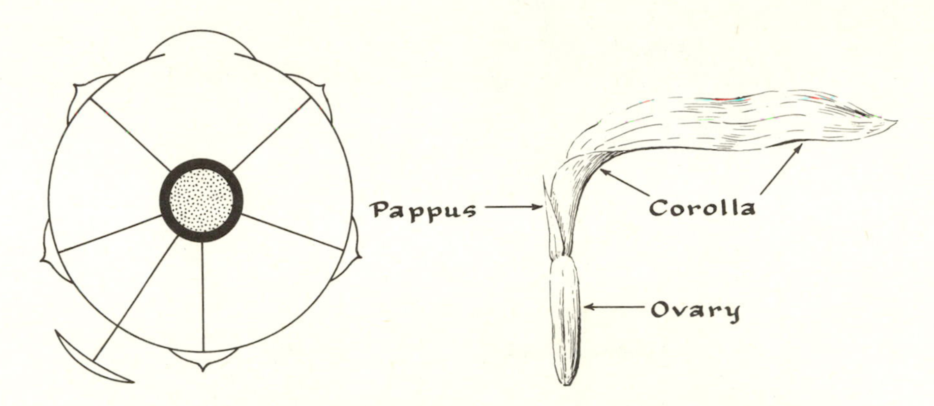

393 *Helianthus annuus*, floral diagram and ray flower, the latter enlarged. Ray flowers are sterile, with a rudimentary pistil.

394 *Balsamorhiza sagittata* (Pursh) Nutt., Balsamroot, of western North America (*Tubuliflorae*).

395 *Balsamorhiza sagittata* (Pursh) Nutt., Balsamroot, three heads, that on the left showing ligulate and tubular flowers, that in the center cut in half to show the typical structure of the inflorescence, and that on the right in side view showing the involucre below the flowers.

Series 2. *Liguliflorae:* The smaller series, including the single tribe *Cichorieae,* and sometimes treated as a separate family. Herbs, rarely shrubs or trees, of various habit. Flowers in the head all ligulate and homogamous. Juice milky.

EXAMPLES: *Cichorium:* Chicory.
Taraxacum: Dandelion.
Lactuca: Lettuce, including 100 species.
Crepis: Hawksbeard, with nearly 200 species.
Hieracium: Hawkweed, with about 400 species.

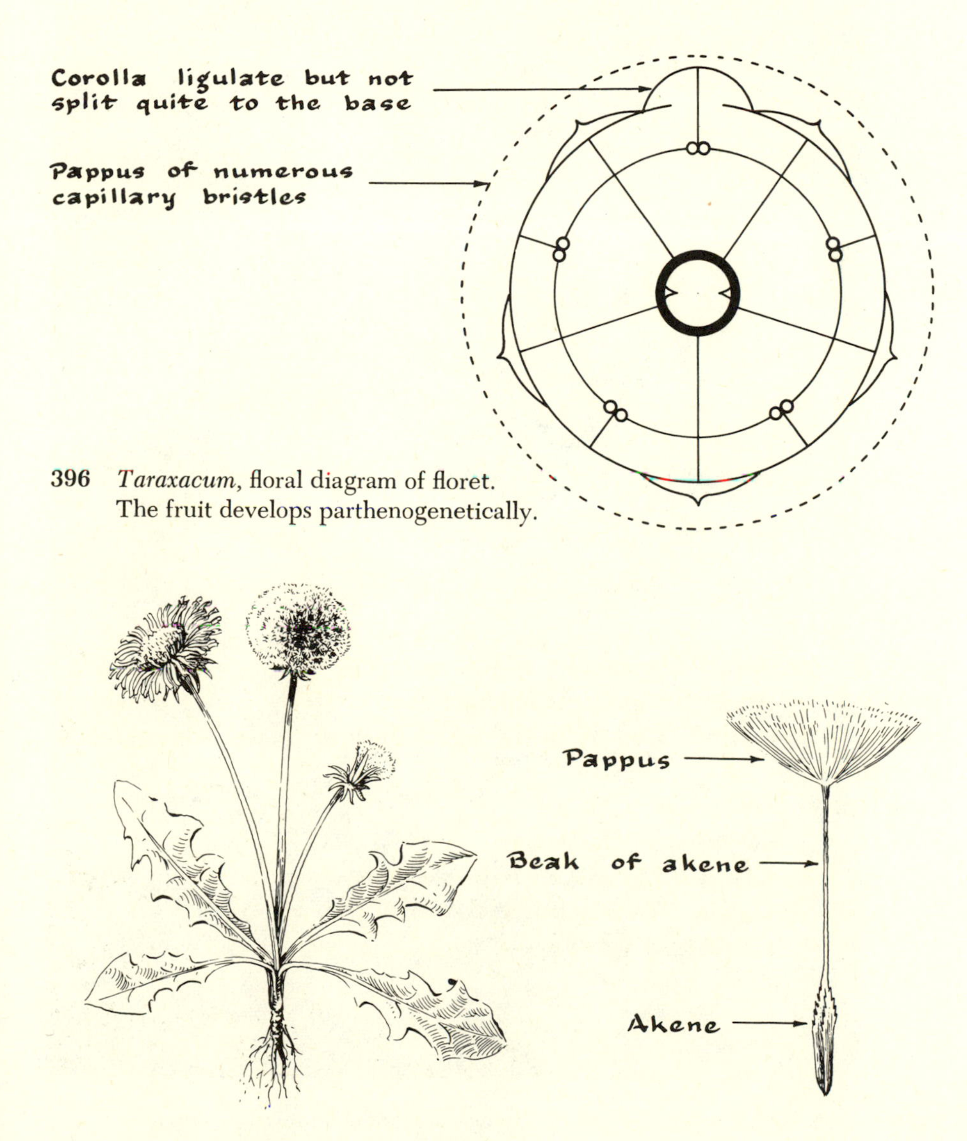

396 *Taraxacum,* floral diagram of floret. The fruit develops parthenogenetically.

397 *Taraxacum,* plant, and fruit, enlarged.

398 *Nothocalais cuspidata* (Pursh) Greene, a Dandelion relative having all the flowers in the head with ligulate corollas (*Liguliflorae*).

399 *Tragopogon dubius* Scop., Goatsbeard, a common weed introduced from Europe. On the left is a flower bud, and in the center is a fruiting head of beaked akenes having a plumose pappus (*Liguliflorae*).

Key to the Tribes of Asteraceae (Compositae)

[Simplified and modified from Arthur Cronquist in *Amer. Midl. Nat.* **53**:494–496 (1955). Used by permission of the author.]

1. Heads radiate or discoid, or with some of the flowers bilabiate, or some or all the flowers tubular; juice not milky, the latex, if any, in isolated cells or pockets

 2. Heads chiefly radiate or not infrequently discoid or disciform; corollas predominantly yellow

  3. Leaves, the lower or all, tending to be opposite; pappus chaffy or of a few awns, or none, rarely capillary; receptacle usually chaffy; involucral bracts tending to be herbaceous and in several series; rays tending to be broad . . . . 1. HELIANTHEAE including HELENIEAE

  3. Leaves very often alternate except in a few genera such as *Arnica;* other characters variable but rarely as given above

   4. Involucral bracts mostly in several series, imbricate or subequal, or the outer bracts larger; pappus variable

    5. Anthers obtuse to strongly sagittate at the base; rays either xanthic like the disk or cyanic to white

     6. Style-branches distinct, flattened; plants not spiny

      7. Involucral bracts herbaceous to chartaceous; plants mostly not aromatic . . . . 2. ASTEREAE

      7. Involucral bracts mostly dry, scarcely herbaceous, becoming hyaline-scarious toward the margins and tips; plants mostly aromatic . . . . 3. ANTHEMIDEAE

6. Style-branches usually connate to near the tip; plants often with the involucre or the leaves spiny . . . . 4. ARCTOTIDEAE

5. Anthers more or less strongly caudate at the base; rays, when present, usually yellow like the disk . . . . 5. INULEAE

4. Involucral bracts usually equal and in nearly or entirely a single series; pappus capillary (sometimes plumose) or none

8. Pappus well developed, rarely absent; akenes mostly all alike; widespread . . . . 6. SENECIONEAE

8. Pappus lacking or poorly developed; akenes mostly heteromorphic; Africa . . . . 7. CALENULEAE

2. Heads discoid or with some bilabiate corollas; pigmentation diverse

9. Anthers obtuse to sagittate at the base; heads usually discoid; plants not spiny; corollas cyanic to sometimes white or cream but never truly yellow

10. Style-branches obtuse and often clavate; leaves mostly opposite . . . . 8. EUPATORIEAE

10. Style-branches gradually attenuate; leaves nearly always alternate . . . . 9. VERNONIEAE

9. Anthers more or less strongly caudate at the base; heads discoid or with some or all the corollas bilabiate; plants often with the leaves or involucre spiny; corollas cyanic or xanthic, or occasionally white

11. Heads usually discoid . . . . 10. CYNAREAE

11. Heads usually with some or all the corollas bilabiate . . 11. MUTISIEAE

1. Heads entirely ligulate; juice milky . . . . 12. CICHORIEAE

COMPARISON OF TRIBES OF ASTERACEAE

	Liguliflorae	**Tubuliflorae**				
	Chichorieae	*Vernonieae*	*Eupatorieae*	*Asterieae*	*Inuleae*	*Heliantheae*
Juice milky	yes	no	no	no	no	no
Corollas *t*ubular, *l*igulate, or *b*ilabiate	l	t	t	t or t & l	t except *Inula*, which has t & l	t or t & l
Nature of pappus	simple or plumose	setose and copious	5-many bristles or scales, or none	various	usually simple or plumose bristles	not capillary, usually 2–3 awns or scales
Nature of anthers	sagittate	sagittate, blunt, or short-tailed	blunt and basifixed	blunt and basifixed	tailed	blunt or acute, not tailed, basifixed
Receptacle chaffy	yes or no	?	?	no (in N. American genera)	yes or no	yes
Heads *hom*ogamous or *het*erogamous	hom	hom	hom	het (rarely hom)	hom or het	het (rarely hom)
Plants aromatic	no	no	no	no	no	no
General range	world-wide, chiefly in Old World, many in Mediterranean region	chiefly in tropical America, also Africa, Asia, and Australia, not in Europe	cosmopolitan, especially in Old World and Brazil; *Eupatorium* in N. America	world-wide, but chiefly in temperate zone	widely distributed	mainly American, few in Old World
Some examples, with approximate numbers of species in each, with other notes.	*Cichorium*, 8 *Crepis*, 200 *Hieracium*, 400 *Lactuca*, 100 *Scorzonera*, 100 *Taraxacum*, 25 *Leontodon*, 50 *Dendroseris* and *Fitchia* of the South Seas are tree genera.	*Vernonia*, 500, includes herbs and some small trees.	*Eupatorium*, 400 *Mikania*, 150 Mainly herbs, some twiners.	*Aster*, *Erigeron*, and *Solidago* have numerous species; *Olearia*, 90, in Australia; *Conyza*, 50, in tropics; *Baccharis*, 300, in tropical America.	*Inula*, 90; *Gnaphalium*, 120; *Leontopodium alpinum* is Edelweiss; *Helichrysum*, 300, species of S. African Everlastings.	Includes 10 subtribes. *Zinnia*, 12 *Helianthus*, 55 *Coreopsis*, 70 *Dahlia*, 9 *Rudbeckia*, 30

Tubuliflorae (Continued)						
Helenieae	*Anthemideae*	*Senecioneae*	*Calenduleae*	*Arctotideae*	*Cynareae*	*Mutisieae*
no	no	no	no	no	no	no
t & l	t or t & l	t or t & l	t & l	t & l	t	b
not capillary, usually 2–3 awns or scales	absent or reduced to a ring or cup	capillary	none	none or not capillary	usually capillary	capillary
blunt or acute, not tailed, basifixed	blunt or acute, not tailed, basifixed	blunt or acute, not tailed, basifixed	pointed at the base	blunt-pointed at the base	usually tailed	long-tailed
no	yes or no	usually no	no	?	yes, many bristles	?
het (rarely hom)	het (rarely hom)	hom or het	het	het	hom (rays may be neuter or female)	hom or het
no	yes	no	no	no	no	no
mainly Mexican and Pacific N. American	some in Old World, some world-wide but northern	world-wide	mainly S. Africa, also in Mediterranean region	mainly S. Africa and mountains of tropical Africa	world-wide, with chief center in Mediterranean region	in Andes, also in Africa, Asia and Tasmania
Helenium, 24 *Gaillardia*, 15	*Anthemis*, 100 *Achillea*, 100 *Chrysanthemum*, 200 *Artemisia*, 200	*Petasites*, 14; *Senecio*, 2, 300, including a few trees and shrubs; *Cineraria*, 25, in S. Africa; *Arnica*, 32	*Calendula*, 20	*Arctotis*, 50 *Gazania*, 30	*Arctium*, 6 *Carduus*, 100 *Cirsium*, 200 *Centaurea*, 500	*Mutisia*, 60 *Gerbera*, 40

GLOSSARY

A-. A prefix meaning without, as in "asepalous," without sepals.

Abaxial. On the side away from the axis.

Acaulescent. Without an evident leafy stem.

Accrescent. Enlarging with age; often applied to the calyx.

Accumbent (cotyledons). Placed with their edges against the radicle.

Achene. See Akene.

Achlamydeous. Without a perianth; naked.

Acicular. Needle-shaped.

Acropetal. Produced in succession toward the apex; usually applied to the order of blooming in an inflorescence. (See Basipetal.)

Actinomorphic. With radial symmetry, the parts similar in size and shape.

Aculeate. Prickly.

Acuminate. Tapering into a long point.

Acute. Pointed; forming less than a right angle.

Adaxial. On the side next to the axis.

Adnate (dissimilar parts). Grown together.

Adventitious. In an unusual place; often applied to roots or buds.

Adventive. Introduced but not well established.

Aestivation. Arrangement of flower parts in the bud.

Aggregate (fruit). A cluster of fruits produced by a single flower, as in the blackberry.

Akene. A small, dry, indehiscent, 1-seeded fruit having a thin pericarp that is free from the seed.

Albumen. An old term largely replaced by the term "endosperm," which see.

Allopatric. Occurring in different areas. (See Sympatric.)

Alternate. Located singly at a node, as leaves on a stem; situated between other parts, as stamens between petals.

Ament (catkin). An elongate, deciduous cluster of unisexual and usually bracteate flowers, as in willows.

Amphitropous (ovule). Attached near its middle; half-inverted.

Amplexicaul. Clasping the stem.

Anatropous (ovule). Inverted, with the micropyle close to the point of attachment.

Androecium. The stamens collectively.

Androgynous. Having staminate flowers above and pistillate flowers below in the same spike.

Androphore. A stalk that supports stamens above the point of attachment of the perianth.

Anemophilous. Pollinated by wind. (See Entomophilous.)

Angiosperm. A plant producing seeds enclosed by an ovary.

Annual. A plant that completes its development in one year or one season and then dies.

Anterior. Away from the axis; in bilabiate flowers the lower lip is anterior and the upper lip is posterior.

Anther. The pollen-containing part of a stamen, usually consisting of two sacs.

Anthesis. The time of blooming.

Anthocarp. A dry, indehiscent fruit composed of an ovary surrounded by a perianth tube or other part, as in *Nyctaginaceae.*

Apetalous. Without petals.

Apical. Concerning the apex or tip.

Apiculate. With a short and abrupt point.

Apocarpous (polycarpous). With separate carpels.

Apochlamydeous (perianth parts). Not united.

Appressed. Lying against, as hairs lying close to the leaf surface.

Approximate. Close to each other but not united.

Aquatic. Growing in water.

Arachnoid. Cobwebby, with fine tangled hairs.

Arborescent. Becoming almost tree-like in size.

Arcuate. Curved or arched.

Areole. A marked space on a surface or beneath it, as spaces between veins in leaves, or spine-bearing areas on cactus plants.

Aril. An appendage or outgrowth from the hilum or funiculus of a seed, often spongy or gelatinous, and sometimes enveloping the seed.

Aristate. With a stiff bristle or awn.

Armed. Having prickles, spines, or thorns.

Article. One of the segments of a jointed fruit such as a loment.

Articulate. With one or more joints or points of separation.

Ascending. Growing upward at an angle, but not erect.

Assurgent. Ascending; growing upward at an angle, but not erect.

Attenuate. Gradually tapering; drawn out into a narrowed portion.

Auricle. An ear-like lobe or appendage.

Austral. Southern. (See Boreal.)

Autophytic. With chlorophyll and therefore independent, as opposed to saprophytic or parasitic.

Awn. A bristle-like appendage such as occurs on the back or at the tip of glumes and lemmas of many grasses.

Axil. The angle formed between two organs, as between a leaf and a stem.

Axillary (*axile*). In the axis; designating flowers borne in the axils of leaves, and ovules or seeds produced in the angles formed by partitions in the ovary of a compound pistil.

Baccate. Like a berry.

Banner (*standard*). The upper petal of a papilionaceous corolla.

Barbed. With rigid points that are reflexed, or directed backward, as in a fish-hook.

Barbellate. Finely barbed.

Basifixed. Attached by the base.

Basipetal. Produced or blooming in succession from the top downward. (See Acropetal.)

Beak. A narrow projection, as on some fruits.

Bearded. With rather long hairs.

Berry. A simple, fleshy, usually indehiscent fruit with one or more seeds, as tomatoes or grapes.

Bi-. A prefix meaning two or twice, as in "bipinnate," twice pinnate.

Biennial. A plant that completes its development in two years, usually blooming the second year.

Bifid. Split about midway into two lobes.

Bilabiate. Two-lipped.

Bipinnate. Twice pinnate.

Bladdery. Thin-walled and inflated.

Blade. The expanded portion of a leaf, sepal, petal, or other part.

Bloom. Whitish powdery covering of a surface, easily rubbed off.

Boreal. Northern. (See Austral.)

Bract. A more or less modified leaf subtending a flower or a flower cluster.

Bracteole (*bractlet*). A secondary bract, often very small.

Bud. An undeveloped or dormant branch, leaf, or flower, usually enclosed by protective scales.

Bulb. A subterranean bud composed of fleshy scales attached to a central or basal stem.

Bulbil (*bulblet*). A little bulb or bulb-like body, often produced above ground on stems or in inflorescences.

Caducous. Falling early; often applied to petals.

Caespitose. See Cespitose.

Calcarate. With a spur.

Callus. A hardened projection; in grasses the hardened base of a lemma at the point of its attachment to the rachilla.

Calyculate. Having bracts or an involucre resembling an outer calyx.

Calyx. The outer part of the floral envelope, composed of sepals.

Campanulate. Shaped like a bell.

Campylotropous (ovule). Curved so as to bring the apex and base close together.

Canaliculate. Grooved lengthwise.

Canescent. With close grayish pubescence.

Capillary. Hair-like.

Capitate. Like a head; in a dense, more or less rounded cluster.

Capitulum. A little head.

Capsule. A simple, dry, dehiscent fruit of two or more carpels, and usually several- to many-seeded.

Carinate. Keeled or creased, with a sharp ridge.

Carpel. A megasporophyll; often regarded as a single, modified, seed-bearing leaf.

Carpophore. A slender stalk between the two carpels of Parsnip Family fruits; from its summit the carpels are suspended at maturity.

Caruncle. An appendage or protuberance adjacent to the hilum of some seeds.

Caryopsis. The fruit of grasses, seed-like, with a thin pericarp adherent to the seed; a grain.

Castaneous. Dark brown, with the color of a chestnut.

Catkin (*ament*). An elongate, deciduous cluster of unisexual, apetalous, and usually bracteate flowers, as in willows.

Caudate. Tailed; with a slender appendage.

Caudex. The thickened and often woody base of a perennial plant.

Caudicle. The thread-like or strap-shaped stalk that connects a mass of pollen to an anther sac in many orchids.

Caulescent. With a leafy stem above ground.

Cauline. Pertaining to the stem, as cauline leaves.

Cell. A unit of which plants and animals are constructed; also a compartment within an ovary.

Ceriferous. Waxy.

Cernuous. Nodding or drooping.

Cespitose. Tufted or matted.

Chaff. Small, dry, scale-like bracts.

Chalaza. The basal portion of an ovule where it joins the funiculus.

Channeled. Grooved lengthwise.

Chartaceous. Papery.

Choripetalous (*polypetalous*). With separate petals.

Chorisepalous. With separate sepals.

Ciliate. With marginal hairs or bristles.

Cincinnus. A type of monochasial inflorescence; a scorpioid cyme.

Cinereous. Ash-colored.

Circinate. Coiled in the bud.

Circumscissile. Opening by a transverse circular split so as to release a lid.

Cladophyll. A modified stem resembling a leaf in form and function.

Clasping. Partly surrounding another structure at the base.

Clavate. Shaped like a club, tapering a little toward the base.

Claw. The slender, stalk-like basal portion of some petals and sepals.

Cleft. Deeply cut.

Cleistogamous. Descriptive of a flower that does not open and is self-pollinated.

Cleistogene. A plant bearing cleistogamous flowers.

Coetaneous. Producing flowers and leaves at about the same time, as in some willows.

Collar. The outer side of the leaf at the place where the blade and sheath come together in grasses.

Column. In orchids the combined style and stamen structure; in certain grasses the basal portion of awns; in the Mallow Family the stamen tube.

Coma. A tuft of fine hairs, as on some seeds.

Commissure. In the Parsnip Family the flat adjacent faces of the two carpels.

Comose. With a tuft of hairs.

Compound. Composed of two or more parts; compound leaves have two or more leaflets; compound pistils have two or more carpels.

Compressed. Flattened.

Conduplicate. Folded in half lengthwise.

Connate. United similar parts, as leaves or anthers.

Connective. The part of a stamen joining the two anther sacs.

Connivent. Coming close together or touching, but not united, as in some stamens.

Contorted. Twisted.

Convolute. Rolled up lengthwise.

Cordate. Heart-shaped, with a basal notch.

Coriaceous. Leathery.

Corm. A modified, usually subterranean stem that is fleshy and thickened, and often bears scale-like leaves.

Corolla. The inner set of floral leaves, consisting of petals.

Corona. A petaloid appendage situated between the corolla and the stamens of some flowers, such as Daffodils and Milkweeds.

Corymb. A rounded or flat-topped inflorescence in which the pedicels or branches are attached at intervals on an elongate axis and are of unequal length, the lower ones longer.

Costate. With longitudinal ribs or veins.

Cotyledon. An embryonic leaf of a seedling or in a seed.

Crenate. With rounded teeth.

Crest. A ridge or elevation on a structure; in some milkweed flowers a horn-like projection from a segment (hood) of the corona.

Crisped. A curled leaf-margin; curled hairs.

Cucullate. Hood-shaped.

Culm. The flowering stem of a grass or sedge.

Cuneate. Wedge-shaped.

Cuspidate. With an abrupt, short, sharp, often rigid point.

Cyathium. The small, cup-like, specialized inflorescence of *Euphorbia.*

Cyclic. In circles or whorls, as opposed to a spiral arrangement.

Cyme. An inflorescence in which the central flower of each group is the oldest; in general a loose term for complex flower clusters that are more or less rounded or flat-topped.

Deca-. A prefix meaning ten.

Deciduous. Falling off, as leaves that are shed in the autumn.

Decompound. Two or more times compound, usually meaning with many small divisions.

Decumbent. With the base prostrate but the upper parts erect or ascending.

Decurrent. Extending downward, as leaves having their bases prolonged downward as wings along the stem.

Definite. Usually meaning ten or less, as opposed to numerous.

Deflexed. Bent downward.

Dehiscence. The method or act of opening or splitting.

Dehiscent (fruit). One that splits open.

Deltoid. Broadly triangular, with the base nearly straight and the sides often a little curved toward the apex.

Dentate. Toothed, the teeth acute and directed outward.

Denticulate. The diminutive of "dentate"; with small teeth.

Depauperate. Stunted or poorly developed.

Determinate (inflorescence). Sometimes applied to those in which the terminal or central flower is the oldest.

Di-. A prefix meaning two.

Diadelphous (stamens). United by their filaments in two groups.

Diandrous. Having two stamens.

Dichasium. An inflorescence having a central older flower and a pair of lateral branches bearing younger flowers.

Dichotomous. Forking; branching by pairs.

Diclinous. Having two sexes; unisexual flowers.

Didymous. Paired or twinned.

Didynamous. With four stamens in two pairs.

Diffuse. With many loose or open branches.

Digitate. With parts diverging from a common base, as fingers of a hand.

Dimorphic. Of two forms.

Dioecious. Having unisexual flowers and these produced on separate plants.

Disk. An enlargement of the floral axis, often fleshy or glandular; in the *Asteraceae,* the central part of the head.

Disk flowers. In *Asteraceae,* those produced in the central part of the head and tubular in shape.

Dissected. Cut into many fine segments.

Distal. Toward the apex.

Distichous. In two rows.

Distinct. Separate from each other.

Divaricate. Widely spreading or divergent.

Divided. Cut deeply to or near the base.

Dolabriform. Shaped like a pick, attached near the middle.

Dorsal. Relating to the back, the side away from the axis.

Drepanium. A flattened and coiled or curved monochasial inflorescence.

Drupaceous. Like a drupe in general appearance but not necessarily with its true structure.

Drupe. A simple, fleshy fruit with a single seed enclosed in a bony endocarp or pit; a stone fruit.

Drupelet. A little drupe, usually found in clusters, as in blackberries.

E-. A prefix meaning without.

Ebracteate. Without bracts.

Echinate. Prickly.

Eglandular. Without glands.

Elliptical. Like an ellipse; longer than wide and with rounded ends.

Emarginate. Notched at the apex.

Embryo. The rudimentary plant within a seed.

Emersed. With parts extending above the water.

Endemic. With a very restricted range; confined to a single geographic area.

Endocarp. The innermost of the three layers forming the wall of the pericarp of a fruit.

Endosperm. Nutritive material or tissue in some seeds outside the embryo.

Ensiform. Sword-shaped.

Entire. With an unbroken or even margin; without teeth or other indentations.

Entomophilous. Pollinated by insects.

Envelope. Surrounding part, as the floral envelope, which consists of the perianth; the sepals and petals.

Ephemeral. Lasting for only a short time, usually less than one day.

Epi-. A prefix meaning upon.

Epicalyx. A set of bracts adjacent to and resembling a calyx.

Epigynous. A flower in which the hypanthium or the perianth is attached to the upper part of the ovary, the ovary then appearing inferior in position.

Epipetalous. Regarding stamens that are attached to the corolla.

Epiphyte. An independent plant growing upon another plant and not connected to the ground.

Equitant. Folded lengthwise and in two flat rows, as the leaves of *Iris.*

Erose. With irregular margin, as though chewed.

Excurrent. Running through and out, as a vein extending beyond a leaf blade into a point.

Exfoliating. Scaling off or shedding in plates, as the bark of some trees.

Exocarp. The outermost of the three layers forming the wall or pericap of a fruit.

Exserted. Projecting beyond, as stamens protruding from the corolla.

Exstipulate. Without stipules.

Extrorse. Facing outward.

Falcate. Curved like a sickle.

Farinose. Mealy; covered with a mealy powder.

Fascicle. A cluster or bundle.

Fastigiate (branches). Erect and closely spaced.

Fertile. Capable of reproducing, as a stamen producing viable pollen or a carpel producing ovules.

Filament. The stalk of a stamen; any thread-like body.

Filiform. Thread-like.

Fimbriate. Fringed.

Fistulose. Hollow and cylindrical, as the leaves of some onions.

Flabellate. Fan-shaped; broadly wedge-shaped.

Flaccid. Limp or flabby.

Flexous. Wavy; curved alternately in opposite directions.

Floccose. With tufts of soft hair.

Floral envelope. The calyx and corolla, or perianth.

Floret. A little flower; in grasses including the lemma.

Floriferous. Flower-bearing.

Foliaceous. Leaf-like, usually meaning with green color.

Foliar. Pertaining to leaves or leaf-like parts.

Follicle. A dry fruit of one carpel that splits on one side.

Free. Not adnate to other parts. (Compare with Distinct.)

Frond. The leaf of a fern or of some other plants such as palms; sometimes applied to the thallus of certain plants such as duckweeds.

Fruit. A ripened ovary, sometimes including other adherent parts.

Frutescent. Becoming shrubby.

Fruticose. Shrubby.

Fugacious. Falling early, as the sepals or petals of some flowers.

Fulvous. Dull yellow.

Funiculus. The stalk of an ovule or seed.

Funnelform. Shaped like a funnel, with gradually widened tube.

Fuscous. Dusky brown or grayish brown.

Fusiform. Spindle-shaped, thickened in the middle and tapering to the ends.

Galea. The upper, usually concave lip of a bilabiate calyx or corolla.

Gamo-. A prefix meaning united.

Geniculate. Bent at a joint; kneed.

Gibbous. Swollen or with a protuberance on one side, usually near the base.

Glabrate. Becoming glabrous or hairless at maturity.

Glabrescent. Same as glabrate.

Glabrous. Without pubescence; smooth.

Gland. A secretory hair or other part that produces nectar or some other liquid.

Glandular. Having glands.

Glaucous. Grayish or bluish in color because of a coating of minute powdery or waxy particles.

Glochid. A finely barbed bristle or hair, especially those forming a tuft at the areolae of stems in *Opuntia.*

Glomerule. A small, compact, more or less rounded cluster.

Glumes. A pair of empty scale-like bracts at the base of a grass spikelet.

Grain (*caryopsis*). The fruit of grasses, seed-like, with a thin pericap adherent to the seed. Also a hardened, seed-like protuberance at the base of inner perianth segments in some species of *Rumex.*

Gynandrous. Having the male and female parts of a flower united, as the stamen attached to the style in orchids.

Gynecandrous. In *Carex,* having pistillate flowers above the staminate flowers of a spike.

Gynobasic style. One that originates between the lobes of a deeply lobed ovary, as in mints and borages.

Gynoecium. The collective term for the female parts of a flower, the pistil or pistils.

Gynophore. A stalk or stipe on which an ovary or fruit is elevated above the floral axis.

Halberd-shaped. Shaped somewhat like an arrowhead but with divergent basal lobes.

Halophyte. A plant usually associated with saline soils.

Hastate. Halberd-shaped; like an arrowhead but with divergent lobes at the base.

Head. A dense inflorescence of sessile or subsessile flowers on a short or broadened axis.

Helicoid. In a spiral like a snail shell.

Herb. A plant that dies completely at the end of the growing season, or one that dies to the ground; not woody-stemmed.

Herbaceous. Like an herb, not woody; or having a green color and a leafy texture.

Hesperidium. A kind of berry having a leathery pericap, as in citrus fruits.

Heterogamous. With two kinds of flowers, as in a daisy. (See Homogamous.)

Hilum. A scar on a seed marking the point of attachment of the funiculus.

Hirsute. With rather stiff or bristly hairs.

Hirtellous. Minutely hirsute.

Hispid. With stiff or rigid, spreading bristles.

Homogamous. With one kind of flower. (See Heterogamous.)

Hood. In the *Asclepiadaceae,* one of the concave segments of the corona.

Hyaline. Thin and translucent.

Hydrophyte. An aquatic plant.

Hypanthium. A saucer-shaped, cup-shaped, tubular, or sometimes rod-shaped expansion of the floral axis that produces floral organs such as sepals, petals, and stamens, from its upper margin.

Hypocotyl. The axis of an embryo or seedling below the cotyledons and above the radicle.

Hypogynous. Having the flower parts attached near the base of the ovary and free from it.

Imbricate. With overlapping edges, as shingles on a roof.

Immersed. Completely submerged in water.

Incised. Cut sharply, irregularly, and rather deeply.

Included. Not protruding beyond the surrounding structure. (See Exserted.)

Incumbent (cotyledons). Placed with their backs to the radicle.

Indefinite (flower parts). Of a number large enough to make an exact count difficult.

Indehiscent. Not splitting open.

Indeterminate (inflorescences). Sometimes applied to those in which the terminal or central flower is the last to open.

Indument. Any covering of a plant surface, especially pubescence.

Indurate. Hardened.

Inequilateral. With sides unequal in length and thus unsymmetrical.

Inferior (ovary). Situated below the point of insertion of the flower parts.

Inflorescence. A flower cluster.

Innate (anther). Attached to the end of the filament.

Innovation. A sterile, basal shoot occurring in some perennial grasses.

Insectivorous. Descriptive of plants that capture insects.

Inserted. Attached to, meaning the point of origin.

Integument. The covering of a body, as the coat of an ovule or seed.

Internode. The portion of a stem between two adjacent nodes.

Interrupted. Having gaps between the parts.

Introrse (anther.) Facing inward.

Involucel. A secondary involucre that subtends a part of an inflorescence.

Involucrate. Having an involucre.

Involucre. A whorl of bracts subtending an inflorescence.

Involute. Rolled lengthwise so as to expose the lower side and conceal the upper side, as in some leaves. (See Revolute.)

Irregular (flower). Having dissimilar parts of the same kind (usually the petals); with bilateral symmetry; zygomorphic.

Jointed. Having swollen or otherwise obvious nodes, as in grass stems.

Keel. A sharp crease or ridge, as in many boats.

Keeled. Sharply creased; with a keel.

Labellum. The lip, or apparently lower petal, of flowers of *Orchidaceae*.

Labiate. Lipped.

Lacerate. With an irregular or ragged margin, as though torn.

Laciniate. Cut deeply into narrow divisions.

Lactiferous. With milky juice.

Lamina. The broad, expanded part of a leaf, sepal, or petal; the blade.

Lanate. Woolly.

Lanceolate. Shaped like the head of a lance, elongate and pointed above, the sides curved, and the broadest part below the middle.

Leaflet. One of the divisions of a compound leaf.

Legume. A simple, dry, dehiscent fruit of one carpel, usually splitting at maturity along two sutures.

Lemma. A bract that usually encloses a flower in the spikelet of grasses.

Lenticel. A corky spot or line on the bark of many woody plants.

Lenticular. Lens-shaped, biconvex with two edges.

Liana. Woody tropical jungle vine.

Ligulate. Tongue-shaped or strap-shaped.

Ligule. A small, often tongue-shaped appendage, as at the junction of blade and sheath of grasses; one of the strap-shaped corollas of *Asteraceae.*

Limb. The upper, expanded portion of a calyx or corolla of united parts, as contrasted with the lower narrow part called the tube.

Linear. Long and narrow with parallel sides, the length generally more than ten times the width.

Lingulate. Strap-shaped.

Lip. One of the two parts of a bilabiate corolla; also the apparently lower and different petal of an orchid.

Lobe. A partial division of a leaf or other organ.

Locule. A cavity within an ovary.

Loculicidal. Splitting along the walls of locules or cavities, as distinct from spliting on the septae or splitting transversely.

Lodicule. One of the two or three minute perianth parts of a grass flower.

Loment. A modified legume having constrictions between the seeds and breaking apart transversely at the constrictions.

Lunate. Shaped like a half-moon, or crescent-shaped.

Lyrate. With relatively small pinnate divisions and a large terminal lobe.

Maculate. Spotted.

Malpighiaceous (hairs). Straight, appressed, and attached near the middle.

Marcescent. Withering but persistent, as flower parts after blooming.

Membranaceous (*membranous*). Thin, soft, flexible, and more or less translucent.

Mericarp. One of the two fruiting carpels of the Parsnip Family.

-merous. A suffix indicating the number of parts or floral organs, as in "3-merous," having the perianth in sets of three.

Mesocarp. The middle layer of the pericarp.

Mesophyte. A plant having medium moisture requirements.

Micropyle. The minute opening into an ovule for the entrance of the pollen-tube, becoming a pit-like mark on the mature seed.

Midrib. The main or central vein of a leaf or other part.

Monadelphous. Stamens that are united into one group by their filaments.

Moniliform. Like a string of beads.

Mono-. A prefix meaning one or once.

Monochasium. A type of inflorescence in which there is a single terminal flower, and below this a single branch bearing one or more younger flowers.

Monoclinous (*perfect*). Having both stamens and pistils in the same flower.

Monoecious. Having separate staminate and pistillate flowers on the same plant.

Mucro. A short, sharp point.

Mucronate. Having a short, sharp point at the apex.

Multi-. A prefix meaning several or many.

Multicipital caudex. A root-crown from which several stems arise.

Multiple (fruit). Derived from several flowers.

Muricate. Having the surface covered with short, sharp projections.

Mycorhiza (*mycorrhiza*). A fungus-root association.

Naked. Lacking organs or parts, a naked flower being one that lacks a perianth.

Nectary. A gland or glands secreting nectar.

Nerve. One of the principal veins of a parallel-veined leaf or other part.

Node. A point on a stem where leaves or branches are attached.

Numerous (stamens or carpels). Usually meaning more than ten. (See Definite.)

Nut. A simple, dry, indehiscent fruit with a bony shell, characteristically derived from a compound pistil, but 1-seeded by abortion.

Nutlet. A small nut. The term is often loosely used to include any small, thick-shelled, seed-like fruit, with or without one or more wings.

Ob-. A prefix meaning inverted, as in "oblanceolate," upside down lanceolate and broadest above the middle.

Oblong. Elongate and with more or less parallel sides, the length usually less than ten times the width.

Obtuse. Blunt, usually forming more than a right angle.

Ocrea. A stipular sheath surrounding the stem.

Odd-pinnate. With a terminal leaflet.

Oligo-. A prefix meaning few.

Oligomerous. Having few parts.

Opposite. (Leaves) in pairs, one on either side of the node; (stamens) inserted in front of petals and thus opposite them.

Orbicular. Circular.

Orthotropous (ovule). Erect, with the micropyle at the upper end.

Oval. Broadly elliptic, the width more than half the length.

Ovary. The bulbous basal portion of a pistil containing one or more ovules.

Ovate. Egg-shaped, the broadest part below the middle.

Ovule. The structure that becomes a seed after fertilization.

Palea (*palet*). The inner and usually smaller of two scaly bracts immediately subtending the grass flower in a spikelet.

Palmate (*digitate*). With parts diverging from a common base, as fingers of a hand.

Panicle. An elongate inflorescence with compound branching.

Papilionaceous. Descriptive of a flower like that of a Sweet Pea, having a standard (banner), two wings, and two keel petals comprising the corolla.

Papillose. Covered with short, rounded projections.

Pappus. The modified and late-maturing calyx of the *Asteraceae,* arising from the summit of the akene, and consisting of hairs, bristles, scales, or awns.

Parasite. A plant that gets its food from another living plant to which it is attached.

Parietal. Produced along the inner side of the ovary wall.

Parted. Cut or lobed more than halfway to the middle or base.

Patent. Spreading.

Pectinate. Like a comb, with many or few, narrow, pinnate divisions.

Pedicel. The stalk of a single flower in an inflorescence.

Peduncle. The stalk supporting a whole inflorescence, or the stalk of a solitary flower.

Peltate. Attached by the lower surface, not by the margin, as the leaves of *Tropaeolum,* the garden Nasturtium.

Perennial. A plant that continues to live year after year.

Perfect (*monoclinous*). A flower having both male and female reproductive parts (stamens and pistils).

Perfoliate. Descriptive of a leaf having the stem apparently passing through it because of a joining of the basal lobes of the blade.

Perianth. The calyx and corolla collectively, or either one when only one is present.

Pericarp. The ovary wall in the fruiting stage.

Perigynous. A type of flower with a hypanthium that arises from the base of the floral axis.

Persistent. Remaining attached rather than falling off.

Petal. One of the parts of the corolla or inner leaf-like parts of a flower.

Petaloid. Resembling a petal in color or texture, usually delicate and not green.

Petiole. The stalk of a leaf.

Phylloclade. A stem that is somewhat broadened and has the function of a leaf.

Phyllode. A broadened petiole without a blade.

Pilose. With rather sparse, soft hairs.

Pinnate. Like a feather; having the parts arranged in two rows along a common axis.

Pinnatifid. Cleft or divided pinnately.

Pistil. The female reproductive part of a flower, occupying a central position; in some flowers single and in others several or many.

Placenta. A point or line of attachment of ovules within an ovary or of seeds within a fruit.

Plumose. Feather-like, having fine, soft hairs along the sides, the hairs divergent from the organ to which they are attached.

Pod. A term often loosely applied to any simple, dry, dehiscent fruit.

Pollen. Microspores; minute spores produced by the anther of a stamen.

Pollinium. A mass of pollen grains adhering together and shed as a unit.

Polycarpous (*apocarpous*). With separate carpels, each a separate pistil.

Polygamo-dioecious. A sexual condition in which two sorts of plants occur: one having some perfect and some staminate flowers, the other having some perfect and some pistillate flowers.

Polygamo-monoecious. A sexual condition in which some perfect and some staminate flowers are produced on the same plant.

Polygamous. Producing some perfect and some unisexual flowers.

Polypetalous. With separate petals.

Polysepalous. With separate sepals.

Pome. A simple, fleshy fruit like an apple, in which the flesh is derived largely from an adnate hypanthium.

Posterior. Next to the axis; in bilabiate corollas the upper lip is posterior and the lower lip is anterior.

Precocious. Blooming before the leaves are expanded.

Procumbent. Lying on the ground but not rooting at the nodes.

Prostrate. Lying flat on the ground.

Pruinose. Covered with a waxy, powdery material and appearing whitish; a "bloom".

Puberulent. Very finely pubescent.

Pubescent. Covered with hairs, especially soft, downy hairs.

Punctate. Covered with dots or pits.

Pyxis. A circumscissile capsule with several seeds.

Raceme. A type of inflorescence having an elongate axis and along this simple pedicels, the order of blooming usually from base to apex.

Racemose. Like a raceme; in general any inflorescence capable of indefinite prolongation, having lateral and axillary flowers.

Rachilla. A little rachis, particularly the axis of a spikelet in grasses.

Rachis. The axis of a spike or of a pinnately compound leaf.

-ranked. Preceded by a number, indicating the number of rows.

Raphe. The part of the funiculus that becomes a part of the seed.

Ray. One of the main branches of a compound umbel; a strap-shaped marginal flower in a head when tubular disk flowers are also present.

Receptacle. The floral axis to which the various flower parts are attached; the enlarged summit of the peduncle of a head to which the flowers are attached.

Recurved. Curved downward or backward.

Reflexed. Bent downward or backward.

Regular (*actinomorphic*). With radial symmetry; the parts of the same sort of similar size and shape.

Reniform. Kidney-shaped; broader than long, with rounded ends, and with a wide basal sinus.

Repand. With a wavy margin.

Replum. The thin partition between the fruiting carpels in mustards.

Reticulate. Like a network.

Retrorse. Directed downward.

Retuse. With a rounded sinus at the apex.

Revolute. Rolled lengthwise so as to expose the upper side and conceal the lower side. (See Involute.)

Rhipidium. A type of monochasial inflorescence that is more or less fan-shaped.

Rhizome. A modified underground stem, usually growing horizontally.

Rib. One of the main veins of a parallel-veined leaf or other organ.

Root. The absorbing, usually underground part of a plant, without nodes.

Rosette. A basal cluster of leaves produced on a very short stem.

Rostrate. With a beak.

Rosulate. With one or more rosettes.

Rotate (corolla). Wheel-shaped; having a short tube and a widely spreading limb, as in a potato flower.

Rugose. Wrinkled.

Runcinate. With jagged lateral indentations, lobes, or teeth that are directed backward.

Runner (*stolon*). A horizontal, aboveground stem that may root at the nodes or apex and develop new plantlets at those places, as in the Strawberry.

Saccate. Shaped like a bag.

Sagittate. Shaped like an arrowhead.

Salverform. A calyx or corolla with a slender tube and widely spreading limb.

Samara. A simple, dry, indehiscent fruit, usually 1-seeded, and with one or more wings.

Saprophyte. A plant without green color that obtains its food from dead organic matter.

Scabrous. Rough to the touch because of minute stiff hairs or other projections.

Scape. A leafless flowering stalk arising from the ground or from a very short stem bearing basal leaves.

Scapose. Bearing a scape or produced on a scape.

Scarious. Thin and dry, like tissue paper, not green.

Schizocarp. A fruit that splits apart into 1-seeded carpels or parts.

Sclerophyllous. With leathery or tough leaves.

Scorpioid. Coiled at the tip.

Scurfy. Covered with minute scales.

Secund. Turned to one side.

Seed. A mature ovule, consisting of an embryo, with or without endosperm, and a surrounding protective coat.

Semi-. A prefix meaning half, as in "semisagittate," shaped like half an arrowhead.

Seminal root. The first or primary root produced by a seedling.

Sepal. One of the parts of the calyx or outer set of floral leaves.

Septate. Having partitions, as many ovaries and some leaves with obvious, internal cross-thickenings.

Septicidal (capsule). Splitting along the septae or partitions.

Septum. A partition.

Seriate. In series, rows, or rings.

Sericeus. Silky with soft hairs.

Serrate. With fine, sharp teeth that are inclined forward or upward.

Serrulate. Finely serrate.

Sessile. Lacking a stalk, as some leaves and flowers.

Seta. A bristle.

Sheath. A tube-like part surrounding another part, as the lower part of a grass leaf that is wrapped round the stem.

Shrub. A plant that is woody and has several main stems, smaller than a tree in height.

Silicle. A little silique, usually not much longer than wide.

Silique. A simple, dry, dehiscent fruit of two carpels that split apart and leave a thin, persistent partition (replum) remaining on the plant; the characteristic fruit of the *Brassicaceae.*

Simple. (Fruit) derived from a single flower and a single pistil; (leaf) having the blade in one piece; (pistil) consisting of a single carpel.

Sinuate. With a wavy margin.

Sinus. The indentation between two lobes.

Solitary. Single.

Sordid. Dirty, not pure white.

Spadix. A spike of flowers on a fleshy axis.

Spathe. A single, large, often showy bract enclosing or subtending an inflorescence, which is commonly a spadix.

Spatulate. Shaped like a spatula, oblong, sometimes a little broader toward the upper end, and with a rounded apex.

Spicate. Produced in a spike.

Spike. An elongated inflorescence of sessile or subsessile flowers.

Spinescent. Bearing spines, or ending in a spine.

Spinulose. Bearing small spines or thorns.

Sporophyll. A reproductive organ, usually thought to be derived from a leaf, that produces spores, as the stamens and carpels of flowering plants.

Spur. A hollow, more or less pointed projection, usually from the calyx or corolla, commonly producing nectar in its tip.

Squarrose. With parts widely spreading or recurved.

Stamen. A pollen-producing organ of a flower, typically consisting of anther and filament; a microsporophyll.

Staminate. Bearing stamens and consequently male; usually used in reference to unisexual flowers or plants.

Staminodium. A sterile stamen.

Standard (*banner*). The uppermost, large petal of a papilionaceous corolla, as in the Sweet Pea.

Stellate. Star-shaped; usually used in reference to hairs.

Stem. The major supporting structure in plants, to which buds, leaves, and flowers are attached at regular intervals at points called nodes.

Stigma. The part of a pistil on which pollen adheres and germinates, generally terminal in position, and often enlarged.

Stipe. A stalk supporting a single organ, particularly an ovary; also the petiole of a fern leaf.

Stipel. A stipule-like appendage at the base of a leaflet of a compound leaf.

Stipitate. Having a stipe or stalk.

Stipular. Pertaining to the stipules.

Stipulate. Possessing stipules.

Stipules. A pair of appendages that may be present at the point of attachment of a leaf to a stem.

Stolon. A modified horizontal stem, aboveground, that may root at the nodes and apex, developing new plantlets at those places, as in a Strawberry.

Stoloniferous. Having stolons.

Strigose. Having the surface covered with straight, appressed hairs that usually are directed forward.

Strophiole. An appendage of the hilum in some seeds.

Style. The stalk-like part of some pistils, connecting the stigma and the ovary.

Sub-. A prefix meaning somewhat, as in "suborbicular," nearly round.

Subtend. To occur immediately below, as a bract subtending a flower.

Succulent. With a fleshy or juicy texture or composition that is usually resistant to drying.

Suffrutescent. Woody or shrubby at the base but not throughout.

Sulcate. Longitudinally grooved.

Super-. A prefix meaning above.

Superior (ovary). Having a position above the point of attachment of the other flower parts, as in hypogynous and perigynous flowers.

Supra-. A prefix meaning above.

Suture. A seam or line along which union has occurred or along which splitting may take place.

Syconium. A hollow, multiple fruit, as that of a Fig.

Sym-. A prefix meaning united.

Symmetrical (flower). Having the same number of each kind of part.

Sympatric. Occurring in one area. (See Allopatric.)

Sympetalous. Having the petals partly or completely united to each other.

Syn-. A prefix meaning united.

Syncarpous. With united carpels.

Syngenesious (stamens or anthers). United by the anthers in a ring.

Tawny. Dull yellowish brown.

Tendril. A part of a stem or leaf modified into a slender, twining, holdfast structure.

Terete. Circular in cross section, cylindrical, rod-shaped.

Ternate. In threes.

Terrestrial. Growing on the land. (See Aquatic.)

Testa. The outer seed coat.

Tetra-. A prefix meaning four.

Tetrad. A group of four, particularly the four pollen grains from one pollen mother cell.

Tetradynamous. With six stamens, four of them longer and two of them shorter, as in flowers of most *Brassicaceae.*

Thalloid. Like a thallus, undifferentiated into stems and leaves.

Thallus. A plant body that is not differentiated into stems and leaves.

Throat. The place in a calyx or corolla of united parts where the tube and limb come together.

Thyrse. A term loosely used to describe a compact panicle; more accurately, a complex group of dichasia resembling a panicle.

Tomentose. Woolly, covered with curly, matted hairs.

Tomentulose. Diminutive of "tomentose," finely woolly.

Tomentum. A covering of woolly, matted hairs.

Torose. Thickened, elongate, and having more or less regular constrictions.

Torulose. Diminutive of "torose."

Torus. The receptacle or floral axis.

Tree. A large, woody plant usually having a single main stem or trunk.

Tri-. A prefix meaning three.

Trigonous. Three-angled.

Triquetrous. Three-angled.

Truncate. Having the base or apex flattened as though cut off.

Tube. The united, more or less cylindrical part of a calyx or corolla of united parts. (See Limb.)

Tuber. A thickened, fleshy, modified stem having functions of food storage and propagation, as the Irish Potato.

Tubercle. A small, swollen structure usually different in color from the part to which it is attached, and often with a hardened texture.

Tubular. Shaped like a tube; also descriptive of corollas that have a well-developed tubular portion but little or no limb portion.

Tufted. Forming clumps. (See Cespitose.)

Tunicate (bulb). Having the leaves arranged in circles when viewed in cross section, as an onion.

Turbinate. Top-shaped, thick at the apex and tapering to a basal point.

Turgid. Swollen.

Turion. A short, scaly branch produced from a rhizome.

Umbel. A flat-topped or rounded inflorescence having flowers on pedicels of nearly equal length and attached to the summit of the peduncle, the characteristic order of blooming being from the outside toward the center.

Umbellet. One of the little umbels of a compound umbel.

Unarmed. Without prickles or spines.

Uncinate. Hooked at the end, as some spines or bristles.

Undershrub. A perennial plant having stems that are woody only in the basal part, the upper part dying back.

Undulate. Having a wavy margin.

Uni-. A prefix meaning one or single.

Uniseriate. In one series or whorl.

Unisexual (*diclinous*). Of one sex only, either male or female, staminate or pistillate.

Urceolate. Urn-shaped; descriptive of a corolla of united petals having a bulbous tube, a narrowed neck, and a very small limb.

Utricle. A 1-seeded fruit with a thin wall, often dehiscent by a lid.

Valvate. With the edges coming together but not overlapping. (See Imbricate.)

Valve. A portion of the wall of a fruit or other part that separates from the remaining part or parts at maturity.

Velutinous. Velvety.

Vein. A bundle of externally visible transporting tissue in a leaf or other organ.

Venation. The system or pattern of veins in an organ.

Ventral. The lower side of a flat organ, or the adaxial side of a carpel. The ventral suture of a carpel bears the seeds along its inner edge.

Ventricose. Enlarged on one side, as some bilabiate corollas.

Vernation. The arrangement of parts in a bud.

Verrucose. Warty.

Versatile. Attached by the middle and free to swing, as some anthers.

Verticil. A whorl.

Villose (*villous*). Covered with fine, long hairs that are not tangled.

Virgate. Wand-like; descriptive of a slender, erect, straight, leafy stem bearing only short or no branches.

Viscid. Sticky, causing foreign particles to adhere to it.

Viviparous. Sprouting from seed or bulblets while still attached to the parent plant; sometimes applied to plants that have flowers modified into bulblets.

Whorl. A group of three or more parts at a node.

Wing. A flat, usually thin appendage on a seed or fruit; also each of the two lateral petals of a papilionaceous corolla, as in the Sweet Pea.

Xerophyte. A plant adapted to very dry situations.

Zygomorphic (flower or corolla). Having the parts of the same kind of different sizes or shapes so as to be bilaterally but not radially symmetrical. (See Irregular.)

INDEX